INTRODUCTION TO
DIGITAL SIGNAL PROCESSING

McGraw-Hill Series in Electrical Engineering

Consulting Editor

Stephen W. Director, Carnegie-Mellon University

CIRCUITS AND SYSTEMS

COMMUNICATIONS AND SIGNAL PROCESSING

CONTROL THEORY

ELECTRONICS AND ELECTRONIC CIRCUITS

POWER AND ENERGY

ELECTROMAGNETICS

COMPUTER ENGINEERING

INTRODUCTORY

RADAR AND ANTENNAS

VLSI

Previous Consulting Editors

Ronald N. Bracewell, Colin Cherry, James F. Gibbons, Willis W. Harman, Hubert Heffner, Edward W. Herold, John G. Linvill, Simon Ramo, Ronald A. Rohrer, Anthony E. Siegman, Charles Susskind, Frederick E. Terman, John G. Truxal, Ernst Weber, and John R. Whinnery

COMMUNICATIONS AND SIGNAL PROCESSING

Consulting Editor

Stephen W. Director, Carnegie-Mellon University

Antoniou: *Digital Filters: Analysis and Design*
Candy: *Signal Processing: The Model-Based Approach*
Candy: *Signal Processing: The Modern Approach*
Carlson: *Communications Systems: An Introduction to Signals and Noise in Electrical Communication*
Cherin: *An Introduction to Optical Fibers*
Cooper and McGillem: *Modern Communications and Spread Spectrum*
Davenport: *Probability and Random Processes: An Introduction for Applied Scientists and Engineers*
Drake: *Fundamentals of Applied Probability Theory*
Guiasu: *Information Theory with New Applications*
Keiser: *Optical Fiber Communications*
Kuc: *Introduction to Digital Signal Processing*
Papoulis: *Probability, Random Variables, and Stochastic Processes*
Papoulis: *Signal Analysis*
Papoulis: *The Fourier Integral and Its Applications*
Peebles: *Probability, Random Variables, and Random Signal Principles*
Proakis: *Digital Communications*
Schwartz: *Information Transmission, Modulation, and Noise*
Schwartz and Shaw: *Signal Processing*
Smith: *Modern Communication Circuits*
Taub and Schilling: *Principles of Communication Systems*

INTRODUCTION
TO DIGITAL
SIGNAL PROCESSING

Roman Kuc

Department of Electrical Engineering
Yale University

McGraw-Hill Book Company

New York   St. Louis   San Francisco   Auckland   Bogotá   Caracas   Colorado Springs
Hamburg   Lisbon   London   Madrid   Mexico   Milan   Montreal   New Delhi   Oklahoma City
Panama   Paris   San Juan   São Paulo   Singapore   Sydney   Tokyo   Toronto

This book was set in Times Roman.
The editors were Alar E. Elken and John M. Morriss;
the cover designer was John Hite
the production supervisor was Leroy A. Young.
Project Supervision was done by Universities Press.
Arcata Graphics/Halliday was printer and binder.

INTRODUCTION TO DIGITAL SIGNAL PROCESSING

234567890   HALHAL   89321098

ISBN 0-07-035570-3

Library of Congress Cataloging-in-Publication Data

Kuc, Roman
    Introduction to digital signal processing / Roman Kuc.
        p.   cm. — (McGraw-Hill series in electrical engineering.
    Communications and signal processing.)
    Includes index.
    ISBN 0-07-035570-3
    1. Signal processing—Digital techniques.   I. Title.   II. Series.
TK5102.5K823 1988
621.38′043—dc19   87-19862
                                                                       CIP

ABOUT THE AUTHOR

 **Roman Kuc** received the B.S.E.E. degree from the Illinois Institute of Technology, Chicago, in 1968, and the Ph.D. degree in electrical engineering from Columbia University, New York, in 1977.

From 1968 to 1975 he was a Member of the Technical Staff of Bell Telephone Laboratories, where he was engaged in the design of audio recording instrumentation and in developing efficient digital speech coding techniques. From 1977 to 1979 he was a postdoctoral Research Associate in the Department of Electrical Engineering, Columbia University, and the Radiology Department of St. Luke's Hospital, New York, where he employed digital signal processing techniques to extract diagnostic information from reflected ultrasound signals. As an Associate Professor with the Department of Electrical Engineering, Yale University, New Haven, CT, he is currently pursuing problems in ultrasonic tissue characterization, speech recognition, and intelligent acoustic sensors for robotic applications.

Dr. Kuc was an Associate Editor of the *IEEE Transactions on Acoustics, Speech, and Signal Processing* and *Ultrasonic Imaging*. He is also the Consulting Editor of Electrical Engineering for *The American Scientist,* a past Chairman of the Instrumentation Section of the New York Academy of Sciences, and a member of the Schevchenko Scientific Society.

DEDICATION

This book is dedicated to my wife, Robin, who took on the lioness' share of coping with our two-year-old son, Alexander Adam, allowing me peace to finish this book.

CONTENTS

Preface   xiii

1 Introduction   1

1-1   What is a Digital Filter?   1
1-2   Anatomy of a Digital Filter   3
1-3   Frequency Domain Description of Signals and Systems   5
1-4   Some Typical Applications of Digital Filters   7
1-5   Replacing Analog Filters with Digital Filters   12
1-6   Overview of the Course   13
1-7   Description of the Computer Projects   17
1-8   Summary   18

2 Discrete-time Description of Signals and Systems   20

2-1   Introduction   20
2-2   Discrete-Time Sequences   20
2-3   Superposition Principle for Linear Systems   26
2-4   Unit-sample Response Sequence   27
2-5   Time-Invariant Systems   31
2-6   Stability Criterion for Discrete-time Systems   38
2-7   Causality Criterion for Discrete-time systems   40
2-8   Linear Constant-Coefficient Difference Equations   41
2-9   Suggested Programming Style   44
2-10 Writing a Digital Filter Program   45
2-11 Summary   55

3 Fourier Transform of Discrete-time Signals   65

3-1   Introduction   65
3-2   Definition of the Fourier Transform   65
3-3   Important Properties of the Fourier Transform   69

ix

3-4	Properties of the Fourier Transform for Real-valued Sequences	72
3-5	Program to Evaluate the Fourier Transform by Computer	80
3-6	Use of the Fourier Transform in Signal Processing	82
3-7	Fourier Transform of Special Sequences	84
3-8	The Inverse Fourier Transform	92
3-9	Fourier Transform of the Product of Two Discrete-time Sequences	94
3-10	Sampling a Continuous Function to Generate a Sequence	96
3-11	Reconstruction of Continuous-time Signals from Discrete-time Sequences	105
3-12	Summary	110

4   The Discrete Fourier Transform 118

4-1	Introduction	118
4-2	The Definition of the Discrete Fourier Transform (DFT)	119
4-3	Computing the Discrete Fourier Transform from the Discrete-time Sequence	121
4-4	Properties of the DFT	128
4-5	Circular Convolution	130
4-6	Performing a Linear Convolution with the DFT	132
4-7	Computations for Evaluating the DFT	133
4-8	Programming the Discrete Fourier Transform	135
4-9	Increasing the Computational Speed of the DFT	136
4-10	Intuitive Explanation for the Decimation-in-time FFT Algorithm	142
4-11	Analytic Derivation of the Decimation-in-time FFT Algorithm	146
4-12	Some General Observations about the FFT	148
4-13	Other Fast Realizations of the DFT	150
4-14	Summary	150

5   The z-transform 157

5-1	Introduction	157
5-2	The Definition of the z-transform	158
5-3	Properties of the z-transform	160
5-4	The System Function of a Digital Filter	162
5-5	Combining Filter Sections to Form More Complex Filters	164
5-6	Digital Filter Implementation from the System Function	167
5-7	The Complex z-plane	169
5-8	The Region of Convergence in the z-plane	174
5-9	Determining the Filter Coefficients from the Singularity Locations	178
5-10	Geometric Evaluation of the z-transform in the z-plane	183
5-11	Relationship Between the Fourier Transform and the z-transform	184
5-12	The z-transform of Symmetric Sequences	188
5-13	The Inverse z-transform	196
5-14	Summary	203

6 Digital Filter Structures 208

6-1	Introduction	208
6-2	System Describing Equations	209
6-3	Filter Catagories	209
6-4	The Direct Form I and II Structures	210
6-5	Cascade Combination of Second-order Sections	214
6-6	Parallel Combination of Second-order Sections	218
6-7	Linear-phase FIR Filter Structures	220
6-8	Frequency-sampling Structure for the FIR Filter	221
6-9	Summary	227

7 From Analysis to Synthesis 230

7-1	Introduction	230
7-2	Interrelationships of Analytic Methods	231
7-3	Practical Magnitude Response Specifications	236
7-4	Log-magnitude Response Curves	238
7-5	Programs to Compute the Log-magnitude Response of Poles and Zeros	239
7-6	Phase Response Considerations	243
7-7	Steps in Performing a Filter Design	246
7-8	Choice of Filter Type	246
7-9	Interactive Filter Design by Intuitive Pole/Zero Placement	247
7-10	Writing an Interactive Design Program	256
7-11	Summary	258

8 Infinite Impulse Response Filter Design Techniques 262

8-1	Introduction	262
8-2	Analog Filter System Function and Frequency Response	263
8-3	Analog Lowpass Filter Design Techniques	263
8-4	Methods to Convert Analog Filters into Digital Filters	280
8-5	Frequency Transformations for Converting Lowpass Filters into other Types	295
8-6	All-pass Filters for Phase Response Compensation	310
8-7	Summary	312

9 Finite Impulse Response Filter Design Techniques 318

9-1	Introduction	318
9-2	Designing FIR Filters with the Windowing Method	319
9-3	The DFT Method for Approximating the Desired Unit-Sample Response	341
9-4	Designing FIR Filters by Combining the DFT and Window Methods	348
9-5	Designing FIR Filters with the Frequency-sampling Method	361

9-6 Use of the FFT Algorithm to Perform FIR Filtering 379
9-7 Summary 384

10 Finite-precision Effects 390

10-1 Introduction 390
10-2 Finite-precision Number Representation within the Computer 391
10-3 Quantization Error in Analog-to-digital Conversion 395
10-4 Expressing Coefficients in Finite Precision 399
10-5 Performing Arithmetic in Finite-Precision Number Systems 404
10-6 Effects of Finite-Precision Arithmetic on Digital Filters 409
10-7 Programs to Simulate Quantization Effects 411
10-8 Summary 416

11 Inverse Filtering 422

11-1 Introduction 422
11-2 Applications of Inverse Filters 423
11-3 Minimum Phase Systems 426
11-4 Formulating the Problem for Applying an Inverse Filter 429
11-5 FIR Filter Approximations to the Inverse Filter 432
11-6 Discrete Hilbert Transform Relationship for Minimum-Phase Systems 433
11-7 Designing Inverse Filters with the Discrete Hilbert Transform 436
11-8 The Effects of Noise on Inverse Filters 441
11-9 Summary 448

Appendix A 455

Appendix B 469

Index 471

PREFACE

This book is intended to be used in the first course covering digital signal processing and filter design, typically offered at the senior or first-year graduate level in electrical engineering. The course is also appropriate for a graduate course in departments other than electrical engineering, such as geophysics and mechanical engineering, in which the analysis of discrete-time data is performed. It is assumed that the student has had a course covering Fourier series and LaPlace transforms on the level of the first linear circuits or control systems course. This text also includes projects that require students to write computer programs to accomplish signal processing projects. The student should be familiar with some programming language such as FORTRAN, BASIC, PASCAL, or C.

This book approaches digital signal processing and filter design in a novel way, by presenting the relevant theory and then having the student apply it by implementing signal processing routines on a computer. This mixture of theory and application has worked successfully for the past six years in teaching this course at Yale University. With this approach, the students receive a deeper and intuitive understanding of the theory, its applications and its limitations.

The text can accommodate a wide variety of courses. Currently, the course on digital filters at the undergraduate level is taught primarily as a theory course, with homework problems and exams to determine the course grade. This book can be used directly in this type of course. However, the course becomes much more interesting and enjoyable when students apply the theory to write programs to perform signal processing tasks. Each chapter describes programming techniques that implement the theory and includes projects that illustrate different applications. Digital filter routines are written from the very beginning of the course, immediately after the time-domain description of discrete-time systems is presented in Chapter 2. The programs can be written on any computer from personal to mainframe, requiring only a standard terminal for displaying the results in graphical form and a printer for generating the hard copy. The programs to be written by the student are not difficult. Each routine performs only one primitive task in the signal processing procedure. For example, the discrete Fourier transform subprogram can be accomplished in only nine lines of code. But once the student has given sufficient thought to the problem to write these nine lines, the *magic* associated with this procedure disappears, being replaced by *understanding*. To minimize the programming effort of the student, the routines are used as building blocks in later projects. Constructing more sophisticated programs from these simpler building blocks aids the student in conceptualizing projects later in the course. By the end of the course, the student has a package of signal processing programs that can be used for a variety of applications.

The graphic outputs for the projects are displayed on the terminal and on a printer, being generated by programs provided in the Appendix. These plots, although having crude resolution, are adequate for illustrating the ideas and also have the advantage of not requiring special plotting hardware. Students having higher-resolution terminal graphic capabilities have readily written their own graphic routines using the programs in the Appendix as guides.

This text differs from others in the area of digital filters and signal processing in the important respect that processing signals and designing filters on a digital computer, using simple programs composed by the students, are an integral part of the course. For this reason, more realistic problems can be assigned and discussed than would be in a course without computers, for which these problems would be mathematically untenable or at least tedious. The organization of the material presented in the text allows the student to start writing meaningful filtering programs from the beginning of the course. Projects included in the text illustrate both digital signal processing and digital filter design concepts.

The book is organized in the following manner. Chapter 1 defines the basic components of a digital filter and provides motivation by describing several common applications of digital filters and signal processing. The material in Chapters 2 through 5 covers the fundamentals of time, frequency, and z-plane analysis. If some of this material has been introduced in previous courses, the respective chapters can be treated as a review. Several common structures that are employed to implement digital filters are described in Chapter 6.

The synthesis of digital filters to meet a desired specification is presented in Chapters 7 through 9. To compare the different filter design procedures, the projects are structured to allow each student to be assigned a *personal* specification to be satisfied. In Chapter 7, the digital filter design is accomplished by interactively and intuitively placing poles and zeros in the z plane. This procedure serves three purposes. First, it develops an understanding of the interaction of the singularities in generating a filter magnitude response or system spectrum. Second, it provides an appreciation for the conventional filter design procedures presented in Chapters 8 and 9. Finally, this intuitive procedure indicates how to correct the locations of the poles and zeros generated by conventional design procedures. The computed positions of these singularities are slightly, but significantly, displaced from the desired positions by the round-off errors that are always present in any computation. These round-off effects become obvious in the filter design projects. The conventional procedures for designing infinite-impulse response (IIR) digital filters are described in Chapter 8, and those for finite-impulse response (FIR) digital filters in Chapter 9.

Two advanced topics are covered in Chapters 10 and 11. Chapter 10 describes the effects of finite precision encountered in practical hardware implementation of digital filters. Some elementary results from probability theory are presented to describe noise effects. Chapter 11 provides an overview of inverse filters. This topic was chosen because it includes and integrates many of the concepts introduced in the previous chapters to illustrate an important practical application.

In addition to the Problems, a set of Computer Projects is included, the programs for which should be written by the student. Instructions for program development are included as an integral part of the text and Project description. The ability to manipulate sequences by a computer program is essential for appreciating the concepts involved for signal processing, which is important when the student is faced with novel problems in his or her own application.

Projects in the analysis part of the book illustrate the practical applications of digital interpolation procedures and the simulation of a frequency-shift keying (FSK) digital communication channel. In the filter synthesis part of the book, the Projects include designing digital filters to satisfy a practical specification. In the chapters describing finite-precision effects and the inverse filter, the Projects illustrate programming techniques for simulating physical hardware and systems.

One main benefit of the Projects is to relate the analytic results to those produced by the computer. At times, the analysis is simple and should then be used to predict and verify the computer results. At other times, the analytic procedure is very complicated mathematically, in which case the computer results should be obtained first to indicate what type of results should be expected. In this latter case, the computer results indicate the direction in which the analysis should proceed. For example, the computer results may indicate that a complex-valued operation in a particular analytic step results in a real-valued sequence. This knowledge may serve to drive the analytic steps. In practice, it is often found that computer simulations demonstrate an effect that can be proven analytically.

An *insider's* knowledge of the signal processing algorithms is attained by having the students write their own programs. This knowledge then allows the student comfortably to make the intellectual move to more sophisticated signal processing and filter design programs and packages, such as those distributed by the IEEE or SIG or offered by many vendors.

This text has the flexibility of being used in a wide variety of digital signal processing courses at the undergraduate and graduate levels. For a conventional undergraduate *theory* course in digital filters, the first nine chapters of this book supply the necessary topics in discrete-time signals and systems. In some universities, such a theory course is followed by a digital filter design laboratory. The Projects in this book can be used as the exercises in such a course. At the graduate level, a faster pace allows topics from all chapters to be covered. Courses stressing the signal processing aspects should emphasize the analytic methods and the discrete Fourier transform and de-emphasize IIR filter design and quantization effects, while courses stressing filter implementation should stress the latter and reduce the emphasis on the former.

To allow adequate space for this treatment of this material, the analytic techniques have been restricted to those that are absolutely necessary for understanding the filter synthesis and signal processing procedures. Topics that are normally covered in a senior/first-year graduate course on digital signal processing, but that are not considered in this text, include the solution of the difference equations with the one-sided z-transform, the numerical methods for optimizing digital filter designs, and the area of multidimensional signal processing. These topics are better treated in a second course in which the students have the familiarity of the material contained in this book.

The author is indebted to the students at Yale University who have taken this course and provided many useful comments regarding the topics and the exercises in this book. Among these are Ata Arjomand, Billur Barshan, Leo Costello, Arnold Englander, Kannan Parthasarathy, Rob Roy, and David Starer. McGraw-Hill would like to thank the following reviewers for their comments and suggestions, J. I. Aunon, Purdue University; Edward Delp, Purdue University; Simon Haykin; McMaster University; M. Kaveh, University of Minnesota; and Paul Prucnal, Columbia University.

The author hopes that the instructors teaching digital signal processing as a combination of theory and computer exercises, as described in this text, have an enjoyable *learning* experience. This learning comes about from the questions posed by the students, who inevitably find some novel approach or interpretation to filter design on the computer. This has certainly been the case every time the author has taught this course.

Roman Kuc
New Haven, Connecticut

INTRODUCTION TO
DIGITAL SIGNAL PROCESSING

CHAPTER
1

INTRODUCTION

In this chapter, we present the concept of a digital filter and define the adder, multiplier and delay elements that are its components. Some typical applications of digital filters for signal processing, system simulation and spectral analysis are described. An overview of the topics covered in the book concludes the chapter.

1.1   WHAT IS A DIGITAL FILTER?

Most broadly defined, a digital filter is a *numerical procedure,* or *algorithm,* that transforms a given sequence of numbers into a second sequence that has some more desirable properties, such as less noise or distortion. As shown in Fig. 1.1, the entire given, or *input,* sequence will be denoted in this book by $\{x(n)\}$, and the entire *output* sequence by $\{y(n)\}$, where n is an index. Typically, the set of index values consists of consecutive integers, which in some cases may take values from minus infinity to plus infinity.

The desired features in the output sequence depend on the application. For example, if the input sequence is generated by a sensing device, such as a microphone, the digital filter may attempt to produce an output sequence having less background noise or interference. In radar applications, digital filters are used to improve the detection of airplanes. In speech processing,

1

Input sequence 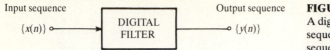 Output sequence

$\{x(n)\}$ $\{y(n)\}$

FIGURE 1-1
A digital filter transforms the input sequence $\{x(n)\}$ into the output sequence $\{y(n)\}$.

digital filters have been employed to reduce the redundancy in the speech signal so as to allow more efficient transmission, and for speech recognition.

Input sequences can be generated in several ways, as shown in Fig. 1.2. One common method is to sample a continuous-time signal at a set of equally-spaced time intervals. If the continuous-time signal is denoted by $x(t)$, then the values of the discrete-time sequence are denoted as

$$x(nT_s) = x(t)\big|_{t=nT_s} \tag{1.1}$$

where T_s is the sampling period. In image processing applications, scanning a line of an image produces an intensity function of position across the image plane, denoted by $i(p)$. This function can be sampled to produce a discrete-position sequence with values given by

$$i(nD_s) = i(p)\big|_{p=nD_s} \tag{1.2}$$

where D_s is the sampling distance interval.

One common application of digital filters is to simulate a physical system. The input sequence is then usually a sequence of numbers generated by the computer and represents the excitation signal $\{e(n)\}$. In this case, $\{e(n)\}$ represents instantaneous values of some physical parameter, such as voltage, torque, temperature, etc. This sequence is then processed to determine the

CONTINUOUS-TIME SIGNAL

DISCRETE-TIME SEQUENCE

$\{x(nT_s)\}$

IMAGE

DISCRETE-TIME SEQUENCE

$\{i(nD_s)\}$

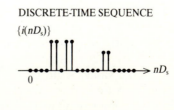

FIGURE 1-2
Deriving one-dimensional discrete-time sequences from analog signals.

response of the system, represented by a sequence $\{r(n)\}$. This application is discussed in more detail later in this chapter.

To allow these different sequence types to be treated with the same notation, we will generalize the element index to be an integer, denoted by n. That is, we will write $x(n)$, rather than $x(nT_s)$, and $i(n)$, rather that $i(nD_s)$.

In addition to sampling, the input sequence could have come in a discrete-time form initially, such as daily sales receipts, $\{m(n)\}$, or weekly measurements of personal weight, $\{w(n)\}$. In these cases, we may be interested in short-term fluctuations or long-term trends.

The above data sequences are indexed in terms of temporal or spatial increments. To simplify the discussion of the various types of input data sequences that can exist, we will refer to all of them as *discrete-time sequences*. This is done with the understanding that the particular sequence under discussion may, in practice, be related to a physical parameter other than time.

Although the theory presented in this text is general, the *implementation* of a digital filter depends on the application and the user's environment. In education and research, a digital filter is typically implemented as a program on a general-purpose computer. The types of computers that can be used vary widely, from personal computers, through the larger minicomputers, to the large time-shared mainframes. In commercial instrumentation and industrial applications, the digital filter program is commonly implemented with a microcomputer that may also be used for control and monitoring purposes as well. For high-speed or large-volume applications, such as use in automobiles for controlling the engine operation, the digital filter may consist of special-purpose integrated circuit chips. These circuits perform the computation and storage functions required for the digital filter operation. These functions, which are common to all digital filter applications, are described in the next section.

1.2 ANATOMY OF A DIGITAL FILTER

A digital filter consists of the interconnection of three simple elements: *adders, multipliers* and *delays*. These are shown in Fig. 1.3. The adder and multiplier are conceptually simple components that are readily implemented in the arithmetic logic unit of the computer. Delays are components that allow access to future and past values in the sequence. Open-headed arrows indicate direction of information flow, while the larger closed-headed arrows indicate multipliers. This convention will be useful for drawing complicated digital filters in the later chapters.

Delays come in two basic "flavors": positive and negative. A positive delay, or simply *delay*, is implemented by a memory register that stores the current value of a sequence for one sample interval, thus making it available for future calculations. A positive delay is conventionally indicated by a box denoted by z^{-1}. The significance of this notation will become clear when we analyze digital filters with the z-transform. A negative delay, or *advance,* is

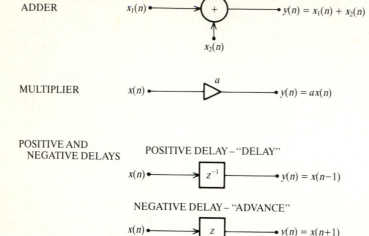

FIGURE 1-3
Elements that are interconnected to implement a digital filter.

used to look ahead to the next value in the sequence, and is indicated by a box denoted by z. Advances are typically used for applications, such as image processing, in which the entire data sequence to be processed is available at the start of processing, so that the advance serves to access the next data sample in the sequence. The availability of the advance will simplify the analysis of digital filters. However, the advance cannot always be implemented in some applications. For example, when the data sequence is obtained by sampling a function of time, each new sample is usually processed as it is acquired. In this case, advances are not allowed, since we cannot gain access to future data values.

A digital filter *design* involves selecting and interconnecting a finite number of these elements and determining the multiplier coefficient values. Examples of interconnecting these elements to form two simple digital filters are considered in the following examples.

Example 1.1. Digital filter used as a three-sample averager. Consider the relationship between the values of the output sequence at time n, denoted by $y(n)$, and the values of the input sequence, $x(n-1)$, $x(n)$ and $x(n+1)$, given by

$$y(n) = \tfrac{1}{3}x(n+1) + \tfrac{1}{3}x(n) + \tfrac{1}{3}x(n-1)$$

As shown in Fig. 1.4, the current output $y(n)$ is equal to the average of the next, current, and previous input values. The advance serves to access the next value of the sequence, while the delay stores the previous value.

Any past or future input data values can be accessed by including the appropriate number of delay or advance elements. Data values can be combined, however, only through an adder.

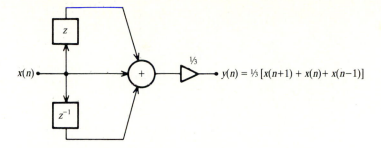

FIGURE 1-4
Nonrecursive digital filter that acts as a three-sample averager.

The term *order* will be used to denote the minimum number of delays and advances that are *necessary* to implement a digital filter. For example, the 3-sample averager is a second-order filter. When the filter output is a function of only the input sequence, as in the 3-sample averager, it is called a *nonrecursive* filter. When the output is also a function of the previous output values, it is called a *recursive* filter. Example 1.2 illustrates a recursive filter.

> **Example 1.2. First-order recursive filter.** Consider the relationship between the input and output sequence values given by
>
> $$y(n) = ay(n-1) + x(n)$$
>
> where a is a constant. As shown in Fig. 1.5, the current value of the output is equal to the sum of the input and a times the value of the previous output. Since only one delay is necessary, this is a first-order filter.

Example 1.2 shows how feedback is implemented in a digital filter. Whereas both positive and negative delays can be applied to the input sequence, *only positive delays* can be applied to the output sequence, since we cannot gain access to output values that have not yet been computed.

1.3 FREQUENCY DOMAIN DESCRIPTION OF SIGNALS AND SYSTEMS

To discuss some typical applications of digital filters in conventional terms, we need to define the frequency domain description of signals and systems. Let us

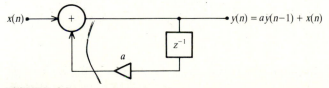

FIGURE 1-5
A first-order recursive digital filter.

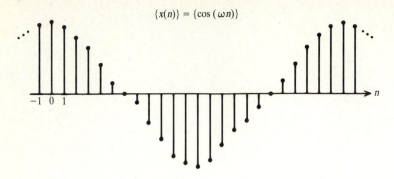

FIGURE 1-6
The cosine sequence $\{\cos(\omega n)\}$ for $\omega = \pi/12$.

DISCRETE-TIME SEQUENCES   MAGNITUDE SPECTRA

$\{x_1(n)\}$

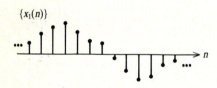

$|X_1(e^{j\omega})|$

$\{x_2(n)\}$

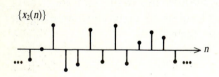

$|X_2(e^{j\omega})|$

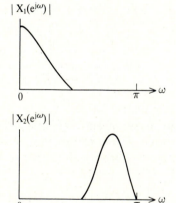

$\{x_3(n)\}$

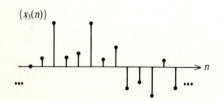

$|X_3(e^{j\omega})|$

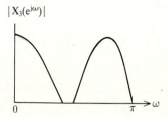

FIGURE 1-7
Discrete-time sequences and their magnitude spectra. The sequence $\{x_3(n)\} = \{x_1(n)\} + \{x_2(n)\}$.

consider the discrete-time sinusoidal sequence $\{x(n)\}$ shown in Fig. 1.6, and given by

$$\{x(n)\} = \{\cos(\omega n)\} \tag{1.3}$$

where ω is the *frequency* of the sequence, having units of *radians*. More interesting signals can be considered to be composed of a sum of such sinusoidal sequences having different frequencies. One example is the Fourier series expansion of a periodic waveform. As described in detail in Chapter 3, the *magnitude spectrum* of a discrete-time sequence, denoted by $|X(e^{j\omega})|$ indicates the magnitude of the various sinusoidal components. Examples of three sequences and their magnitude spectra are shown in Fig. 1.7. The sequence $\{x_1(n)\}$ contains mostly low-frequency components, $\{x_2(n)\}$ contains mostly high-frequency components, and $\{x_3(n)\} = \{x_1(n)\} + \{x_2(n)\}$ contains both.

One common application of digital filters is to separate signals on the basis of their spectral content. For example, in Fig. 1.8, we start with the sequence $\{x_3(n)\}$. To retrieve the low-frequency signal $\{x_1(n)\}$, we pass $\{x_3(n)\}$ through a filter whose *magnitude response,* denoted by $|H_{LP}(e^{j\omega})|$, passes the low-frequency components of the input, while eliminating the high-frequency ones. Such a filter is called a *lowpass* filter. To retrieve the high-frequency signal $\{x_2(n)\}$, we pass $\{x_3(n)\}$ through a *highpass* filter, whose response is shown as $|H_{HP}(e^{j\omega})|$. Two other important filters are the *bandpass* filter, which passes the input components that lie only within a desired band of frequencies, and the *bandreject* filter, which eliminates the input components that lie within a specified band of frequencies.

1.4   SOME TYPICAL APPLICATIONS OF DIGITAL FILTERS

Studying the field of digital filtering provides a fundamental understanding of processing of discrete-time data. This is essential for manipulating data by digital computer and for applying the signal processing procedures to such applications as robotics, intelligent sensors and seismic signal processing. Digital filter techniques are also important for modeling linear systems, such as speech production models and transmission channels. Often, such modeling provides insights into the underlying principles of the problem, from which creative solutions can stem. We now consider some typical applications.

SPEECH PROCESSING. Digital filters are commonly used for speech recognition systems. Let us consider the sequences $\{x_1(n)\}$ and $\{x_2(n)\}$, shown in Fig. 1.9, which represent samples taken from two speech signals. The first is the /a/ sound, as in "father", and the second is the /i/ sound, as in "see". These two periodic signals are most easily differentiated by computing the magnitude spectrum of one period. We note that the two spectra, $|X_1(e^{j\omega})|$ and $|X_2(e^{j\omega})|$, consist of peaks that correspond to the resonances of the vocal tract shape that

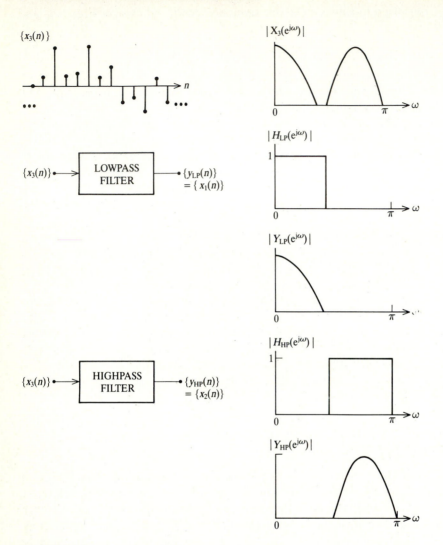

FIGURE 1-8
Recovering sequences by filtering. The sequence $\{x_1(n)\}$ is recovered from $\{x_3(n)\}$ by lowpass filtering, and $\{x_2(n)\}$ is recovered by highpass filtering.

was used to generate the sound. In speech recognition systems, various speech sounds can be differentiated on the basis of the frequency locations of these peaks.

COMPUTER SIMULATION OF A PHYSICAL SYSTEM. When the physical system, such as a robot, is very complicated or expensive to build, often it is first simulated on a digital computer in its digital filter form. This simulation is employed to determine the performance and the sensitivities of the analog

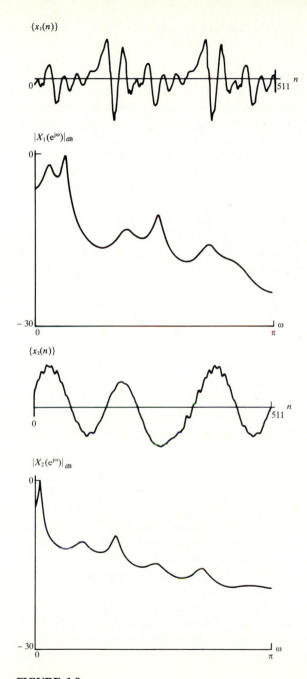

FIGURE 1-9

Speech signals and their spectra. The sequence $\{x_1(n)\}$ consists of 512 samples of the speech sound /a/ and $\{x_2(n)\}$ of /i/. The sample values are connected by straight lines. The corresponding spectra $|X_1(e^{j\omega})|$ and $|X_2(e^{j\omega})|$, given in logarithmic units of dB, exhibit peaks that can be used to differentiate these sounds.

system before it is actually built. The frequency-dependent transformations, such as those produced by the inertia and compliance of a robot arm, can be mimicked by a digital filter.

The advantages of performing a digital filter simulation include the following. First, researchers can obtain data without performing expensive and time-consuming experiments. Second, the simulation provides a flexible means of varying the input and system parameters and observing the effect of these changes. A third advantage of simulations is that any desired signal in the model can be observed. In the physical system, it may be very costly or even impossible to observe the value of certain parameters that are important in the operation of the system. The following cases illustrate these advantages.

Simulating a computer communication system. One common digital data communication system that allows computers to transfer data uses the *frequency-shift keying* (FSK) scheme. In the FSK system, the binary digits, 1 or 0, are transformed into sections of a sinusoidal waveform. As shown in Fig. 1.10, the frequency of a particular section is determined by the binary value to be transmitted. A simulation program including a digital filter to model the transmitter, detector and the transmission channel may attempt to answer the following questions: What frequencies should be used? How long should the duration of each segment be? What detector is suitably reliable but reasonably expensive or fast? To investigate the complicated effects of interference, random noise can be added to the transmitted signal and the resulting characteristics of a given detector can be examined. All these procedures can be performed before any physical equipment is manufactured or installed.

Such a simulation program is developed by the student in the projects at the end of the chapters. In one project, the transmitter is simulated by a program that generates the waveform that corresponds to a given binary sequence. Other projects consider different digital filters that operate on the received sinusoidal sequence to retrieve the value of the binary digit transmitted.

Simulating a robot arm. Another advantage of simulation programs is that the various parameters acting in a physical system can each be examined in detail. For example, consider the simulation of a robot arm, shown in Fig. 1.11. The actual arm is constructed of some material that has mass and elasticity. The input to such an arm is shown as a rotational displacement that is modeled as a step function at one joint of the arm. The desired output may be the location of the manipulator at the end of the arm. The time response of the output may be important for ensuring that the arm construction satisfies the specification for quickness or overshoot.

One simple digital filter model of the arm is a second-order filter whose coefficient values are related to the mass and elasticity of the material. The physical input signal is converted into an input sequence, as shown in the figure, and the output sequence of the filter indicates the response to be

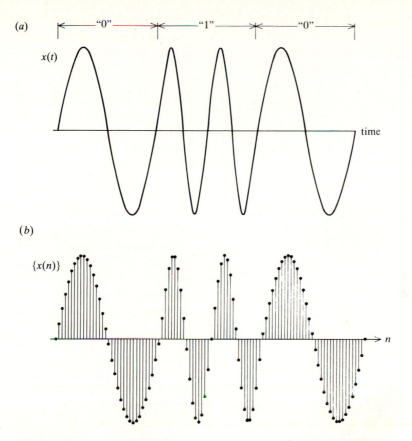

FIGURE 1-10
Frequency-shift keying (FSK) signals. a) Example of continuous-time signal from actual system. b) Simuated sequence produced by computer.

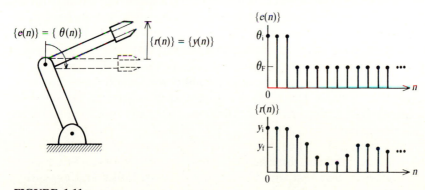

FIGURE 1-11
Simulating a robot arm by applying an excitation sequence and computing the response sequence.

expected if the model is accurate. The value of this simulation model is that the coefficient values of the digital filter can be adjusted to produce the desired response. Then the final values can be employed to determine the physical dimensions of the robot arm.

1.5 REPLACING ANALOG FILTERS WITH DIGITAL FILTERS

One of the first applications of digital filters was to replace analog filters, composed of resistors R, capacitors C and inductors L. This replacement is usually performed to overcome some of the inherent limitations of analog components, including the fluctuation of the component values with temperature and age. Another disadvantage of analog components is their large physical size, especially in the case of large-valued inductors and capacitors.

A further disadvantage of analog filters is the difficulty encountered in changing the component values in an existing filter. In many cases, the equivalent digital filter is implemented as a sequence of instructions in a computer, which can easily be changed with an editor, while soldering may be required to change a physical component in the analog filter. When the digital filter is implemented in hardware, the multiplier coefficients are often stored in a read-only memory (ROM), which can readily be replaced if a filter modification is desired. These coefficients can also be stored in read/write memory and be changed automatically as the nature of the input signal sequence changes. This added flexibility of digital filters has created an area of research in so-called *adaptive filters* that is currently very active.

Replacing an analog filter with a digital filter in a given application involves several considerations. For illustration, let us examine the situation in Fig. 1.12, which shows a continuous-time input signal $x(t)$, a simple analog *RC* lowpass filter and the output signal $y(t)$. To replace this analog filter with the equivalent digital filter, analog-to-digital conversion (ADC) must be performed to produce a discrete-time input sequence $\{x(n)\}$ from $x(t)$. The input sequence $\{x(n)\}$ is then processed with a digital filter, implemented with special-purpose hardware or as a computer program, to produce the output sequence $\{y(n)\}$. Finally, a digital-to-analog conversion (DAC) must be performed to reconstruct the continuous-time signal $y(t)$. For simple analog filters, the added expense involved in the ADC, DAC and the hardware to implement the digital filter does not usually make this substitution economically feasible. But with the proliferation of microcomputers and the reduced cost of analog-to-digital converters, performing a filter operation with digital circuits is becoming common.

In some applications, the analog filters themselves may be costly. In geophysical and biomedical applications, signals with very low-frequency components, typically less that one Hertz, require very bulky and expensive inductors and capacitors. In these cases, digital filters are commonly employed as replacements. In some applications, such as those involving very high

ANALOG FILTER

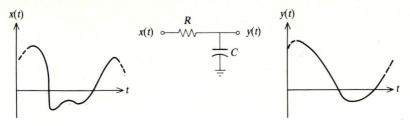

DIGITAL FILTER

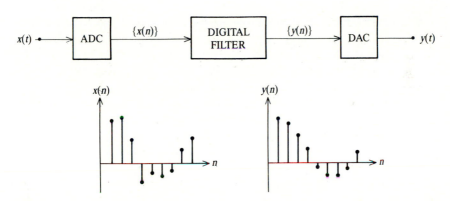

FIGURE 1-12
Replacing an analog filter with a digital filter requires the additional operations of an analog-to-digital conversion (ADC) and a digital-to-analog conversion (DAC).

frequencies, analog filters still have the advantage. But with decreasing cost and increasing speed of digital circuits, more and more analog filters are being replaced with their digital counterparts.

1.6 OVERVIEW OF THE COURSE

The digital filter design and signal processing concepts described in this book can be divided into five categories: analysis, synthesis, the fast Fourier transform, finite-accuracy effects, and inverse filters.

ANALYSIS. We analyze digital signal processing algorithms by determining their time and frequency domain characteristics. This allows us to attain an intuitive understanding of their capabilities and limitations. The time-domain performance of digital filters is described in terms of the filter's *unit-sample* *response sequence,* denoted by $\{h(n)\}$. This sequence is analogous to the impulse response of analog filters. A convolutional equation allows us to determine the output sequence $\{y(n)\}$ from the input sequence $\{x(n)\}$ and the unit-sample response $\{h(n)\}$. The unit-sample response sequence also permits us to determine whether a filter is stable. *Linear difference equations* provide an alternate time-domain description that is be useful for implementing digital filter structures.

Filter specifications are commonly expressed in the frequency domain. Hence, we examine the relevant aspects of the Fourier domain that define the frequency domain properties of signals and filters. The Fourier transform of the unit-sample response $\{h(n)\}$ is the *transfer function* $H(e^{j\omega})$ of the filter and it describes the gain of the filter at different frequencies. The Fourier transform of a data sequence $\{x(n)\}$ is called the *spectrum* $X(e^{j\omega})$ and it defines the frequency content of the signal. We explore the problems encountered when generating a discrete-time sequence by sampling a continuous-time signal and when reconstructing the continuous-time signal from the discrete-time sequence. The discrete Fourier transform (DFT) allows us to evaluate the Fourier transform by computer. The consequences of employing the DFT for signal processing and spectral analysis applications are investigated.

An alternate and more general analytic technique than the Fourier transform is the *z-transform.* The *system function* $H(z)$ is defined as the z-transform of the unit-sample response $\{h(n)\}$ and is used for the analysis and synthesis of digital filters. The complex *z plane* is the domain of interest for the z-transform. The representation of a digital filter as a collection of poles and zeros in the z plane will provide a useful interpretation of the filter frequency response.

SYNTHESIS. After having a firm and intuitive understanding of the time and frequency behavior of digital filters, the goal is then to design and implement a digital filter to perform a desired task. The synthesis procedure involves combining the digital filter elements and specifying the values of the multiplier coefficients to satisfy a desired specification.

Most often, the transformation that is to be performed by the digital filter is specified in terms of the *magnitude response* $|H(e^{j\omega})|$. The magnitude response for an ideal lowpass filter, for which the frequency components in the input sequence from zero to some cutoff frequency ω_c are passed through the filter unaffected, while all the others are completely eliminated from the output sequence, is shown in Fig. 1.13(a). This ideal cannot be achieved in practice, but must be approximated. A practical specification that can be satisfied by a digital filter is shown in Fig. 1.13(b). This desired specification allows slight deviations from the ideal response: ε in the passband, δ in the stopband, and a nonzero frequency interval for the transition between the passband and stopband. A successful digital filter design produces a magnitude response that falls within the allowed regions.

We will consider two different digital filter types to satisfy a given

(a) $|H_{ideal}(e^{j\omega})|$

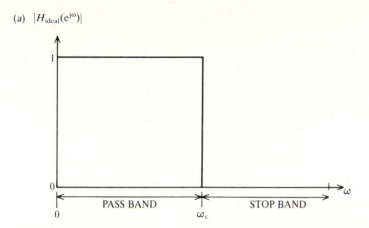

(b) $|H_{desired}(e^{j\omega})|$

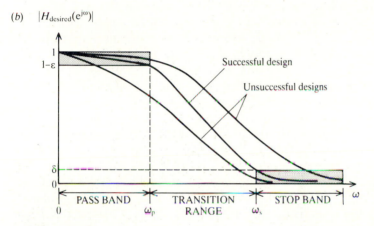

FIGURE 1-13
Magnitude response curves for a digital low pass filter. (a) Ideal low pass filter response. (b) Practical low pass filter specification allows derivations from the ideal response. A successful digital filter has a magnitude response that falls within the allowed regions.

specification: *infinite impulse response (IIR)* filters and *finite impulse response (FIR)* filters. IIR digital filters are usually designed by extending classical analog filter design procedures. An IIR filter design is typically accomplished in three steps. First, an analog lowpass filter is designed to meet the desired passband specification. The most commonly employed are the Butterworth, Chebyshev and elliptic analog lowpass filter design procedures. Second, an analog-to-digital filter transformation is employed to obtain a digital lowpass filter. The *impulse-invariance* method and the *bilinear z-transform* method are described for transferring these analog designs into their digital counterparts. In the final step, a *frequency transformation* is employed to obtain a highpass, bandpass or bandreject filter.

The design procedures for FIR filters have been developed especially for digital filters and do not have analog filter counterparts. Two FIR filter structures will be investigated. The first involves implementing the desired filter unit-sample response in a tapped-delay line structure. Since the desired response often has an infinite number of nonzero elements, a finite-duration window must be applied to eliminate all but a finite number of these elements. We investigate the effect of this truncation on the resulting filter's transfer function. The second structure is the frequency-sampling structure, which employs a comb-filter followed by a bank of resonators and implements the filter by specifying the magnitude response at a finite number of frequency points. A comparison of the two design procedures is performed to determine under which conditions one is more efficient than the other.

FAST FOURIER TRANSFORM (FFT). In addition to digital filters, we also explore the use of the fast Fourier transform (FFT) algorithm for processing data and performing spectral analysis. The FFT is an elegant and efficient method for computing the discrete Fourier transform that allows the filtering operation to be performed in the frequency domain. We present both an intuitive description of the FFT, as well as a more rigorous mathematical one, to indicate how this efficiency is achieved. The utility of employing the FFT for filtering applications is described and compared with the methods of direct digital filter implementation.

FINITE-ACCURACY EFFECTS. An important concern for implementing practical digital filters involves finite-accuracy effects caused by the limited precision to which the sequence and coefficients values are represented within the computer. Similar errors are also introduced in the numerical round-off routinely performed in tedious hand calculations. In the analysis and synthesis discussions above, we dealt with the ideal, or infinite-precision, case. To process data on a computer, the data values must first be quantized to a finite number of bits, usually equal to or smaller than the word size on the computer. The effects of this *quantization error* will be investigated. The coefficients used in the digital filter multipliers must also be quantized. The slight change in these coefficient values modifies the frequency response of the filter. A third source of error occurs when several numbers are added or multiplied and the result exceeds the limits of the computer. In this case, the result must be scaled and truncated, introducing still another source of error that can produce unwanted oscillations at the output of some IIR filters. We consider analytic methods of characterizing these three common quantization problems.

INVERSE FILTERS. In the final chapter, we discuss the interesting and important topic in signal processing called *inverse filtering*. This topic integrates many of the digital signal processing concepts presented in the previous chapters to solve the problem of removing distortions from observed signals that are introduced by measuring instruments or by physical transmission media. It is hoped that this topic whets the student's appetite for pursuing further study in this most interesting area of research and development.

1.7 DESCRIPTION OF THE COMPUTER PROJECTS

Along with the standard problem sets at the end of each chapter, a set of projects to be performed on a computer is also included. The pedagogical value of these projects cannot be overstressed, since they are indicative of the problems encountered in practice. The computer is a powerful tool for signal processing and the projects given in this book illustrate its flexibility. Most of the projects in Chapters 2 through 6 can be analytically determined with little effort. The analytic results are meant to verify the computer results and to instill some faith in the algorithms. As the analysis becomes more difficult, the computer results can help guide the analytic procedure. The design projects in Chapters 7 through 11 are almost impossible to do analytically, as are most practical filter designs. By that stage in the course, the intellectual insights are achieved almost exclusively from the computer results. Using the computer in this fashion is the basis for *computer simulations,* which are often performed to explore the performance of complicated systems before they are implemented.

The subroutines for the projects are meant to be written by the student, in FORTRAN, the current universal language for signal processing, or in the increasingly popular C or PASCAL (BASIC is slightly less desirable, because of its slow speed, but acceptable). To minimize the amount of programming, a structured approach is used in which the routines written for the initial projects are used again in subsequent projects. Each routine performs a simple, and hence easily verifiable, operation. Constructing more sophisticated programs from these simpler building blocks aids the student in conceptualizing the digital filter operation and for trouble-shooting the more complicated projects later in the course. By the end of the course, the student should have an intuitive understanding of digital signal processing that is invaluable for applying it to novel situations. As an added benefit, students will also have a flexible library of signal processing programs that they have written themselves, and hence understand intuitively.

Graphic outputs produced by the projects are displayed on the terminal and on a printer, being generated by programs provided in Appendix A. These plots, although having crude resolution, are adequate for illustrating the ideas. They also have the advantage of not requiring special plotting hardware. The plotting subprograms can also easily be modified to display the results on other types of digital plotters and display terminals.

The experience gained through this course should whet the appetite for the study of advanced topics in filter design and signal processing. References to material describing advanced topics are provided at the end of each chapter. The student will also be well prepared to apply digital filtering techniques to practical problems.

After this course, the student can easily become facile with the various programs that are available for automatically designing optimal digital filters. Examples are the digital filter program packages offered by the IEEE, or in SIG, the software package from the Lawrence Livermore National Laboratory at the University of California. An important prerequisite for employing these programs successfully is the intuitive understanding of the tradeoffs and subtlety involved in the methods for digital signal processing. This is the material stressed in this book.

1.8 SUMMARY

In this chapter, we have provided the motivation for digital filtering, defined the components of a digital filter, and indicated some common applications. An overview of the course material has described the inter-relationship of the topics that describe the theory and application of digital filters to signal processing.

FURTHER READING

IEEE Acoustics, Speech and Signal Processing Magazine. This bimonthly magazine of the Acoustics, Speech and Signal Processing Society of the Institute of Electronics and Electrical Engineers (IEEE) provides tutorial articles on various topics covering digital signal processing and digital filters.

IEEE Transactions on Acoustics, Speech and Signal Processing. This bimonthly journal publishes articles describing the most recent results in the development and implementation of digital filters and digital signal processing techniques.

Rabiner, L. R. and C. M. Rader (eds.): *Digital Signal Processing,* IEEE Press, 1972; and *Digital Signal Processing II,* IEEE Press, 1975. [These books contain reprints from various journals that provide an organized and historical treatment of digital filter theory and applications.]

REFERENCES TO TOPICS

Pascal

Cooper, Doug, and Michael Clancey: *Oh! Pascal!,* W. W. Norton, 1982.

Jensen, Kathleen, and Niklaus Wirth: *Pascal User Manual and Report,* Springer-Verlag, Berlin, 1985.

Mallozzi, John, and Nicholas De Lillo: *Computability with Pascal,* Prentice-Hall, Englewood Cliffs, NJ, 1984.

C language

Harbison, Samuel, and Guy Steele Jr.: *C, A Reference Manual,* Prentice-Hall, Englewood Cliffs, NJ, 1984.

FORTRAN

Didday, Richard, and Rex Page: *FORTRAN for Humans,* West, 1981.

McCracken, Daniel: *FORTRAN,* Wiley, New York, 1967.

BASIC

Maratek, Samual: *BASIC,* Academic Press, New York, 1975.

Computer communications

Schwartz, Mischa: *Telecommunication Networks,* Addison-Wesley, Reading, Mass. 1987.
Tanenbaum, Andrew: *Computer Networks,* Prentice-Hall, Englewood Cliffs, NJ, 1981.
Lathi, B. P.: *Modern Digital and Analog Communication Systems,* Holt, New York, 1983.

Digital signal processing program packages

Programs for Digital Signal Processing, IEEE Press, New York, 1979.
Lager, Darrel L., and Stephen G. Azevedo: *SIG – A General-Purpose Signal Processing Program.* Proceedings of the IEEE, Vol. 75, No. 9, pp. 1322–1332, 1987.

Robot simulation

IEEE Transactions on Pattern Analysis and Machine Intelligence.

Speech processing

IEEE Transactions on Acoustics, Speech and Signal Processing.

Medical signal processing

IEEE Transactions on Biomedical Engineering.

CHAPTER
2

DISCRETE-TIME DESCRIPTION OF SIGNALS AND SYSTEMS

2.1 INTRODUCTION

In this chapter we investigate procedures to describe the time-domain behavior of discrete-time systems, of which digital filters are a special class. After establishing the discrete-time signal notation, we define several sequences that are important for the analysis of digital filters. The superposition principle for linear systems is then employed to derive useful relationships governing digital filters. The unit-sample response is presented as a time-domain descriptor of a discrete-time system. The linear convolution operation relates the output sequence of the filter to the input sequence and the unit-sample response. Linear constant-coefficient difference equations are employed for the analysis and synthesis of digital filters. A procedure to write a digital filter program directly from these time-domain equations is described.

2.2 DISCRETE-TIME SEQUENCES

In this book, we will be dealing with discrete-time sequences, which will be denoted either by $\{x(n)\}$, or by $x(n)$; for $-N_1 \leq n \leq N_2$, where N_1 and N_2 may possibly be infinite. As mentioned in Chapter 1, the term *discrete-time* can also refer to sequences of other physical parameters, such as discrete-location for image processing applications, discrete-distance for sonar ranging systems, or discrete-position for robotic arm sensors. The particular value of the sequence at time k will be denoted simply by $x(k)$, with no additional qualifiers on the index. In this book, we are primarily concerned with sequences that are produced by sampling a function of a continuous-valued variable at equally spaced intervals. These, in fact, are the most common types of sequences in digital filter applications. We now consider several important sequences that are used throughout this book for analyzing digital filters.

UNIT-SAMPLE SEQUENCE. The unit-sample sequence contains only one nonzero valued element and is defined by

$$d(n) = \begin{cases} 1 & \text{for } n = 0 \\ 0 & \text{otherwise} \end{cases} \tag{2.1}$$

This sequence is shown in Fig. 2.1 and will play an important role as the input sequence to a digital filter. Since the unit-sample sequence contains only one nonzero element, the form of the output sequence then directly indicates the effect produced by the digital filter.

The *delayed* unit-sample sequence, denoted by $\{d(n-k)\}$, has its nonzero element at sample time k:

$$d(n-k) = \begin{cases} 1 & \text{for } n = k \\ 0 & \text{otherwise} \end{cases} \tag{2.2}$$

The unit-sample sequence is also an important analytic tool because of its *sifting property*, by which a particular element can be extracted from the entire sequence. To demonstrate this property, we note that the value of $\{x(n)\}$ at time k can be written as

$$x(k) = \sum_{n=-\infty}^{\infty} d(n-k)x(n) \tag{2.3}$$

as shown in Fig. 2.2. This follows because $d(n-k)$ is nonzero only when its argument is zero, or when $n = k$. When multiplying the two sequences together term by term, only the single element $x(k)$ remains nonzero and the sum evaluates to $x(k)$. Hence, the unit-sample sequence can be used to select, or sift, a single element from the entire sequence. This sifting property will be important below in the derivation of the convolution relationship between the digital filter input and output sequences. A more convenient, but equivalent, form of Eq. (2.3) that is used below employs the time-reversed version of the

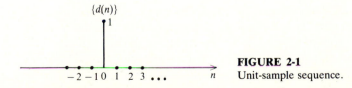

FIGURE 2-1
Unit-sample sequence.

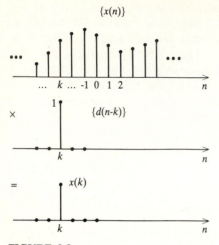

FIGURE 2-2
Sifting property of the unit-sample sequence.

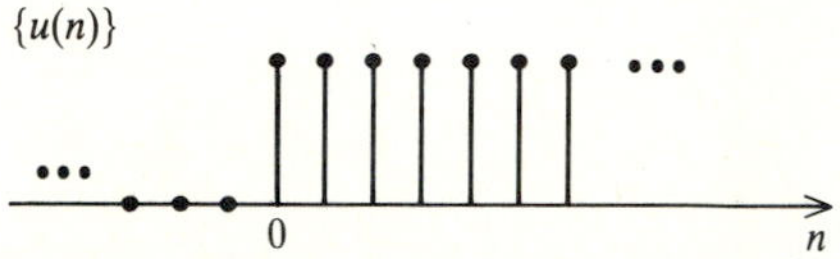

FIGURE 2-3
Unit-step sequence.

unit-sample sequence having its nonzero element at $n = k$:

$$x(k) = \sum_{n=-\infty}^{\infty} d(k - n) x(n) \tag{2.4}$$

UNIT-STEP SEQUENCE. The unit-step sequence is defined as

$$u(n) = \begin{cases} 1 & \text{for } n \geq 0 \\ 0 & \text{otherwise} \end{cases} \tag{2.5}$$

and is shown in Fig. 2.3. The unit-step sequence is used for defining the starting point of a sequence in analytic expressions. For example, the sequence

$$x(n) = \begin{cases} a^n, & \text{for } n \geq 0 \\ 0, & \text{otherwise} \end{cases} \tag{2.6}$$

can be written simply as

$$x(n) = a^n u(n) \qquad \text{for all } n \tag{2.7}$$

SINUSOIDAL SEQUENCES. Sinusoidal sequences play an important role in the frequency-domain analysis of digital filters. Given some radian frequency ω, we can form the discrete-time sinusoidal sequences $\{\cos(\omega n)\}$ and $\{\sin(\omega n)\}$ by evaluating these functions at the corresponding value of the argument. Examples are shown in Fig. 2.4. Strictly speaking, the units of ω are *radians per sampling interval*. By convention, the sampling interval is taken as a dimensionless quantity, allowing ω to be specified in *radians*. In the next chapter, we will consider the sampling of continuous-time signals and relate ω to continuous-time frequency values.

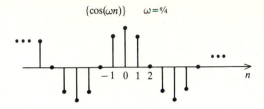

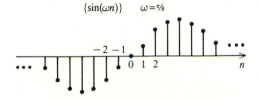

FIGURE 2-4
Two discrete-time sinusoidal sequences.

Discrete-time sinusoids exhibit an interesting property that is important for evaluating the Fourier transform of a discrete-time sequence: The set of all distinct values that a discrete-time sinusoidal sequence can have occurs for values of ω lying in the range $-\pi \le \omega < \pi$.

To illustrate this property, let us consider two frequency values, ω_0 is in the interval $[-\pi, \pi)$, and ω_1 outside this range. For example, let

$$\omega_1 = \omega_0 + 2m\pi \qquad (2.8)$$

where m is a nonzero integer. The values of the cosine sequence at frequency ω_1 are given by

$$\{\cos(\omega_1 n)\} = \{\cos((\omega_0 + 2m\pi)n)\} = \{\cos(\omega_0 n)\} \qquad (2.9)$$

This last result follows because the cosine function is periodic with period 2π and the argument is $\omega_0 n$ plus an integral number ($= mn$) of 2π. In Fig. 2.5, this effect is shown for $\omega_0 = 0.1\pi$ and $\omega_1 = 2.1\pi$. This result is also true for the sine sequence.

COMPLEX EXPONENTIAL SEQUENCE. Before we consider the complex exponential sequence, we will establish the notation for complex numbers and the complex plane used throughout this book. Let c be a complex number. Then it can be written as

$$c = a + jb \qquad (2.10)$$

where a is the *real part*, b is the *imaginary part*, and $j = (-1)^{1/2}$. The representation of c on the complex plane is shown in Fig. 2.6. An alternate form of a complex number is the *complex vector* notation:

$$c = |c|\, e^{j\,\text{Arg}[c]} \qquad (2.11)$$

where $|c|$ is the *magnitude* and $\text{Arg}[c]$ is the *argument* or *phase*, which is

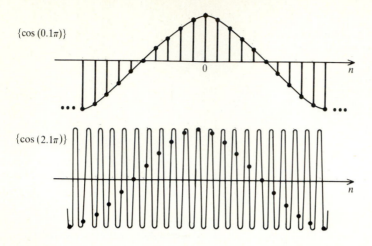

$\{\cos(0.1\pi)\}$

$\{\cos(2.1\pi)\}$

FIGURE 2-5
Two analog signals having different frequencies can produce the same values for a discrete-time sinusoidal sequence.

defined by the angle with respect to the real axis in the complex plane. The following geometric relationships can be derived from Fig. 2.6:

$$|c|^2 = a^2 + b^2 \tag{2.12}$$

and

$$\text{Arg}[c] = \arctan[b/a]. \tag{2.13}$$

The complex exponential sequence is important for a compact frequency analysis of digital filters, and is given by

$$\{e^{(\sigma+j\omega)n}\} = \{e^{\sigma n}\, e^{j\omega n}\} \tag{2.14}$$

When both σ and ω are real numbers, each element of the sequence is expressed in complex vector notation with *magnitude* $e^{\sigma n}$ and *phase* ωn.

A special case of this sequence occurs when $\sigma = 0$. The exponential sequence can then be related to the sinusoidal sequences by the geometric

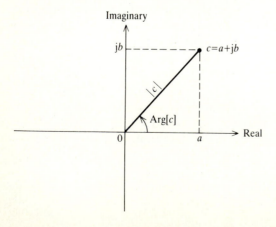

FIGURE 2-6
The complex plane.

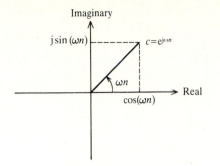

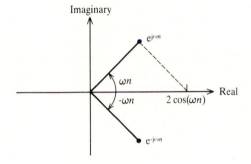

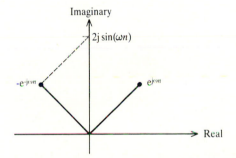

FIGURE 2-7
Interpretation of the Euler identities in the complex plane.

relationships shown in Fig. 2.7, which are known as the *Euler identities*. These are given by

$$e^{j\omega n} = \cos(\omega n) + j\sin(\omega n) \tag{2.15}$$

$$\cos(\omega n) = \frac{e^{j\omega n} + e^{-j\omega n}}{2} \tag{2.16}$$

$$\sin(\omega n) = \frac{e^{j\omega n} - e^{-j\omega n}}{2j} \tag{2.17}$$

The property of sinusoidal sequences, that all values can be represented with ω in the range $[-\pi, \pi)$, is also true for complex exponentials. To illustrate this, consider $\omega_1 = \omega_0 + 2m\pi$, where m is any integer. Then

$$e^{j\omega_1} = e^{j(\omega_0 + 2m\pi)} = e^{j\omega_0} e^{j2m\pi} = e^{j\omega_0} \tag{2.18}$$

since $e^{j2m\pi} = 1$, when m is any integer. This identity can also be demonstrated in the complex plane. Increasing the phase of a complex number by α is equivalent to rotating the complex vector by angle α, counter-clockwise if $\alpha > 0$ and clockwise if $\alpha < 0$. If $\alpha = 2m\pi$, where m is any integer, then the complex vector makes m complete revolutions about the origin and returns to its original orientation.

2.3 SUPERPOSITION PRINCIPLE FOR LINEAR SYSTEMS

The digital filters considered in this book are a special class in the general category of linear systems. Any discrete-time system can be thought of as transforming an input sequence $\{x(n)\}$ into an output sequence $\{y(n)\}$ through a transformation denoted by $G\{\cdot\}$. However, two characteristics define a linear system: (1) for a given input sequence, a change in the amplitude scale of the input sequence results in the same amplitude scale change in the output sequence, and (2) if two input sequences are added and the sum is applied to the system, the resulting output is the same as the sum of the responses to the individual inputs. These ideas can be formalized in the statement of the superposition principle.

SUPERPOSITION PRINCIPLE. Given a system producing a transformation $G\{\cdot\}$, two arbitrary input sequences $\{x_1(n)\}$ and $\{x_2(n)\}$ that result in $\{y_1(n)\} = G\{x_1(n)\}$ and $\{y_2(n)\} = G\{x_2(n)\}$, and two constants α and β, then the system is defined to be *linear* when

$$G\{\alpha x_1(n) + \beta x_2(n)\} = \alpha G\{x_1(n)\} + \beta G\{x_2(n)\}$$
$$= \alpha\{y_1(n)\} + \beta\{y_2(n)\} \qquad (2.19)$$

The following example illustrates the application of the superposition principle.

> **Example 2.1.   Linearity of the 3-sample averager.** The transformation produced by the 3-sample averager is given by
>
> $$y(n) = \tfrac{1}{3}[x(n+1) + x(n) + x(n-1)] = G\{x(n)\}$$
>
> To test the linearity, we apply the scaled sum of two arbitrary sequences, or $\{\alpha x_1(n) + \beta x_2(n)\}$, to the averager. The output is then equal to
>
> $$G\{\alpha x_1(n) + \beta x_2(n)\} = \tfrac{1}{3}[\alpha x_1(n+1) + \beta x_2(n+1) + \alpha x_1(n) + \beta x_2(n)$$
> $$+ \alpha x_1(n-1) + \beta x_2(n-1)]$$
>
> $$= \frac{\alpha}{3}[x_1(n+1) + x_1(n) + x_1(n-1)]$$
>
> $$+ \frac{\beta}{3}[x_2(n+1) + x_2(n) + x_2(n-1)]$$
>
> $$= \alpha\{y_1(n)\} + \beta\{y_2(n)\}$$

This last equation is simply the sum of the two outputs that would have been observed if only $\{\alpha x_1(n)\}$ or $\{\beta x_2(n)\}$ had been applied. Hence, the 3-sample averager is a linear system.

Example 2.2.   A nonlinear system. Let us consider a square-law device, in which the output is the square of the input:

$$\{y(n)\} = G\{x(n)\} = \{x^2(n)\}$$

By applying two scaled inputs into this system, we have at the output:

$$G\{\alpha x_1(n) + \beta x_2(n)\} = [\alpha x_1(n) + \beta x_2(n)]^2$$
$$= \alpha^2 x_1^2(n) + \beta^2 x_2^2(n) + 2\alpha\beta x_1(n)x_2(n)$$

which is not equal to $\alpha x_1^2(n) + \beta x_2^2(n)$, the output required for the system to be linear. Hence, the square-law device is a nonlinear system.

2.4   UNIT-SAMPLE RESPONSE SEQUENCE

One important consequence of the superposition principle is that it allows us to derive a convenient relationship between the input and output of a linear system in the time domain. Applying the sifting property of the unit-sample sequence given by Eq. (2.4), we can write

$$\{y(n)\} = G\{x(n)\} = G\left\{ \sum_{k=-\infty}^{\infty} d(n-k)\,x(k) \right\}$$
$$= G\{\cdots + x(-1)d(n+1) + x(0)d(n) + x(1)d(n-1) + \cdots\} \quad (2.20)$$

Since the system operates through its transformation $G\{\cdot\}$ on *sequences*, we can view the output of a system as the superposition of responses to delayed and scaled unit-sample sequences, as shown in Fig. 2.8. The unit-sample sequence having its nonzero element at $n = k$, $d(n-k)$, is scaled by $x(k)$. Applying the superposition principle, we can write the output as

$$\{y(n)\} = \cdots + x(-1)\,G\{d(n+1)\} + x(0)\,G\{d(n)\} + x(1)\,G\{d(n-1)\} + \cdots$$
$$= \sum_{k=-\infty}^{\infty} x(k)\,G\{d(n-k)\} \quad (2.21)$$

$G\{d(n-k)\}$ is the output sequence produced by the system when the input is a unit-sample sequence having its nonzero element at $n = k$. This output is called the *unit-sample response*, and is sufficiently important to merit its own symbol:

$$G\{d(n-k)\} = \{h(n, k)\}. \quad (2.22)$$

In words, $\{h(n, k)\}$ is the response of a linear system whose input is a unit-sample sequence with nonzero element at position $n = k$. The following examples illustrate the concept of the unit-sample response.

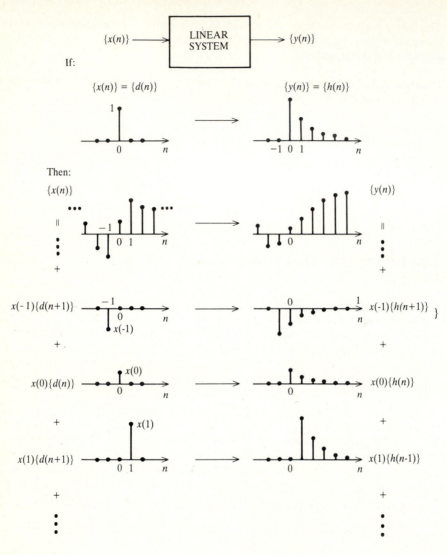

FIGURE 2-8
The input sequence to a linear system can be viewed as the superposition in time of a set of scaled unit-sample sequences. The output sequence is then the superposition of the scaled unit-sample responses.

Example 2.3.   Unit-sample response of the 3-sample averager. Let

$$y(n) = \tfrac{1}{3}x(n+1) + \tfrac{1}{3}x(n) + \tfrac{1}{3}x(n-1)$$

To find the unit-sample response $\{h(n, 0)\}$, we set $\{x(n)\} = \{d(n)\}$ and solve for the values of $y(n)$ for each n:

For $n = -2$, $y(-2) = \tfrac{1}{3}d(-1) + \tfrac{1}{3}d(-2) + \tfrac{1}{3}d(-3) = 0.$

For $n \leq -2$, $y(n) = 0$, since $d(n) = 0$ for $n < 0$.

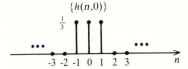

FIGURE 2-9
Unit-sample response of nonrecursive three-sample averager.

For $n = -1$, $y(-1) = \frac{1}{3}d(0) + \frac{1}{3}d(-1) + \frac{1}{3}d(-2) = \frac{1}{3}$.

For $n = 0$, $y(0) = \frac{1}{3}d(1) + \frac{1}{3}d(0) + \frac{1}{3}d(-1) = \frac{1}{3}$.

For $n = 1$, $y(1) = \frac{1}{3}d(2) + \frac{1}{3}d(1) + \frac{1}{3}d(0) = \frac{1}{3}$.

For $n = 2$, $y(2) = \frac{1}{3}d(3) + \frac{1}{3}d(2) + \frac{1}{3}d(1) = 0$.

For $n > 2$, $y(n) = 0$, since the nonzero value of $\{d(n)\}$ has moved out of the *memory* of this filter. Since $\{d(n)\}$ is the input, the output is defined to be $\{h(n, 0)\}$. Hence

$$h(n, 0) = \begin{cases} \frac{1}{3}, & \text{for } -1 \leq n \leq 1 \\ 0, & \text{otherwise} \end{cases}$$

This sequence is shown in Fig. 2.9.

The unit-sample response is computed by starting at a point in time, n, for which all the previous inputs and the contents of the delays and advances are all equal to zero. This state of the filter is known as the *zero initial condition*.

Example 2.4. Unit-sample response of a first-order recursive filter. Let

$$y(n) = \begin{cases} ay(n-1) + x(n) & \text{for } n \geq 0 \\ 0 & \text{otherwise.} \end{cases}$$

To find $\{h(n, 0)\}$, we let $\{x(n)\} = \{d(n)\}$ and apply the zero initial condition.

For $n < 0$, $y(n) = 0$, because $d(n)$ is zero for $n < 0$ and $y(-1) = 0$.

For $n = 0$, $y(0) = ay(-1) + d(0) = 1$.

For $n = 1$, $y(1) = ay(0) + d(1) = a$.

For $n = 2$, $y(2) = ay(1) + d(2) = a^2$.

$\cdots$

For $n = k$, $y(k) = ay(k-1) + d(k) = a^k$.

Hence,

$$h(n, 0) = a^n u(n) \qquad \text{for all } n$$

Some typical sequences are shown for different values of a in Fig. 2.10.

This concept of a unit-sample response can also be applied to time-varying systems, in which the multiplier coefficients vary with n, as illustrated in the next example.

$$h(n,0) = a^n u(n) \quad \text{for all } n$$

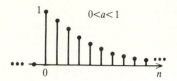

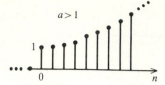

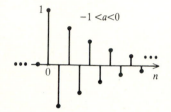

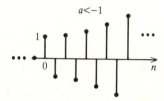

FIGURE 2-10
Four possible unit-sample response sequences that can be produced by first-order recursive filter.

Example 2.5. Unit-sample response of a time-varying system. Let

$$y(n) = \begin{cases} \dfrac{1}{n+1} y(n-1) + x(n) & \text{for } n \geq 0 \\ 0 & \text{otherwise.} \end{cases}$$

Applying the zero initial condition and using the same approach as in the previous example, we have

$$h(0, 0) = (1)(0) + 1 = 1$$
$$h(1, 0) = (\tfrac{1}{2})(1) + 0 = \tfrac{1}{2}$$
$$h(2, 0) = (\tfrac{1}{3})(\tfrac{1}{2}) = \tfrac{1}{6}$$
$$h(3, 0) = (\tfrac{1}{4})(\tfrac{1}{6}) = \tfrac{1}{24}$$
$$\cdots$$

If the unit-sample is applied at time $n = 1$, i.e. $\{x(n)\} = \{d(n-1)\}$, then $\{y(n)\} = \{h(n, 1)\}$, where

$$h(0, 1) = (1)(0) + 0 = 0$$
$$h(1, 1) = (\tfrac{1}{2})(0) + 1 = 1$$
$$h(2, 1) = (\tfrac{1}{3})(1) + 0 = \tfrac{1}{3}$$
$$h(3, 1) = (\tfrac{1}{4})(\tfrac{1}{3}) = \tfrac{1}{12}$$
$$\cdots$$

Note that $\{h(n, 1)\}$ is a different-valued sequence from $\{h(n, 0)\}$. This example illustrates that, for a *time-varying filter* (whose coefficient values vary with time n), the values of the unit-sample response depend upon the time of application of the unit-sample sequence.

The general time-domain relationship between the system output value at time n and the input values can be obtained by combing Eqs. (2.21) and (2.22), to give

$$y(n) = \sum_{k=-\infty}^{\infty} x(k)\, h(n, k) \qquad (2.23)$$

This form is called a *convolutional relationship*. This equation can be simplified when it is applied to time-invariant systems, which are discussed in the next section. The references describe adaptive filters whose coefficient values change to meet various conditions.

2.5 TIME-INVARIANT SYSTEMS

We will further restrict our attention in this book to the important class of *linear time-invariant* (LTI) systems, or those whose transformations do not change with time. In simplest terms, for an LTI system, a shift of n_0 samples in the input produces *only* a corresponding shift in the output. That is, if $\{y(n)\} = G\{x(n)\}$, then

$$\{y(n - n_0)\} = G\{x(n - n_0)\} \qquad \text{for all } n_0 \qquad (2.24)$$

One characteristic of an LTI system is that the form of the unit-sample response does not change with the time of application of the unit-sample sequence. Then, rather than being a function of two variables, n and k, the unit-sample response becomes a function of only $n - k$, or the time as measured from the application of the unit-sample. Analytically, we then write

$$G\{d(n - k)\} = \{h(n, k)\} = \{h(n - k)\} \qquad (2.25)$$

This equation implies that a shift in the time that the unit-sample sequence is applied to the system results *only* in a corresponding time shift in its unit-sample response, as shown in Fig. 2.11. A time-invariant digital filter can easily be recognized, since its coefficient values do not vary with time.

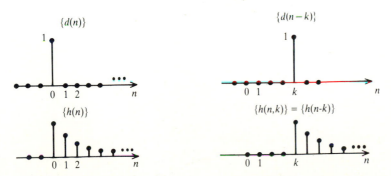

FIGURE 2-11
Shift-invariance property of the unit-sample response for a time-invariant system.

Applying Eq. (2.25), the convolution relationship of a LTI system can be written as

$$y(n) = \sum_{k=-\infty}^{\infty} x(k) h(n-k) \tag{2.26}$$

or, more concisely,

$$\{y(n)\} = \{x(n)\} * \{h(n)\} \tag{2.27}$$

where $*$ will denote the convolution operation. In evaluating a convolution expression, it is often helpful to note that convolution is a *commutative* operation, i.e.

$$\{y(n)\} = \{x(n)\} * \{h(n)\} = \{h(n)\} * \{x(n)\} \tag{2.28}$$

To demonstrate this, we let $m = n - k$ in Eq. (2.26) to get

$$y(n) = \sum_{m=\infty}^{-\infty} x(n-m) h(m) = \sum_{m=-\infty}^{\infty} h(m) x(n-m) \tag{2.29}$$

The desired result is obtained when m is replaced by k. The limits of the first sum in Eq. (2.29) appear in reverse order, but can be reordered without consequence because the same terms appear in the evaluation of the sum.

The convolution operation is also *distributive*, i.e.

$$\{h(n)\} * \{x_1(n) + x_2(n)\} = \{h(n)\} * \{x_1(n)\} + \{h(n)\} * \{x_2(n)\} \tag{2.30}$$

This result can be shown directly by applying the superposition principle.

COMPUTING A DISCRETE-TIME CONVOLUTION. Since the convolution operation is one of the most common and important in digital signal processing, we present a step-by-step procedure for the evaluation of (2.26).

Step 1. Choose an initial value of n, the starting time for evaluating the output sequence. If $\{x(n)\}$ starts at $n = n_x$ and $\{h(n)\}$ starts at $n = n_h$, then $n = n_x + n_h$ is a good choice. The values of $x(n)$ for $n < n_x$ and of $h(n)$ for $n < n_h$ are then assumed to be zero.

Step 2. Both sequences are expressed in terms of the index k. Note that $x(k)$ is identical to $x(n)$, while $h(n-k)$ is a time-reversed version of $h(n)$. If $h(n)$ starts at $n = n_h$, then the last point of $h(n-k)$ occurs at $k = n - n_h$.

Step 3. The two sequences are multiplied element by element and the products are accumulated over all values of k. This sum of products then becomes the value of the output for index n, or $y(n)$.

Step 4. The index n is incremented and steps 2, 3 and 4 are repeated until the sum of products computed in Step 3 is zero for all the remaining values of n.

These four steps are illustrated in the following examples.

Example 2.6.   Convolution of two finite-duration sequences. Let

$$x(n) = \begin{cases} 1 & \text{for } -1 \le n \le 1 \\ 0 & \text{otherwise} \end{cases}$$

and

$$h(n) = \begin{cases} 1 & \text{for } -1 \le n \le 1 \\ 0 & \text{otherwise} \end{cases}$$

The evaluation of $\{y(n)\} = \{h(n)\} * \{x(n)\}$ by Eq. (2.26) using the four steps given above is shown in Fig. 2.12.

Step 1. The evaluation will start at $n = -2$.

Step 2. Both sequences are shown indexed by k, with $\{h(n - k)\}$ being the time-reversed version of $\{h(n)\}$. Since $\{h(n)\}$ starts at $n = -1$, the last point of $\{h(n - k)\}$ occurs at $k = n + 1$. When $n = -2$, $k = -1$.

Step 3. Computing the product of the pair of sequence values for each k, only the $k = -1$ terms yield a nonzero product. This result then becomes the value of $y(-2)$.

Step 4. The value of n is incremented and the process above is repeated until $n = 3$. The relative positions of the two sequences are shown for the values of n that produce nonzero values for $y(n)$.

Example 2.7.   Convolution of a finite-duration sequence with an infinite-duration sequence. Let

$$x(n) = \begin{cases} n + 1, & \text{for } 0 \le n \le 2 \\ 0, & \text{otherwise} \end{cases}$$

and

$$h(n) = a^n u(n) \qquad \text{for all } n$$

We will perform the convolution in two ways. First, using Eq. (2.29), we have

$$y(n) = \sum_{k=-\infty}^{\infty} h(k) x(n - k) = \sum_{k=-\infty}^{\infty} a^k u(k) x(n - k)$$

From Fig. 2.13, note that

$$x(n - k) = \begin{cases} 1 & \text{for } n - k = 0, & \text{or } k = n \\ 2 & \text{for } n - k = 1, & \text{or } k = n - 1 \\ 3 & \text{for } n - k = 2, & \text{or } k = n - 2 \\ 0 & \text{otherwise} \end{cases}$$

Hence, the only nonzero terms in the sum are

$$y(n) = a^n u(n) + 2a^{n-1} u(n - 1) + 3a^{n-2} u(n - 2)$$

The second way of evaluating the convolution is by using Eq. (2.26). The evaluation, of course, produces the same result, but is simpler because of the

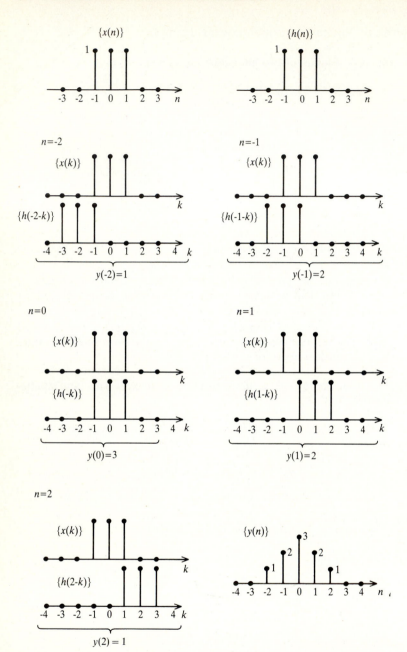

FIGURE 2-12
The convolution of two finite-duration sequences.

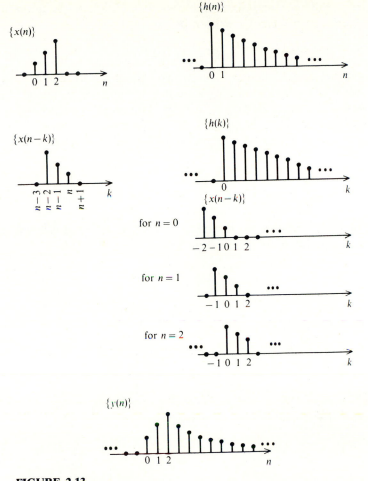

FIGURE 2-13
The convolution of a finite-duration sequence with an infinite-duration sequence.

finite duration of $\{x(n)\}$:

$$y(n) = \sum_{k=-\infty}^{\infty} x(k)\, h(n-k) = \sum_{k=0}^{2} (k+1)a^{n-k}u(n-k)$$

$$= a^n u(n) + 2a^{n-1}u(n-1) + 3a^{n-2}u(n-2)$$

Let us define a *finite-duration* sequence as one that contains only a finite number of nonzero elements. When both $\{h(n)\}$ and $\{x(n)\}$ are finite-duration sequences, then the convolution produces a finite-duration sequence, as illustrated by Example 2.6. It is left as an exercise to show that if the input sequence $\{x(n)\}$ contains N_X elements and the unit-sample response sequence $\{h(n)\}$ contains N_H elements, then the output sequence $\{y(n)\}$ contains $N_Y = N_X + N_H - 1$ elements.

A *geometric series* is a series in which consecutive elements differ by a constant ratio. Such a series can be written in the form

$$x(n) = r^n, \qquad \text{for } -N_1 \leq n \leq N_2 \qquad (2.31)$$

where r is a constant and N_1 and N_2 are any two numbers. In evaluating the sum of products in the convolution, we commonly encounter sequences that form such a series. The sum can then be easily computed by employing the following two geometric sum formulas. To derive these formulas, let us consider the following Taylor series expansion: Given a complex number c, where $|c| < 1$, then

$$\frac{1}{1-c} = 1 + c + c^2 + c^3 + \cdots = \sum_{n=0}^{\infty} c^n \qquad (2.32)$$

Applying this result in the reverse direction yields the *infinite geometric sum formula:* Given a complex number c, where $|c| < 1$, then

$$\sum_{n=0}^{\infty} c^n = \frac{1}{1-c} \qquad (2.33)$$

We can also compute the sum of a finite number of elements in a geometric series. Let us consider the following sum

$$1 + c + c^2 + \cdots + c^{N-1} = \sum_{n=0}^{N-1} c^n \qquad (2.34)$$

The sum of this finite-duration sequence can be expressed as the difference between the sums of two infinite-duration sequences as

$$\sum_{n=0}^{N-1} c^n = \sum_{n=0}^{\infty} c^n - \sum_{n=N}^{\infty} c^n$$

$$= \sum_{n=0}^{\infty} c^n - c^N \sum_{n=0}^{\infty} c^n \qquad (2.35)$$

Applying the infinite geometric sum formula to the two summations, we obtain the desired *finite geometric sum formula:* Given a complex number c, then

$$\sum_{n=0}^{N-1} c^n = \frac{1 - c^N}{1 - c} \qquad (2.36)$$

While the infinite geometric sum formula requires that the magnitude of c be *strictly less than unity*, the finite geometric sum is valid for *any* value of c.

Example 2.8. Convolution of two infinite-duration sequences. Let

$$h(n) = a^n u(n) \qquad \text{for all } n,$$
$$x(n) = b^n u(n) \qquad \text{for all } n$$

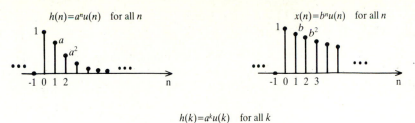

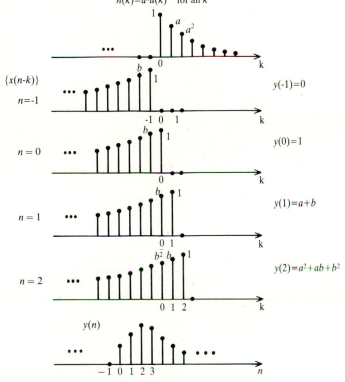

FIGURE 2-14
The convolution of two infinite-duration sequences.

as shown in Fig. 2.14. Then the output sequence values are equal to

$$y(n) = \sum_{k=-\infty}^{\infty} a^k u(k) b^{n-k} u(n-k) = \sum_{k=0}^{n} a^k b^{n-k}$$

$$= b^n \sum_{k=0}^{n} (a/b)^k$$

The finite geometric sum formula can be applied to yield

$$y(n) = b^n \frac{1 - (a/b)^{n+1}}{1 - (a/b)}$$

2.6 STABILITY CRITERION FOR DISCRETE-TIME SYSTEMS

For practical applications, we are interested in designing only stable digital filters, or those whose outputs do not become infinite. In this section, we define a stability criterion and provide a necessary and sufficient condition for a filter to meet this condition. In later chapters, we will describe alternate means of determining whether or not a digital filter is stable.

Let us define a *bounded* sequence to be one in which the absolute value of every element is less than some finite number M, i.e. $\{x(n)\}$ is bounded if $|x(n)| < M$ for all n, where M is some finite number. Then, a system is said to be *stable* when every possible bounded input sequence produces a bounded output sequence. Since the output values of a digital filter should not exceed the number range of the computer, it is appropriate to consider this bounded output criterion.

STABILITY CRITERION. An LTI system is stable if, and only if, the stability factor, denoted by S, and defined by

$$S = \sum_{k=-\infty}^{\infty} |h(k)| \tag{2.37}$$

is finite, i.e. $S < \infty$.

The stability of a digital filter is expressed in terms of the absolute values of its unit-sample response. We now show that this criterion is both *sufficient* and *necessary*. To prove that it is *sufficient*, we show that if S is finite, then the system is stable. Let $\{x(n)\}$ be a bounded input sequence. We must show that the output is bounded when S is finite. Taking the absolute value of the output, expressed by the convolution equation, we have

$$|y(n)| = \left| \sum_{k=-\infty}^{\infty} h(k)\, x(n-k) \right| \tag{2.38}$$

Recall that the magnitude of the sum of terms is less than or equal to the sum of the magnitudes. Hence,

$$|y(n)| \le \sum_{k=-\infty}^{\infty} |h(k)\, x(n-k)| \tag{2.39}$$

Finally, since all the input values are bounded, say by M, we have for all n:

$$|y(n)| \le M \sum_{k=-\infty}^{\infty} |h(k)|$$

or

$$|y(n)| \le MS \tag{2.40}$$

Hence, since both M and S are finite, the output is also bounded.

To show that it is *necessary* for S to be finite for the filter to be stable, we present a counter-example. We need to find a bounded input sequence that produces an unbounded output. For our input sequence, let us choose

$$x(n) = \begin{cases} h(-n)/|h(-n)| & \text{for all } n \text{ for which } h(-n) \neq 0 \\ 1 & \text{otherwise} \end{cases} \qquad (2.41)$$

This sequence consists of plus and minus ones and is obviously bounded. Let us now consider the output value for $n = 0$, or $y(0)$. From the convolution equation, we have

$$y(0) = \sum_{k=-\infty}^{\infty} h(k)x(-k) = \sum_{k=-\infty}^{\infty} h^2(k)/|h(k)|$$

$$= \sum_{k=-\infty}^{\infty} |h(k)| = S \qquad (2.42)$$

Hence, for $y(0)$ to be bounded, it is necessary for S to be finite. We have thus proved the necessary and sufficient conditions. The following examples illustrate the application of the stability criterion.

Example 2.9.   Stability of the generalized 3-sample averager. Let us generalize the 3-sample averager by letting the coefficients be arbitrary constants b_{-1}, b_0 and b_1. The output is then equal to

$$y(n) = b_{-1}x(n+1) + b_0 x(n) + b_1 x(n-1)$$

To test whether this filter is stable, we compute the value of S:

$$S = \sum_{n=-\infty}^{\infty} |h(n)| = |b_{-1}| + |b_0| + |b_1|$$

Hence, S is finite for finite values of b_{-1}, b_0 and b_1.

Example 2.9 can be easily extended to show that any filter with a unit-sample response having a finite number of nonzero elements is always stable.

Example 2.10.   Stability of first-order recursive filter. Let

$$y(n) = ay(n-1) + x(n)$$

Recall that $h(n) = a^n u(n)$, for all n. Checking for stability, we must calculate

$$S = \sum_{n=-\infty}^{\infty} |h(n)| = \sum_{n=0}^{\infty} |a|^n$$

It is obvious that S is unbounded for $|a| \geq 1$, since then each term in the series is ≥ 1. For $|a| < 1$, we can apply the infinite geometric sum formula, to find

$$S = \frac{1}{1-a} \qquad \text{for } |a| < 1$$

Since S is finite for $|a| < 1$, the system is stable for $|a| < 1$.

Example 2.11.   Stability of second-order filter. Let

$$y(n) = 2r \cos(\omega_0) y(n-1) - r^2 y(n-2) + x(n) - r \cos(\omega_0) x(n-1)$$

To test the stability of this filter, we must first determine its unit-sample response and then compute the value of S. To determine $\{h(n)\}$, we assume zero initial conditions, or $y(n) = 0$, for $n < 0$ and let the input sequence be $\{d(n)\}$. We then get the following output values.

$$y(0) = 1$$
$$y(1) = 2r \cos(\omega_0) - r \cos(\omega_0) = r \cos(\omega_0)$$
$$y(2) = 2r^2 \cos^2(\omega_0) - r^2 = r^2(2 \cos^2(\omega_0) - 1) = r^2 \cos(2\omega_0)$$
$$\cdots$$
$$y(n) = r^n \cos(n\omega_0)$$

Hence,

$$h(n) = r^n \cos(n\omega_0) u(n)$$

Checking for stability, we have

$$\sum_{n=0}^{\infty} |h(n)| = \sum_{n=0}^{\infty} |r^n \cos(n\omega_0)| \leq \sum_{n=0}^{\infty} |r|^n = \frac{1}{1-|r|} \qquad \text{for } |r| < 1$$

Hence, the filter is stable for $|r| < 1$. The inequality of the sums in the above stability calculation comes about because the magnitude of the cosine function is less than or equal to unity.

2.7 CAUSALITY CRITERION FOR DISCRETE-TIME SYSTEMS

A system is said to be *causal* when the response of the system does not precede the excitation. All physical systems are causal in that they do not react until a stimulus is applied. An LTI system is causal if, and only if,

$$h(n) = 0 \qquad \text{for } n < 0 \tag{2.43}$$

This is true because, by definition, $\{h(n)\}$ is the response of a system to a unit-sample sequence, whose nonzero element occurs at $n = 0$. For a causal system, the output cannot have nonzero values for $n < 0$, the time before the nonzero-valued element is applied.

Causal filters are typically employed in applications in which the data samples are processed as they are received. Clearly, in this case we cannot have access to future sample values. In this book, we will not restrict ourselves to causal digital filters. When filters are used to process data for which the entire data record is available at the start of processing, there is no restriction against looking ahead to future values. One example of such data is that obtained from images. As described in the previous chapter, this look-ahead feature is implemented by the advance, or negative delay, component of the digital filter. For noncausal filters, $\{h(n)\}$ has nonzero elements for $n < 0$, as illustrated with the 3-sample averager in Example 2.3. Since the unit-sample response of the first-order recursive filter in Example 2.4 obeys the condition given in Eq. (2.43), it is a causal filter.

2.8 LINEAR CONSTANT-COEFFICIENT DIFFERENCE EQUATIONS

In this section, we formalize the notation that we have been using above for describing the time-domain behavior of digital filters. In general, the output of a finite-order LTI system at time n can be expressed as a linear combination of the inputs and outputs in the following form:

$$y(n) = \sum_{k=1}^{M} a_k \, y(n-k) + \sum_{k=-N_F}^{N_P} b_k \, x(n-k) \qquad (2.44)$$

In words, the current output $y(n)$ is equal to the sum of past outputs, from $y(n-1)$ to $y(n-M)$, which are scaled by the delay-dependent *feedback* coefficients a_k, plus the sum of future, present and past inputs, which are scaled by delay-dependent *feedforward* coefficients b_k. The output value at time n is determined by the input values from N_F samples in the future $x(n + N_F)$, through the current value $x(n)$, to the sample N_P time units in the past $x(n - N_P)$. For a causal filter $N_F \leq 0$. When $N_F > 0$, then the future values of the input determine the current output value, and the filter is noncausal.

> **Example 2.12. Difference equation of a simple filter.** We can combine the equations of the nonrecursive 3-sample averager and the first-order recursive filter, to get
>
> $$y(n) = a \, y(n-1) + \tfrac{1}{3}x(n+1) + \tfrac{1}{3}x(n) + \tfrac{1}{3}x(n-1)$$
>
> This filter has both a recursive and a nonrecursive part and is also noncausal.

The difference equation is useful not only for analyzing digital filters but can be employed for implementing them. A digital filter can be implemented from the general difference equation by performing the following four steps, as shown in Fig. 2.15.

Step 1. Two points are drawn, one corresponding to the current input $x(n)$ and the other corresponding to the current output $y(n)$.

Step 2. To gain access to the past values of the input and output sequences, delay elements are connected to the output and input points. Future values of the input sequence are obtained by connecting advances to the input point.

Step 3. Multipliers are connected to the outputs of the delay elements to produce the required products. Multipliers with gains of unity need not be included, but replaced by a simple wire.

Step 4. Finally, the digital filter is accomplished by connecting the outputs of the multipliers to two adders. The first generates the sum of

STEP 1

$$x(n) \bullet\!\!-\!\!-\qquad -\!\!-\!\!\bullet y(n)$$

STEP 2

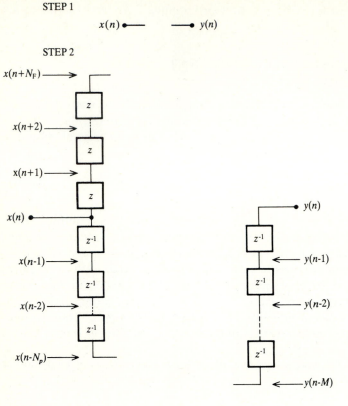

STEP 3

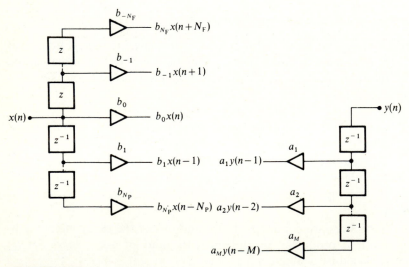

FIGURE 2-15

Steps to produce a digital filter structure from the general difference equation. This filter structure will be implemented in the computer programs.

STEP 4

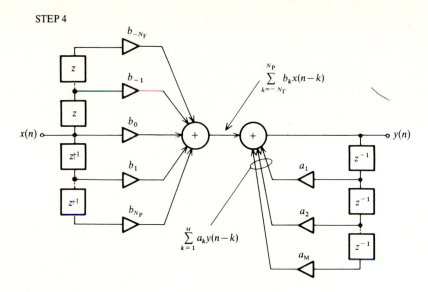

products of the inputs and the feedforward coefficients, and the other that of the outputs and the feedback coefficients.

The following example illustrates the application of these four steps.

Example 2.13. Implementation of the second-order digital filter from the difference equation. The filter implementation procedure from the difference equation given in Example 2.11 is shown in Fig. 2.16.

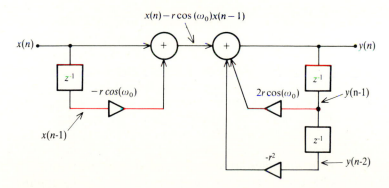

FIGURE 2-16
Digital filter structure that implements the difference equation given by

$$y(n) = 2r\cos(\omega_0)\,y(n-1) - r^2\,y(n-2) + x(n) - r\cos(\omega_0)\,x(n-1)$$

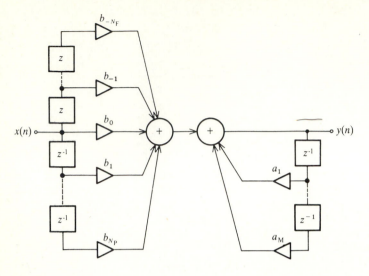

FIGURE 2-17
General digital filter structure implemented in the computer programs.

The general digital filter structure shown in Fig. 2.17 is convenient for describing the logic behind the computer programs described below. However, it is not the most efficient structure in terms of the number of delays that are used. In a later chapter, we will consider other filter structures that have a smaller number of delays but for which the corresponding computer programs are more complicated. Conceptually, however, these more memory-efficient programs perform the same types of operations as the simple programs described below.

2.9 SUGGESTED PROGRAMMING STYLE

Before describing the programs, we provide the motivation for the particular programming style suggested below. The goal of the programs is to yield the greatest intuitive understanding of the digital filter operation with the least programming effort. Techniques to improve the efficiency of these programs are suggested throughout the text. The logical steps in the operation of a digital filter are best illustrated by writing simple subprograms that perform the primitive tasks that are most commonly employed in signal processing. These simple subprograms are then combined to form the digital filter program.

To provide a FORTRAN example of the recommended programming style, one common task is to initialize an array to zero. Even though this can be done quite easily in a simple loop in the main proram, the programming style suggested above would use the following subroutine:

```
      SUBROUTINE NULL(X,NX)
      DIMENSION X(NX)
      DO 1 I=1,NX
  1   X(I)=0.
      RETURN
      END
```

An array B(N), for $1 \leq N \leq NB$, can be zeroed in the main program with the following statement:

```
      CALL NULL(B,NB)
```

Two additional subroutines, described below, compute the sum of products of two arrays and implement a shift register. Use of such simple subroutines will simplify the main program by making it more *readable*, which is important for debugging and trouble shooting. This programming style also allows the student to reason in terms of these primitive tasks.

In trying to explain the programs in a manner that is not specific to a particular programming language, a problem immediately arises with array indexing. The various computer languages used for signal processing can be divided into two classes: those that allow arrays to be indexed starting with zero, such as BASIC and C, and those that do not, such as FORTRAN and PASCAL. In the first set, $x(0)$ is the first element of an array, in the second. the first element is $x(1)$. Since all languages can start indexing from 1, we will use this convention in the programs below. For computer languages that allow the array index to start from zero, a slight modification in the indexing of the programs should be made. The advantage of starting the array with $x(0)$ is that it then conforms to the notation used for sequences in the analytic sections of this book, i.e. $x(n)$, for $0 \leq n \leq N - 1$.

2.10 WRITING A DIGITAL FILTER PROGRAM

In this section, we describe the computer programs to implement a digital filter from a difference equation. These programs will be used in the following chapters for analyzing and implementing digital filters to meet some desired specification. To implement a digital filter as a flexible program, it is most convenient to write separate subprograms for the nonrecursive part, to be called NONREC, and the recursive part, to be called REC. The general filter can then be implemented by first using NONREC followed by REC, as described below.

First, it is useful to establish the notation for the arrays to be used in the programs. We are typically given an input array of real values, denoted by

X(N), for $1 \leq N \leq NX$, where NX is a known value. This array represents the input data to be processed. In this book, the most common input will be the unit-sample sequence. We are usually asked to design a digital filter that is specified by an array of feedback coefficients ACOEF(N), for $1 \leq N \leq NA$, to be used in REC and an array of feedforward coefficients BCOEF(N), for $1 \leq N \leq NB$, to be used in NONREC.

Since we are implementing nonrecursive and recursive filters separately, it is helpful to define two output arrays. The output of a nonrecursive filter is denoted by YN(N) and will have a duration $NY = NX + NB - 1$. This duration follows from the convolution of two finite-duration sequences. The output array of the recursive filter is denoted by YR(N) and has a duration NY. For the recursive filter, however, the value of NY must be specified. Because of the feedback loop, the output of an ideal recursive filter contains an infinite number of non-zero elements. However, for a stable filter, the output values eventually approach zero when the input has a finite duration. Hence, there will be only a finite number of output elements whose amplitudes are significantly greater than zero. For all the filters in this book, NY will be less than or equal to 128.

Since most programming languages require that the size of all arrays be specified at the beginning of the program, a length of 128 *floating-point,* or *real,* elements for all arrays will be adequate for most of the projects in this book. In the descriptions of the variables below, *real* is used to specify a floating-point variable.

We first consider the implementation of a causal filter, for which $N_F = 0$ in Eq. (2.44), and then describe the simple modification that can be added to implement a noncausal filter.

PROGRAMMING A CAUSAL FILTER. The general causal filter is implemented by combining two causal components: a *strictly nonrecursive causal* part with a *strictly recursive* part.

The general difference equation (2.44) of a *causal nonrecursive* filter, for which $N_F = 0$ and $a_k = 0$ for all k reduces to

$$y(n) = \sum_{k=0}^{N_P} b_k \, x(n-k) \tag{2.45}$$

The feedforward coefficients b_k for $0 \leq k \leq N_P$ are stored in BCOEF(N) for $1 \leq N \leq NB$ (NB $= N_P + 1$). The value of b_0 is stored in BCOEF(1), b_1 in BCOEF(2), etc. Notice that Eq. (2.45) has the form of the following convolution

$$y(n) = \sum_{k=0}^{N_P} h(k) \, x(n-k) \tag{2.46}$$

By computing the unit-sample response of the nonrecursive filter, we find that it has a finite duration and is equal to the feedforward coefficient sequence, or

$$h(k) = b_k \qquad \text{for } -N_F \leq k \leq N_P \tag{2.47}$$

Hence, a nonrecursive filter can be implemented directly from its unit-sample response. For the causal filter considered here, $N_F = 0$.

The difference equation for a *strictly recursive* filter, which employs the past outputs and only the current input value, for which $b_k = 0$ for $k \neq 0$, reduces to

$$y(n) = \sum_{k=1}^{M} a_k \, y(n-k) + x(n) \tag{2.48}$$

The feedback coefficients a_k for $1 \leq k \leq M$ will be stored in ACOEF(N) for $1 \leq N \leq NA$, where $NA = M$. The value of a_1 is stored in ACOEF(1), a_2 in ACOEF(2), etc.

In both Eqs. (2.45) and (2.48), a sum-of-products calculation employing an array of coefficients and an array of data is required. We implement this primitive task as a separate subprogram, called SUMP (**sum** of **p**roducts), that can be used in both NONREC and REC. To simplify the programming logic, we will use two auxiliary buffer arrays. The first, denoted XBUF(N), for $1 \leq N \leq NB$, stores the current and past NB values of $\{X(N)\}$. Only NB values are needed since that is the number of coefficients in BCOEF(N). The second buffer array, denoted YBUF(N), for $1 \leq N \leq NA$, stores the past NA values of $\{Y(N)\}$. A second subprogram, called SHFTRG (**sh**ift **r**e**g**ister), acts as a shift register to maintain the proper time sequence of values in these arrays.

SUMP Subprogram. The SUMP subroutine computes the sum of products of two arrays, a coefficient array, denoted COEF(N), for $1 \leq N \leq NC$, and a buffer array BUF(N), of the same size, to produce the output value SUM. The call to the subprogram is then

    SUMP(COEF,BUF,NC,SUM)

where

COEF   (real input array) is an array of coefficients;
BUF   (real input array) is an array of data values;
NC   (integer input) is the duration of the COEF array;
SUM   (real output) is the sum of products equal to

$$SUM = \sum_{I=1}^{NC} COEF(I) * BUF(I) \tag{2.49}$$

as shown in Fig. 2.18. When SUMP is used in NONREC, BUF is equal to XBUF, COEF corresponds to BCOEF and NC is equal to NB ($= N_P + 1$). When used in REC, BUF is equal to YBUF, COEF corresponds to ACOEF and NC is equal to NA ($= M$).

SHFTRG Subroutine. To update the buffer arrays, we implement a shift register routine, called SHFTRG, that shifts the contents of the buffer array by one element at each iteration of the filter operation. The shift is performed

COEFFICIENT ARRAY   BUFFER ARRAY

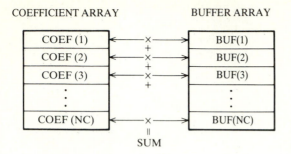

FIGURE 2-18

Operation performed by the SUMP routine.

such that the oldest element in the buffer is lost and the current value of the time sequence is put into the first element of the buffer. That is, at time L, $X(L)$ is put into XBUF(1) and only the current and past $N_B - 1$ values are retained in the buffer.

In general, given a value VAL and buffer array BUF(N), for $1 \leq N \leq$ NBUF, we want to write the subroutine called

```
SHFTRG(VAL,BUF,NBUF)
```

where

   VAL   (real input) is the value to be inserted into the buffer;

   BUF   (real input and output array) is the buffer array;

   NBUF   (integer input) is the duration of the buffer array.

This subroutine shifts BUF(K) into BUF(K + 1), for $1 \leq K \leq$ NBUF, and puts VAL into BUF(1), as shown in Fig. 2.19. When used in NONREC, the buffer BUF is equal to XBUF, NBUF is equal to NB, and at time L, VAL is equal to $X(L)$. When used in REC, BUF is equal to YBUF, NBUF is equal to NA and at time L, VAL is equal to YR(L). We now describe how to use these subroutines to implement the nonrecursive filter subprogram NONREC and the recursive filter filter subprogram REC.

NONREC Subprogram. For the nonrecursive filter, the input array X(N), for $1 \leq N \leq$ NX and the feedforward coefficient array BCOEF(N), for $1 \leq N \leq$ NB

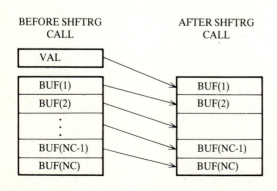

FIGURE 2-19

Operation performed by the SHFTRG routine.

($NB = N_P + 1$), are specified and the output array YN(N), for $1 \leq N \leq NX + NB - 1$, is to be generated. The nonrecursive filter subroutine is denoted by

    NONREC(X,NX,BCOEF,NB,YN)

where

X	(real input array) is the input array to the nonrecursive filter;
NX	(integer input) is the duration of the input array;
BCOEF	(real input array) is the feedforward coefficient array;
NB	(integer input) is the duration of the coefficient array;
YN	(real output array) is the output array of duration $NY = NX + NB - 1$.

The procedure to be performed within NONREC is indicated in the logic diagram shown in Fig. 2.20. The NONREC routine is accomplished using the three basic steps described below.

Step 1. The buffer array XBUF(N), for $1 \leq N \leq NB$, is defined and initialized to zero.

Step 2. In a loop with index I, for I equal from one to NX, VAL is set equal to the Ith element of the input, or X(I). VAL is then entered into the shift register XBUF. The sum of products of XBUF and BCOEF are computed to produce SUM. The Ith value of the output sequence for the nonrecursive filter, YN(I), is then set equal to SUM.

Step 3. When $I > NX$, all the elements of X(N) have been used. VAL is then set to zero and then entered into the buffer XBUF. The sum of products is computed and the result SUM becomes the value of the output YN(I). This loop is executed until no more elements of the input sequence $X(N)$ remain in the buffer, or until $I > NX + NB - 1$.

After this routine is completed, the nonrecursive filter operation has been accomplished and the output values will reside in the YN array. The combination of the SHFTRG and SUMP routines results in a convolution operation.

REC Subroutine. For the recursive filter, the input array X(N), for $1 \leq N \leq NX$, and the feedback coefficient array ACOEF(N), for $1 \leq N \leq NA$ ($NA = M$), are specified and the output array YR(N), for $1 \leq N \leq NY$, is to be generated. Unlike the nonrecursive filter, the duration of a recursive filter can be infinite because of the feedback. Hence, the value of NY must be sufficiently large so that the important features of the output are observed. In the case of determining the unit-sample response, the value of NY must be chosen large enough such that the values of $h(n)$ for $n > NY$ are approximately zero. In producing a plot of the sequence, it is sufficient that the omitted magnitudes be smaller than the resolution of the plotting device. Typical values for NY used in the projects will be 55 and will not exceed 128.

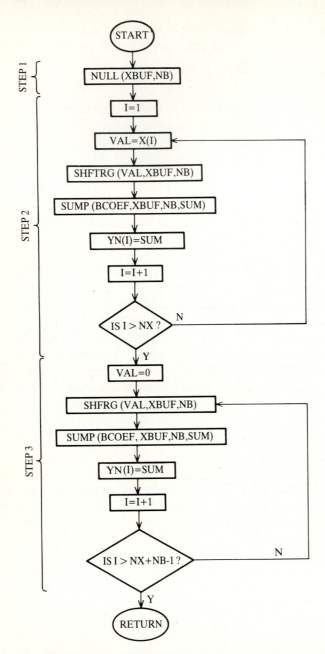

FIGURE 2-20

Logic diagram for the NONREC filter routine. The values for the input array X(N), the size of the array NX, the feed-forward coefficient array BCOEF(N), and the size of the coefficient array NB are given; the output array of the nonrecursive filter YN(N) is to be computed.

The recursive filter is implemented by the subroutine

```
REC(X,NX,ACOEF,NA,YR,NY)
```

where

X	(real input array) is the input array to the recursive filter;
NX	(integer input) is the duration of the input array;
ACOEF	(real input array) is the feedback coefficient array;
NA	(integer input) is the duration of the coefficient array;
YR	(real output array) is the output array;
NY	(integer input) is the duration of the output array.

The logic diagram for the REC routine is shown in Fig. 2.21. Within REC, the following steps are performed:

Step 1. The buffer array YBUF(N), for $1 \leq N \leq NA$, is defined and initialized to zero.

Step 2. In a loop with index I, for $1 \leq I \leq NX$, the output at time I, YR(I), is equal to the sum of the input X(I) and SUM, the product of the past outputs stored in YBUF and ACOEF. For $I = 1$, SUM is equal to zero. VAL is then set equal to the output YR(I) and is sequenced into YBUF.

Step 3. When $I > NX$, all the elements of $X(N)$ have been used. At that point the output YR(I) is then determined from only the past output values. The recursive filter operation is accomplished after the desired number of output points NY have been computed.

GENERAL FILTER IMPLEMENTATION. The general filter can be implemented by calling the NONREC followed by the REC routines, as shown in Fig. 2.22. The input sequence X(N), for $1 \leq N \leq NX$, to NONREC produces the output sequence YN(N), for $1 \leq N \leq NX + NB - 1$. This sequence then becomes the input sequence to the REC routine. The REC routine will then produce the output sequence YR(N), for $1 \leq N \leq NY$. Usually NY will be greater than $NX + NB - 1$. The array YR is then the output of the general filter.

If the REC routine is called first and then followed by NONREC, care must be taken in choosing the value of NY, the duration of the output sequence YR produced by REC. This sequence then becomes the input to NONREC. The output of NONREC, YN, then contains $NY + NB - 1$ elements, a duration that must not exceed the dimension of YN.

NONCAUSAL FILTER IMPLEMENTATION. By definition, a filter is noncausal when future values of the input sequence are used in the computation of the current output value. Noncausal filters can be identified easily from the difference equation by N_F being greater than zero. Since arrays with negative indices are not generally allowed in programming languages, the simplest way to implement a noncausal filter is to change the difference equation in such a

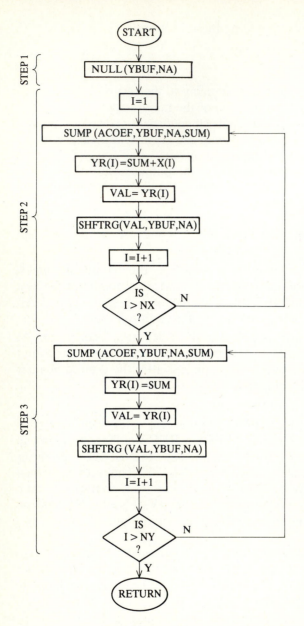

FIGURE 2-21

Logic diagram for the REC filter routine. The values for the input array X(N), the size of the array NX, the feedback coefficient array ACOEF(N), and the size of the coefficient array NA are given; the output array of the recursive filter YR(N) is to be computed.

way as to produce a causal filter and then shift the final output array *into the future* by N_F index units, as described below. Using this procedure, the causal filter can then be implemented in the manner described above, and only an additional shift routine is needed, to obtain the noncausal result.

The noncausal property of the filter is due to the N_F feedforward coefficients having negative index values in the nonrecursive part of the filter. Hence, a noncausal filter can be made causal by shifting the coefficient

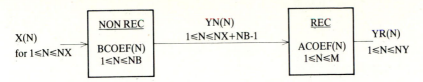

FIGURE 2-22
Combining NONREC and REC to form the general filter having both nonrecursive and recursive parts.

sequence $\{b_k\}$ to produce $\{b'_k\}$, where

$$b'_k = b_{k-N_F} \qquad \text{for } 0 \le k \le N_F + N_P$$

as shown in Fig. 2.23. This new set of coefficients then generates the causal output, denoted by $\{y'(n)\}$, whose elements are equal to

$$y'(n) = \sum_{k=1}^{M} a_k \, y'(n-k) + \sum_{k=0}^{N_P+N_F} b'_k \, x(n-k) \qquad (2.50)$$

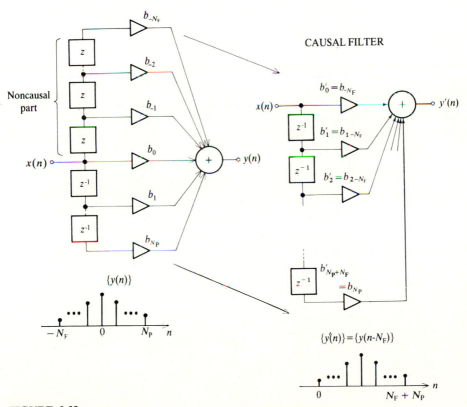

FIGURE 2-23
Transforming a noncausal filter into a causal filter.

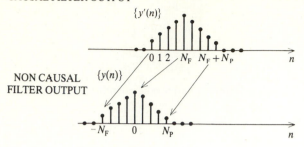

CAUSAL FILTER OUTPUT

NON CAUSAL FILTER OUTPUT

FIGURE 2-24
The noncausal filter output is obtained by shifting the output of the causal filter *to the left*.

This causal filter can be implemented by using the routines described above. The values of $\{y'(n)\}$ are exactly the same as those of $\{y(n)\}$, but only shifted by N_F index values. To obtain the noncausal output sequence $\{y(n)\}$, a time shift *to the future* is required, i.e.

$$y(n) = y'(n + N_F), \qquad \text{for } -N_F \le n \le NY - N_F.$$

as shown in Fig. 2.24. This time shift is performed by the subprogram

    SHIFT(Y,NY,NF)

where
- Y   (real input and output array) is the array to be shifted;
- NY   (integer input) in the duration of the array;
- NF   (integer input) is the number of elements that the input array is to be shifted to produce the output.

After the shift is performed, the first NF samples of the input array are lost and the last NF samples of the output array are set to zero, as shown in Fig. 2.25.

BEFORE SHIFT CALL		AFTER SHIFT CALL
Y(1)		Y(1)
Y(2)		Y(2)
⋮		⋮
Y(N_F)		Y($N_Y - N_F - 1$)
Y($N_F + 1$)		Y($N_Y - N_F$)
Y($N_F + 2$)		Y($N_Y - N_F + 1$)
⋮	0	
Y($N_Y - 1$)		Y($N_Y - 1$)
Y(N_Y)	0	Y(N_Y)
	0	

FIGURE 2-25
Operation performed by the SHIFT routine.

In programming this shift, we still have a problem, because the shifted array may require negative index values. To avoid this problem and still observe the noncausal components, if any, in the filter unit-sample response, the input to the filter can be the delayed unit-sample sequence, whose nonzero component occurs at $N = N_F$. Using this as the input sequence, the noncausal components of the filter can then be observed in the output sequence YN(1) through YN(NF − 1).

2.11   SUMMARY

This chapter has presented the time-domain analysis techniques for describing linear time-invariant digital filters and discrete-time systems, from basic definitions to programming techniques. A particularly useful sequence that describes the behavior of a digital filter is its unit-sample response, since from it we can determine whether a filter is stable and whether it is causal. Further, for any input sequence to a digital filter, we can determine the output sequence by performing convolution with the unit-sample response.

An alternate time-domain description of a digital filter is the linear difference equation. The unit-sample response can be determined from the difference equation by simply making the input equal to the unit-sample sequence and computing the output sequence. Of practical importance, it was found to be a rather straightforward task to implement a digital filter structure from the difference equation. If the unit-sample response has a finite duration, a nonrecursive filter can be implemented by identifying the feedforward coefficients with the corresponding elements of the unit-sample response.

The steps for implementing a digital filter as a computer program were presented. By writing simple routines to perform a sum-of-products computation and a shift-register operation, the implementation of nonrecursive and recursive filters becomes conceptually easy to grasp. The general filter structure containing both nonrecursive and recursive parts is implemented by employing both routines. Noncausal filters can be implemented by employing the causal filter routines and applying a subprogram that shifts the output sequence "into the future".

PROBLEMS

2.1. Express the following in complex vector notation:

(a) $1 + e^{-j\theta}$

(b) $1 - a\,e^{-j\theta}$

(c) $1 + 2e^{-j\theta} + e^{-j2\theta}$

(d) $\dfrac{1 + ja}{1 + jb}$

(e) $\dfrac{1 + a\,e^{j\theta}}{1 + b\,e^{j\theta}}$

(f) $(a + jb)^{1/2}$

2.2. Express the following in real plus imaginary notation:

(a) $2\cos(\theta)e^{-j\theta}$

(b) $\dfrac{1}{1+ja}$

(c) $\dfrac{1}{1+a\,e^{j\theta}}$

(d) $\dfrac{1}{1+e^{-j\theta}+e^{-j2\theta}}$

(e) $(20+16\cos\theta)^{1/2}\exp\left[-j\arctan\left(\dfrac{-\sin\theta}{2+\cos\theta}\right)\right]$

2.3. Use the Euler identities to derive the formula for the cosine of the sum of two angles:

$$\cos(\alpha+\beta)=\cos(\alpha)\cos(\beta)-\sin(\alpha)\sin(\beta)$$

2.4. Demonstrate that the first-order recursive filter is linear. (Hint: Express $G\{x(n)\}$ as an infinite sum of scaled terms containing current and past values of $x(n)$.)

2.5. Evaluate the following summations and give the result in complex vector notation:

(a) $\displaystyle\sum_{n=0}^{\infty} r^n\,e^{-j\omega n}$ for $|r|<1$

(b) $\displaystyle\sum_{n=0}^{N-1} e^{-j\omega n}$

(c) $\displaystyle\sum_{n=0}^{N-1} \cos(\omega n)$

(d) $\displaystyle\sum_{n=0}^{N-1} a^n\,e^{-j\omega n}$

2.6. This problem considers the case of determining the unit-sample response of a filter having nonzero initial conditions. This condition can occur when the delay element contents are not initialized to zero.

(a) *The 3-sample averager.* Determine $\{h(n,0)\}$ when the content of the delay element is equal to 2 and that of the advance is equal to 1 at $n=0$.

(b) *The first-order recursive filter.* Determine $\{h(n,0)\}$ when the content of the delay element is equal to 3 at $n=0$.

2.7. Given two finite-duration causal sequences, $\{x(n)\}$ containing N_X elements and $\{h(n)\}$ containing N_H elements, show that $\{y(n)\}=\{x(n)\}*\{h(n)\}$ produces a causal, finite-duration sequence containing $N_Y=N_X+N_H-1$ elements.

2.8. Given the following two sequences, determine an analytic expression for $\{y(n)\}=\{x(n)\}*\{h(n)\}$. Let

$$x(n)=\begin{cases}1 & \text{for } 0\le n\le N-1\\ 0 & \text{otherwise}\end{cases}$$

and

$$h(n)=\begin{cases}1 & \text{for } 0\le n\le N-1\\ 0 & \text{otherwise}\end{cases}$$

Hint: First try $N = 2$ and then $N = 3$, to obtain the general analytic form of the sequence.

2.9. Prove that any realizable filter having a finite number of elements in its unit-sample response is always stable.

2.10. For the following difference equations, determine and sketch the unit-sample response sequence and implement the digital filter structure:

(a) $y(n) = x(n) - x(n - N)$
(b) $y(n) = ay(n - 1) + x(n) + x(n - 1)$

Let $a = 0.9$.

2.11. For the digital filter structures shown in Fig. P2.11, determine the difference equation, compute and sketch the unit-sample response.

2.12. The steady-state condition of a discrete-time system is defined as that in which the output value does not change with time, i.e. $y(n + 1) = y(n)$ for n very large. Determine this steady-state value when the input sequence is a constant-valued step sequence, $x(n) = Ku(n)$, for the following system:

$$y(n) = ay(n - 1) + x(n) + x(n - 1)$$

Let $a = 0.9$.

2.13. Implement the digital filter that generates $\cos(\omega_0 n)$, for $n \geq 0$. (Hint: see Example 2.11).

2.14. Demonstrate that the sequence $\{x(n)\} = \{\sin(\omega n)\}$ produces identical values when $\omega = \omega_0$ and $\omega = \omega_0 + 2m\pi$, where m is an integer. Sketch the results for a particular value of ω_0, with $m = 0$ and 1.

2.15. Demonstrate that the convolution operation is commutative by performing the convolution given in Example 2.8 in the reverse order to obtain the same result.

2.16. Restructure the logic diagram of the nonrecursive filter such that only one loop is required.

2.17. Restructure the logic diagram of the recursive filter such that only one loop is required.

2.18. For each of the following transformations, determine whether the system is stable, causal, linear and shift-invariant:

(a) $G_1\{x(n)\} = ax(n - n_0) + bx(n - n_1)$, for constants a, b, n_0 and n_1
(b) $G_2\{x(n)\} = x(n)x(n - n_0)$
(c) $G_3\{x(n)\} = x(2n)$

2.19. That a system is unstable does not imply that the output at time n, or $y(n)$, approaches infinity for every input sequence. Consider the system whose unit-sample response is given by

$$h(n) = \begin{cases} 1 & \text{for } n \geq 0 \\ 0 & \text{otherwise} \end{cases}$$

(a) Show that this system is unstable.
(b) Let the input sequence be given by

$$x(n) = e^{j\omega n} \qquad \text{for all } n$$

Using the convolution equation, determine the values of ω for which $y(n)$ is finite, and the values of ω for which $y(n)$ is infinite.

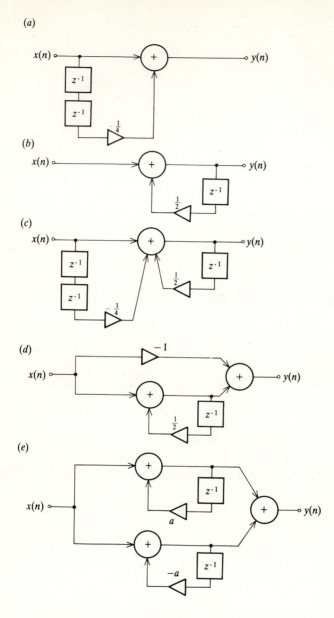

FIGURE P2-11
Digital filter structures.

2.20. Find the output sequence produced for the system having the unit-sample response given by

$$h(n) = \begin{cases} 1 & \text{for } n = 0 \\ -1 & \text{for } n = 4 \\ 0 & \text{otherwise} \end{cases}$$

when the input sequence is given by

$$x(n) = \begin{cases} 1 & \text{for } n \geq 0 \\ 0 & \text{otherwise} \end{cases}$$

2.21. A common input to an LTI system used in control situations is the unit-step sequence, $\{u(n)\}$. The response of the system is the *unit-step response,* denoted by $\{g(n)\}$. Determine the unit-step response of the first-order recursive system, whose unit-sample response is given by $h(n) = a^n u(n)$.

COMPUTER PROJECTS

Preface

Most of the projects contain the following parts:

Object, that is a concise statement of the purpose of the project.

Analytic results, which describe the required analysis to be performed for later comparison with the computer results. In some cases, references are made to a problem at the end of the chapter.

Computer results, which describe the programs to be written and the computations to be performed.

Definitions for the following terms will be helpful for performing the projects.

Analytically determine means to apply the appropriate mathematical formulas and evaluate the expression. All of the summations in the analytic expressions should be evaluated and complex quantities should be expressed in complex vector notation. The analysis is usually not difficult when requested. If the student finds that the procedure is becoming very complicated, it means that either a mistake has been made, or that a simpler approach can be taken. In this case, the programming should be performed first to provide the answer towards which the analysis should lead.

Sketch means to provide a graphical interpretation of the analytic results. If the results were not analytically determined, the graphical interpretation is to indicate a qualitative understanding of the operation being discussed. In all cases, the time- or frequency-domain scales and landmarks are to be indicated on the sketches. In many cases, only relative amplitudes are necessary.

Compute means that the procedure is to be performed numerically by the computer and compared with the analytic results.

Plot means that the computed data are to be presented in graphical form and compared with the sketches obtained from analysis. The graphs produced by the computer should be labeled by hand to indicate the correct time scale (n) or frequency scale (ω). The amplitudes and abscissa locations of features of interest in the plots, such as peaks, valleys or discontinuities, should be noted and related to the sketches.

The project report should be submitted with the material for each project grouped together. The analysis should be presented first, followed by the corresponding computer plot for easy comparison. Lists of numerical printouts should not be included. The programs used in the project should be appended to the end of the report, to facilitate evaluation of the programming style and for easy reference in the future.

2.1. Sinusoidal sequences

Object. Demonstrate that the values of a sinusoidal sequence are unique for frequencies only in the range $-\pi \leq \omega < \pi$.

Analytic results. Demonstrate that $\{x(n)\} = \{\sin(\omega n)\}$ produces identical values when $\omega = \omega_0$ and $\omega = \omega_0 + 2m\pi$, where m is any integer and $-\pi \leq \omega_0 < \pi$. Sketch the results for a particular value of ω_0, with $m = 0$ and 1.

Computer results. Using the program TEST0.FOR in Appendix A, enter the two values of frequency used above to produce the sinusoidal sequences. The program asks for the frequency in Hertz per sample interval F, where $F = \omega/2\pi$. The range of F that produces unique values is then $-0.5 \leq F < 0.5$.

Use the PLOT.FOR subroutine given in Appendix A to plot the data on the terminal and on the printer. Since 55 lines fit on one piece of paper, limit the number of points plotted to 55 whenever possible.

2.2. Exponential sequences

Object. Generate an exponential sequence and to determine the value number range of the computer.

(1) *Real-valued exponential*

Analytic results. Determine and sketch the exponential sequence

$$x(n) = r^n \qquad \text{for } 0 \leq n \leq 8.$$

for some positive value of $0.9 < r < 1$. Repeat for some negative value of $-1 < r < -0.9$.

Computer results

(a) Write a program to accept a real-valued constant r and compute 55 elements of the exponential sequence. Plot the sequences for the two values of r used above. Compare your results with analytically determined sequences and sketches.

(b) Determine an approximate range for the largest floating-point number that can be expressed on your computer by accepting increasingly larger values of r until an overflow error is encountered in computing the sequence. Determine the smallest magnitude number that is still nonzero by accepting decreasingly smaller values of r, until either an underflow error occurs, or the sequence values become zero.

(2) *Complex-valued exponential*

Generate and plot 55 elements of the real and imaginary parts for the complex exponential sequence given by

$$x(n) = (r\, e^{j\omega_0})^n$$

for $\omega_0 = \pi/8$ and for the two values of r used above.

2.3. Determination of the effective duration of discrete-time sequence

Object. Determine the effective duration of an infinite duration unit-sample response sequence.

For the infinite-duration exponential sequences, the magnitude of the elements typically decreases as r^n. Since the plotting routines have a resolution of approximately 1 percent, we can define an *effective duration* of the sequence, as the number of sequence values that are greater than 1 percent of the maximum value of the sequence. Beyond this effective duration, all the elements can be considered as being zero-valued.

Analytic results. For the real-valued exponential sequence $x(n) = r^n u(n)$, determine an analytic expression for the effective duration of the sequence as a function of r.

Computer results. Verify this result by plotting several representative sequences.

2.4. Verification of the digital filter programs

Object. Write and verify the subroutines to implement nonrecursive and recursive filters.

Write the subroutines NONREC and REC as described in the text. To implement the noncausal feature, write the subroutine SHIFT. To verify the filter operation, let the input array be a delayed unit-sample sequence, whose nonzero element occurs at $N = 6$, i.e., $X(6) = 1$, $X(N) = 0$, for $1 \le N \le NX$, $N \ne 6$. With this definition, the sixth element of the output sequence will correspond to $n = 0$ and the noncausal components will occur in $Y(1)$ through $Y(5)$.

For each of the filters below, perform the following:
  (i) draw the filter structure;
  (ii) determine an analytic expression for the unit-sample response sequence and sketch the sequence;
  (iii) compute and plot the unit-sample response;
  (iv) compare the analytic and computed results, indicating the agreement or explaining any differences.

Computer results

(1) *Causal nonrecursive filters.* Implement the following difference equations as causal digital filters. Let $NX = 20$.

(a) $y(n) = x(n) + 2x(n-1) + 3x(n-2)$
(b) $y(n) = x(n) - x(n-10)$

(2) *Noncausal nonrecursive filter.* Implement the 3-sample averager given by $y(n) = \frac{1}{3}[x(n+1) + x(n) + x(n-1)]$.
(3) *Causal recursive filters.* Implement the following difference equations. Let $NY = 55$.

(a) $y(n) = a_1 y(n-1) + x(n)$ for $a_1 = -0.95$

(b) $y(n) = a_1 y(n-1) - a_2 y(n-2) + x(n)$ for $a_1 = 2r \cos(\omega_0)$ and $a_2 = r^2$.

Let $r = 0.95$ and $\omega_0 = \pi/8$. Hint:

$$h(n) = \frac{r^n}{\sin(\omega_0)} \sin[(n+1)\omega_0] \qquad \text{for } n \geq 0$$

2.5. Application of a noncausal filter: a linear interpolator

Object. Implement a digital interpolator as a noncausal filter.

Consider the following sequence of values

$$x(n) = \begin{cases} 0.95^{(n-6)} & \text{for } 6 \leq n \leq 36 \text{ and } n \text{ even} \\ 0 & \text{otherwise} \end{cases}$$

In practice, this sequence could have been obtained by placing observed data values into the even-indexed elements of an array. The idea is to determine the values of $\{x(n)\}$ that are zeros by employing an interpolation sequence.

To perform the interpolation, let us consider the noncausal filter whose unit-sample response is given by

$$h(-1) = 0.5$$
$$h(0) = 1$$
$$h(1) = 0.5$$

Analytic results. If $\{y(n)\} = \{x(n)\} * \{h(n)\}$, prove that

$$y(n) = x(n) \qquad \text{for } n \text{ even}$$

and

$$y(n) = \frac{x(n-1) + x(n+1)}{2} \qquad \text{for } n \text{ odd}.$$

Computer results. Implement the causal filter to obtain the intermediate array $\{y'(n)\}$, then shift $\{y'(n)\}$ to obtain the desired array $\{y(n)\}$. Compare the plots of $\{y(n)\}$ with $\{y'(n)\}$ to verify the shift and with $\{x(n)\}$ to verify the interpolation.

Notice the strange values at the ends of the interpolated sequence. These are artifacts produced by assuming the values of the sequence are zero outside the specified range.

2.6. Cascading filters

Object. To implement more complicated digital filters by cascading simpler filter sections.

Write a program to cascade NONREC and REC. To verify the filter operation, let the input array be the delayed unit-sample sequence given by

$$X(N) = \begin{cases} 1 & \text{for } N = 6, \\ 0 & \text{otherwise}. \end{cases}$$

For each of the filters below, perform the following:

(i) draw the filter structure and determine the difference equation;

(ii) analytically determine and sketch the unit-sample response;

(iii) compute and plot the unit-sample response;

(iv) compare the analytic and computed results.

(1) *Cascade of nonrecursive filters.* Cascade

$$yn(n) = x(n) + 2x(n-1) + 3x(n-2)$$

and

$$y(n) = \tfrac{1}{3}[yn(n+1) + yn(n) + yn(n-1)]$$

Determine the number of nonzero elements in the unit-sample response sequence.

(2) *Cascade of nonrecursive and recursive filters.* Cascade

$$yn(n) = x(n) - b_1 x(n-1) \qquad \text{for } b_1 = r \cos(\omega_0)$$

and

$$yr(n) = a_1 yr(n-1) - a_2 yr(n-2) + yn(n) \qquad \text{for } a_1 = 2r \cos(\omega_0) \text{ and } a_2 = r^2.$$

Let $r = 0.95$, $\omega_0 = \pi/8$ and $NY = 55$.

(3) *Cascade of recursive filters.* Cascade

$$yr(n) = a \, yr(n-1) + x(n)$$

and

$$yr'(n) = b \, yr'(n-1) + yr(n)$$

for $a = 0.8$ and $b = -0.9$ and $NY = 55$. (Compare this result with that of Example 2.8.)

2.7. Digital tone generators for frequency-shift keying (FSK) system.

Object. Simulate a frequency-shift keying (FSK) system by implementing two digital tone generators at different frequencies and an appropriate switching scheme.

In the FSK system, each bit value (1 or 0) will be transformed into 32 elements of a sinusoid at a different frequency: a *zero* will be transmitted as $\cos(\omega_0 n)$, for $0 \le n \le 31$, and a *one* will be transmitted as $\cos(\omega_1 n)$, for $0 \le n \le 31$.

The two frequencies will be personalized to each student: let K be the sum modulo-8 of the last four digits of your social security number, then

$$\omega_0 = 2\pi K/32 \qquad \text{and} \qquad \omega_1 = 2\pi(K+8)/32$$

(if K is equal to zero, make $K = 1$).

The 32 sinusoidal values are then inserted into the sequence $\{x(n)\}$ to be transmitted, the first bit sequence going into $x(0)$ to $x(31)$, the second value into $x(32)$ to $x(63)$, etc.

Analytic results. Draw a second-order recursive digital filter that acts like a digital oscillator and generates $\cos(\omega n)$, for $n \ge 0$, for each of the two frequencies above. Then, 32 values from the unit-sample response can then be inserted into the transmitted sequence at the appropriate time from the desired oscillator.

Computer results

(a) Generate and plot the two 32 element sequences that correspond to the two binary values. On the computer plot, sketch the continuous-time sinusoid from which the samples were derived.

(b) To produce an FSK sequence for future projects, write the subroutine

$$FSK(X)$$

where X (real output array) is an array of duration 96 that contains the discrete-time sequence that results for the binary pattern 010.

Plot the output sequence. Sketch directly on the plot the continuous-time sinusoidal function that corresponds to the samples.

FURTHER READING

Gold, Bernard, and Charles M. Rader: *Digital Processing of Signals,* McGraw-Hill, 1969. [This classic is one of the earliest books in the field.]

Oppenheim, Alan V., and Ronald W. Schafer: *Digital Signal Processing,* Prentice-Hall, Englewood Cliffs, NJ, 1975. [This book was probably the most popular book in digital signal processing, before the availability of the present text, and is suitable for advanced study of many of the topics covered here.]

Rabiner, Lawrence R., and Bernard Gold: *Theory and Application of Digital Signal Processing,* Prentice-Hall, Englewood Cliffs, NJ, 1975. [This book provides many practical aids for digital filter design and implementation.]

REFERENCE TO TOPIC

Time-varying filters

Cowan, C. F. N. and P. M. Grant: *Adaptive Filters,* Prentice-Hall, Englewood Cliffs, NJ, 1985.

Goodwin, C. G. and K. S. Sin: *Adaptive Filtering, Prediction and Control,* Prentice-Hall, Englewood Cliffs, NJ, 1984.

Haykin, Simon: *Adaptive Filter Theory,* Prentice-Hall, Englewood Cliffs, NJ, 1986.

Honig, M. L. and D. G. Messerschmitt: *Adaptive Filters,* Kluwer Academic, Boston, 1984.

CHAPTER
3

FOURIER TRANSFORM OF DISCRETE-TIME SIGNALS

3.1 INTRODUCTION

In many signal processing applications, the distinguishing features of signals and filters are most easily interpreted in the frequency domain. This chapter describes the frequency-domain properties of discrete-time signals and the frequency-domain behavior of discrete-time systems. The main analytic tool is the *Fourier transform* of a discrete-time sequence, which can be applied either to the unit-sample response of a filter to define the *filter transfer function*, or to a data sequence to define the *signal spectrum*. The important properties of the Fourier transform applied to discrete-time sequences are itemized and discussed. The magnitude and phase responses of a digital filter are examined. In practice, discrete-time sequences are commonly obtained by sampling continuous-time signals. The relationships between the spectra of continuous-time signals and the corresponding discrete-time sequences are investigated. Interpolation functions to convert discrete-time sequences into continuous-time functions are also described.

3.2 DEFINITION OF THE FOURIER TRANSFORM

The frequency-domain analysis of signals and systems usually employs sinusoidal sequences, given by $\{\sin(\omega n)\}$ and $\{\cos(\omega n)\}$, or more concisely $\{e^{j\omega n}\}$, where ω is the radian frequency. Sinusoidal sequences have the important property that when a sinusoid of a particular frequency is applied at the input of a linear time-invariant (LTI) system, then the output is also a sinusoid at the same frequency, although it may have a different amplitude and phase. The same statement cannot be made about other functions, such as a square-wave, for example.

To demonstrate this form of invariance, let the input to an LTI system be the complex sinusoid $\{x(n)\} = \{e^{j\omega_0 n}\}$. Then, as shown in the previous chapter, the output is equal to the convolution of the input and the unit-sample response of the system:

$$
\begin{aligned}
y(n) &= \sum_{k=-\infty}^{\infty} h(k)\, x(n-k) \\
&= \sum_{k=-\infty}^{\infty} h(k)\, e^{j\omega_0(n-k)} \\
&= e^{j\omega_0 n} \sum_{k=-\infty}^{\infty} h(k)\, e^{-j\omega_0 k}
\end{aligned}
\tag{3.1}
$$

We can interpret this last result by using complex-vector notation. Let us define a complex-valued coefficient $H(e^{j\omega_0})$ that is equal to the evaluation of the above sum, or

$$
H(e^{j\omega_0}) = \sum_{k=-\infty}^{\infty} h(k)\, e^{-j\omega_0 k} = |H(e^{j\omega_0})| \exp\{j \operatorname{Arg}[H(e^{j\omega_0})]\}
\tag{3.2}
$$

Including this coefficient in Eq. (3.1), we get

$$
\begin{aligned}
y(n) &= H(e^{j\omega_0})\, e^{j\omega_0 n} \\
&= |H(e^{j\omega_0})| \exp\{j(\omega_0 n + \arg[H(e^{j\omega_0})])\}
\end{aligned}
\tag{3.3}
$$

Hence, the output sequence is a complex exponential at the same frequency ω_0, but has been multiplied by a complex-valued time-invariant coefficient $H(e^{j\omega_0})$. This coefficient has introduced both gain and phase factors, as shown in Fig. 3.1.

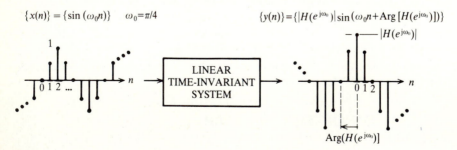

FIGURE 3-1
Magnitude and phase effects produced by a linear time-invariant system.

This result is true for any frequency ω and hence this coefficient can be generalized to be a function of frequency. We will denote this frequency function by $H(e^{j\omega})$, where

$$H(e^{j\omega}) = \sum_{n=-\infty}^{\infty} h(n)\, e^{-j\omega n} \qquad (3.4)$$

This equation defines the *Fourier transform* of the discrete-time sequence $\{h(n)\}$.

The Fourier transform of a sequence is said to *exist* if it can be expressed in a valid functional form. Since the computation involves summing a possibly infinite number of terms, the Fourier transform exists only for sequences that are *absolutely summable*. That is, given a sequence $\{h(n)\}$, $H(e^{j\omega})$ exists only when

$$\sum_{n=-\infty}^{\infty} |h(n)| < \infty \qquad (3.5)$$

When the sequence defines the unit-sample response of a digital filter, then its Fourier transform is called the *transfer function*. It defines the frequency transmission characteristics of the filter. When the sequence is a discrete-time signal $\{x(n)\}$, then its Fourier transform is given by

$$X(e^{j\omega}) = \sum_{n=-\infty}^{\infty} x(n)\, e^{-j\omega n} \qquad (3.6)$$

Then $X(e^{j\omega})$ defines the frequency content of $\{x(n)\}$, and will be called the *signal spectrum*.

Example 3.1. Transfer function of the 3-sample averager. Recall that the unit-sample response for the noncausal 3-sample averager is given by

$$h(n) = \begin{cases} \frac{1}{3} & \text{for } -1 \le n \le 1 \\ 0 & \text{otherwise} \end{cases}$$

Then its transfer function is given by

$$H(e^{j\omega}) = \sum_{n=-\infty}^{\infty} h(n)\, e^{-j\omega n} = \sum_{n=-1}^{1} \tfrac{1}{3} e^{-j\omega n}$$

$$= \tfrac{1}{3}(e^{j\omega} + 1 + e^{-j\omega})$$

This last result can be simplified by applying the Euler identity

$$e^{j\alpha} + e^{-j\alpha} = 2\cos\alpha$$

to get

$$H(e^{j\omega}) = \tfrac{1}{3}(1 + 2\cos\omega)$$

Example 3.2. Transfer function of the first-order recursive filter. Recall that the unit-sample response is given by

$$h(n) = a^n u(n) \qquad \text{for all } n$$

Because we have an infinite number of elements in $\{h(n)\}$, we must check to determine whether it is absolutely summable:

$$\sum_{n=-\infty}^{\infty} |h(n)| = \sum_{n=0}^{\infty} |a|^n$$

If $|a| \geq 1$, the sum becomes infinite and the Fourier transform does not exist. For $|a| < 1$, we can apply the infinite geometric sum formula

$$\sum_{n=0}^{\infty} |a|^n = \frac{1}{1 - |a|} \qquad \text{for } |a| < 1$$

from which we note that the sum is finite for $|a| < 1$. Then

$$H(e^{j\omega}) = \sum_{n=0}^{\infty} a^n e^{-j\omega n} = \sum_{n=0}^{\infty} (a e^{-j\omega})^n$$

Since $|a e^{-j\omega}| < 1$ when $|a| < 1$, we can apply the infinite geometric sum formula to obtain

$$H(e^{j\omega}) = \frac{1}{1 - a e^{-j\omega}} \qquad \text{for } |a| < 1.$$

The necessary condition to apply the infinite geometric sum formula in evaluating the Fourier transform is the same for the sequence to be absolutely summable. Hence, it is not necessary to determine whether the sequence is absolutely summable as a separate step.

Example 3.3. Transfer function of the second-order recursive filter. From Example 2.11, we have

$$h(n) = r^n \cos(\omega_0 n) u(n) \qquad \text{for all } n$$

Then

$$H(e^{j\omega}) = \sum_{n=0}^{\infty} r^n \cos(\omega_0 n) e^{-j\omega n}$$

Applying the Euler identity to the cosine, we have

$$H(e^{j\omega}) = \frac{1}{2} \left[\sum_{n=0}^{\infty} r^n e^{j\omega_0 n} e^{-j\omega n} + \sum_{n=0}^{\infty} r^n e^{-j\omega_0 n - j\omega n} \right]$$

$$= \frac{1}{2} \left[\sum_{n=0}^{\infty} (r e^{-j(\omega - \omega_0)})^n + \sum_{n=0}^{\infty} (r e^{-j(\omega + \omega_0)})^n \right]$$

For $|r| < 1$, we can apply the infinite geometric sum formula to get

$$H(e^{j\omega}) = \tfrac{1}{2}[(1 - r e^{-j(\omega - \omega_0)})^{-1} + (1 - r e^{-j(\omega + \omega_0)})^{-1}]$$

Combining over a common denominator, we have

$$H(e^{j\omega}) = \frac{1}{2} \frac{2 - r(e^{j\omega_0} + e^{-j\omega_0}) e^{-j\omega}}{1 - r(e^{j\omega_0} + e^{-j\omega_0}) e^{-j\omega} + r^2 e^{-j2\omega}}$$

$$= \frac{1 - r \cos(\omega_0) e^{-j\omega}}{1 - 2r \cos(\omega_0) e^{-j\omega} + r^2 e^{-j2\omega}}$$

In the examples above, the sequences were written in the form of the unit-sample response of a system $\{h(n)\}$. It is to be understood that the same results would have been obtained for a signal sequence $\{x(n)\}$, having the same element values. In this case, we would be calculating the spectrum, rather than the transfer function.

3.3 IMPORTANT PROPERTIES OF THE FOURIER TRANSFORM

We now present some important properties of the Fourier transform. These are summarized in Table 3.1.

LINEARITY. To demonstrate that the Fourier transform is a linear transformation, we apply the superposition principle to the definition. Let

$$\{y(n)\} = \{a\,x_1(n)\} + \{b\,x_2(n)\} \tag{3.7}$$

Then its Fourier transform is equal to

$$Y(e^{j\omega}) = \sum_{n=-\infty}^{\infty} y(n)\,e^{-j\omega n}$$

$$= \sum_{n=-\infty}^{\infty} [a\,x_1(n) + b\,x_2(n)]\,e^{-j\omega n}$$

$$= a\sum_{n=-\infty}^{\infty} x_1(n)\,e^{-j\omega n} + b\sum_{n=-\infty}^{\infty} x_2(n)\,e^{-j\omega n}$$

$$= aX_1(e^{j\omega}) + bX_2(e^{j\omega}) \tag{3.8}$$

Since $X_1(e^{j\omega})$ is the Fourier transform of $\{x_1(n)\}$ and $X_2(e^{j\omega})$ is that of $\{x_2(n)\}$, we have the desired result.

PERIODICITY. The Fourier transform is a periodic function of the continuous-valued variable ω, with period 2π. This can be demonstrated by increasing the frequency in the definition of the Fourier transform by $2k\pi$, where k is any integer. We then have

$$H(e^{j(\omega+2k\pi)}) = \sum_{n=-\infty}^{\infty} h(n)\,e^{-j(\omega+2k\pi)n}$$

$$= \sum_{n=-\infty}^{\infty} h(n)\,e^{-j\omega n}\,e^{-j2kn\pi} \tag{3.9}$$

For integer values of k and n, the case here, $e^{-j2kn\pi} = 1$. Hence, we have $H(e^{j(\omega+2k\pi)}) = H(e^{j\omega})$.

This periodicity can also be noted because the variable ω appears in the function only as $e^{j\omega}$, which is itself periodic with period 2π. If the argument of a function is periodic, it follows that the function itself is also periodic.

TABLE 3.1
Properties of the Fourier transform

1. Definition:

$$H(e^{j\omega}) = \sum_{n=-\infty}^{\infty} h(n) e^{-j\omega n} \quad \left(\text{exists when } \sum_{n=-\infty}^{\infty} |h(n)| < \infty\right)$$

2. Linearity: If

$$\{y(n)\} = \{x_1(n)\} + \{x_2(n)\} \quad \text{then} \quad Y(e^{j\omega}) = X_1(e^{j\omega}) + X_2(e^{j\omega})$$

3. Periodicity:

$$H(e^{j\omega}) = H(e^{j(\omega + 2k\pi)}) \quad \text{for any integer } k$$

4. Magnitude and phase functions: If

$$H(e^{j\omega}) = H_R(e^{j\omega}) + jH_I(e^{j\omega})$$

then

$$|H(e^{j\omega})|^2 = H_R^2(e^{j\omega}) + H_I^2(e^{j\omega})$$
$$\text{Arg}[H(e^{j\omega})] = \arctan[H_I(e^{j\omega})/H_R(e^{j\omega})]$$

5. Fourier transform of delayed sequence: For any value of n_0, if

$$\{y(n)\} = \{x(n - n_0)\} \quad \text{then} \quad Y(e^{j\omega}) = X(e^{j\omega}) e^{-j\omega n_0}$$

6. Fourier transform of the convolution of two sequences: If

$$\{y(n)\} = \{h(n)\} * \{x(n)\} \quad \text{then} \quad Y(e^{j\omega}) = H(e^{j\omega})X(e^{j\omega})$$

7. Fourier transform of the product of two sequences: If

$$\{y(n)\} = \{h(n)x(n)\} \quad \text{then} \quad Y(e^{j\omega}) = H(e^{j\omega}) \circledast X(e^{j\omega}).$$

$$\text{(periodic convolution)}$$

8. For the *real-valued* sequence $\{h(n)\}$:

$$H_R(e^{j\omega}) = \sum_{n=-\infty}^{\infty} h(n) \cos(\omega n) \qquad H_I(e^{j\omega}) = -\sum_{n=-\infty}^{\infty} h(n) \sin(\omega n)$$

   (a) Complex-conjugate symmetry: $H(e^{j\omega}) = H^*(e^{-j\omega})$
   (b) Real component is even function: $H_R(e^{j\omega}) = H_R(e^{-j\omega})$
   (c) Imaginary component is odd function: $H_I(e^{j\omega}) = -H_I(e^{-j\omega})$
   (d) Magnitude spectrum is even function: $|H(e^{j\omega})| = |H(e^{-j\omega})|$
   (e) Phase spectrum is odd function: $\text{Arg}[H(e^{j\omega})] = -\text{Arg}[H(e^{-j\omega})]$
   (f) If $\{h(n)\} = \{h(-n)\}$ (even sequence), then $H(e^{j\omega})$ is purely real.
   (g) If $\{h(n)\} = \{-h(-n)\}$ (odd sequence), then $H(e^{j\omega})$ is purely imaginary.

REAL AND IMAGINARY COMPONENTS OF THE FOURIER TRANSFORM. Since the Fourier transform is a complex function, it is helpful to define alternate forms that can be plotted on a pair of two-dimensional graphs. Let us consider the general case of a complex sequence $\{h(n)\}$, which can be written as

$$h(n) = h_R(n) + jh_I(n) \tag{3.10}$$

where $h_R(n)$ is the real component and $h_I(n)$ is the imaginary component. By applying Euler's identity to the definition of the Fourier transform, we have

$$H(e^{j\omega}) = \sum_{n=-\infty}^{\infty} h(n)\, e^{-j\omega n} = \sum_{n=-\infty}^{\infty} [h_R(n) + jh_I(n)][\cos(\omega n) - j\sin(\omega n)]$$

$$(3.11)$$

This function can be expressed explicitly in terms of its real and imaginary parts as

$$H(e^{j\omega}) = H_R(e^{j\omega}) + jH_I(e^{j\omega}) \qquad (3.12)$$

where

$$H_R(e^{j\omega}) = \sum_{n=-\infty}^{\infty} [h_R(n)\cos(\omega n) + h_I(n)\sin(\omega n)] \qquad (3.13)$$

and

$$H_I(e^{j\omega}) = \sum_{n=-\infty}^{\infty} [h_I(n)\cos(\omega n) - h_R(n)\sin(\omega n)] \qquad (3.14)$$

MAGNITUDE AND PHASE COMPONENTS OF THE FOURIER TRANSFORM. Being a complex quantity, we can also express $H(e^{j\omega})$ as a complex vector:

$$H(e^{j\omega}) = |H(e^{j\omega})| \exp\{j\, \text{Arg}[H(e^{j\omega})]\} \qquad (3.15)$$

If $H(e^{j\omega})$ denotes the transfer function of a system, it is conventional to call $|H(e^{j\omega})|$ the *magnitude response*, and $\text{Arg}[H(e^{j\omega})]$ the *phase response*. For signal sequences, $H(e^{j\omega})$ is the spectrum, $|H(e^{j\omega})|$ is called the *magnitude spectrum*, and $\text{Arg}[H(e^{j\omega})]$ is the *phase spectrum*. The magnitude response is equal to

$$|H(e^{j\omega})| = (H(e^{j\omega})H^*(e^{j\omega}))^{1/2} \qquad (3.16)$$

where $H^*(e^{j\omega})$ is the *complex-conjugate* of $H(e^{j\omega})$. An equivalent expression for the magnitude is given by

$$|H(e^{j\omega})| = (H_R^2(e^{j\omega}) + H_I^2(e^{j\omega}))^{1/2} \qquad (3.17)$$

The phase response is equal to

$$\text{Arg}[H(e^{j\omega})] = \arctan[H_I(e^{j\omega})/H_R(e^{j\omega})] \qquad (3.18)$$

The phase is an indeterminate quantity when $|H(e^{j\omega})| = 0$, which occurs at values of ω for which $H_R(e^{j\omega}) = H_I(e^{j\omega}) = 0$.

Before giving some examples of computing the magnitude and phase responses, we show that we can usually simplify the calculation of the Fourier transform for *real-valued sequences* by considering the special properties of their transforms.

3.4 PROPERTIES OF THE FOURIER TRANSFORM FOR REAL-VALUED SEQUENCES

If $\{h(n)\}$ is a sequence of real numbers, then its Fourier transform can be written as

$$H(e^{j\omega}) = \sum_{n=-\infty}^{\infty} h(n)\, e^{-j\omega n} = \sum_{n=-\infty}^{\infty} h(n) \cos(\omega n) - j \sum_{n=-\infty}^{\infty} h(n) \sin(\omega n)$$

$$= H_R(e^{j\omega}) + jH_I(e^{j\omega}) \tag{3.19}$$

The Fourier transform of a real sequence has the following useful symmetry properties.

SYMMETRY OF THE REAL AND IMAGINARY PARTS. Let us consider the effect of reversing the sign of the frequency variable in the argument of the transform. We can then write

$$H(e^{-j\omega}) = H_R(e^{-j\omega}) + jH_I(e^{-j\omega}) \tag{3.20}$$

Evaluating $H(e^{-j\omega})$, we find

$$H(e^{-j\omega}) = \sum_{n=-\infty}^{\infty} h(n)\, e^{j\omega n} = \sum_{n=-\infty}^{\infty} h(n) \cos(\omega n) + j \sum_{n=-\infty}^{\infty} h(n) \sin(\omega n) \tag{3.21}$$

Comparing Eqs. (3.19) and (3.21), we note that

$$H_R(e^{j\omega}) = \sum_{n=-\infty}^{\infty} h(n) \cos(\omega n) = H_R(e^{-j\omega}) \tag{3.22}$$

indicating that the real part is an even function of ω. Comparing the imaginary parts, we find

$$H_I(e^{j\omega}) = - \sum_{n=-\infty}^{\infty} h(n) \sin(\omega n) = -H_I(e^{-j\omega}) \tag{3.23}$$

indicating that the imaginary part is an odd function of ω. Also, since the complex conjugate is obtained by changing the sign of the j terms, we find

$$H(e^{-j\omega}) = H^*(e^{j\omega}). \tag{3.24}$$

Since $H(e^{j\omega})$ is a periodic function of ω with period 2π, an important consequence of these results is that, for real-valued discrete-time sequences, all the information in the Fourier transform is contained in the frequency range $0 \le \omega \le \pi$. The unspecified portion of the Fourier transform in the range $-\pi \le \omega < 0$ can be determined from these symmetry properties.

SYMMETRIES OF THE MAGNITUDE AND PHASE FUNCTIONS. The magnitude response is also an even function of frequency, since

$$|H(e^{j\omega})| = (H(e^{j\omega})H^*(e^{j\omega}))^{1/2} = (H^*(e^{-j\omega})H(e^{-j\omega}))^{1/2} = |H(e^{-j\omega})| \tag{3.25}$$

The phase response can be written

$$\text{Arg}[H(e^{-j\omega})] = \arctan[H_I(e^{-j\omega})/H_R(e^{-j\omega})]$$
$$= -\arctan[H_I(e^{j\omega})/H_R(e^{j\omega})]$$
$$= -\text{Arg}[H(e^{j\omega})] \tag{3.26}$$

indicating that the phase response is an odd function of frequency. To illustrate the above properties let us consider a few examples.

Example 3.4. Magnitude and phase spectra of the delayed unit-sample sequence. Let

$$d(n-k) = \begin{cases} 1 & \text{for } n = k \\ 0 & \text{otherwise} \end{cases}$$

Since the unit-sample sequence is commonly used as the input to a filter, it is considered to be a signal sequence and it is proper the call its Fourier transform a spectrum. Let us denote the spectrum of $\{d(n-k)\}$ by $D_k(e^{j\omega})$. Then

$$D_k(e^{j\omega}) = \sum_{n=-\infty}^{\infty} d(n-k)\, e^{-j\omega n} = e^{-j\omega k}$$

The real and imaginary parts are most easily determined by applying the Euler identities:

$$D_{k,R}(e^{j\omega}) = \cos(\omega k) \qquad \text{and} \qquad D_{k,I}(e^{j\omega}) = -\sin(\omega k)$$

The magnitude spectrum is equal to

$$|D_k(e^{j\omega})| = [\cos^2(\omega k) + \sin^2(\omega k)]^{1/2} = 1$$

or, alternately,

$$|D_k(e^{j\omega})| = (e^{-j\omega k})(e^{j\omega k})^{1/2} = 1$$

The phase spectrum is equal to

$$\text{Arg}[D_k(e^{j\omega})] = \arctan[-\sin(\omega k)/\cos(\omega k)] = \arctan[-\tan(\omega k)] = -\omega k$$

The magnitude and phase spectra could also have been obtained by inspection by noting that $D_k(e^{j\omega})$ is already in the complex vector form. The magnitude and phase functions are shown in Fig. 3.2 for different values of k.

In Fig. 3.2, for $k = 0$, the magnitude spectrum is constant with frequency and the phase spectrum is identically zero. It is for this reason that $\{d(n)\}$ is desirable as a filter input, since the spectrum of the output sequence can then be attributed to the filter transfer function. For the unit-sample sequence delayed by k samples, the magnitude response is still constant with frequency, but the phase response is linear with frequency, equal to $-k\omega$.

It is the accepted convention to plot the *principal value* of the phase lying in the range $[-\pi, \pi]$. Even though the phase values of $+\pi$ and $-\pi$ are equivalent, we will allow both to occur to accommodate the plotting conventions described below. Note from Fig. 3.2, that when the phase exceeds

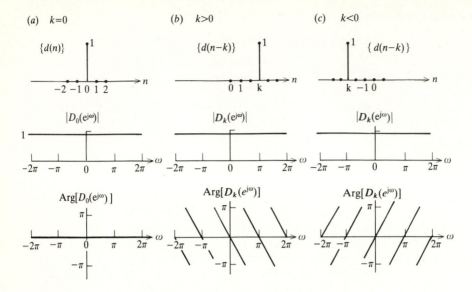

FIGURE 3-2
Magnitude and phase spectra of the unit-sample sequence. (*a*) Standard unit-sample sequence ($k = 0$); (*b*) Delayed sequence ($k > 0$); (*c*) Delayed sequence ($k < 0$).

this range, a jump of $\pm 2\pi$ is needed to bring the phase back into this range. This principal value range is also convenient, since it is the value returned by the commonly available ATAN2(Y, X) computer function, where Y is the imaginary component and X is the real component.

> **Example 3.5. Magnitude and phase response of the 3-sample averager.** From Example 3.1, we have
>
> $$H(e^{j\omega}) = \tfrac{1}{3}(1 + 2\cos(\omega))$$
>
> which is a real-valued function of ω. The magnitude response is equal to
>
> $$|H(e^{j\omega})| = \left|\frac{1 + 2\cos(\omega)}{3}\right|$$
>
> and the phase response is equal to
>
> $$\text{Arg}[H(e^{j\omega})] = \begin{cases} 0 & \text{when } H(e^{j\omega}) > 0 \quad \text{or for } -2\pi/3 < \omega < 2\pi/3 \\ \pm\pi & \text{when } H(e^{j\omega}) < 0 \quad \text{or for } -\pi \le \omega < -2\pi/3 \end{cases}$$
>
> $$\text{and } 2\pi/3 < \omega < \pi.$$
>
> The magnitude and phase responses are shown in Fig. 3.3. The appropriate sign of π is chosen to make the $\text{Arg}[H(e^{j\omega})]$ an odd function of frequency.

As a filter, the magnitude response indicates that the 3-sample averager is a lowpass filter. If the same discrete-time sequence were a signal, then the

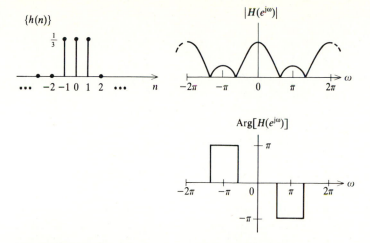

FIGURE 3-3
Magnitude and phase responses of the 3-sample averager.

magnitude spectrum indicates that it would have more energy at low frequencies than at high frequencies.

The previous example illustrates that precautions must be taken when determining the phase response of a filter having a real-valued transfer function, because negative real values produce an additional phase of π radians. For example, let us consider the following *linear-phase* form of the transfer function that will occur repeatedly throughout the book:

$$H(e^{j\omega}) = B(e^{j\omega})\, e^{-jk\omega} \tag{3.27}$$

where $B(e^{j\omega})$ is a real-valued function of ω, called the *amplitude* function, that can take on both positive and negative values. When expressing this transfer function in the complex vector notation,

$$H(e^{j\omega}) = |H(e^{j\omega})|\, \exp\{j\, \text{Arg}[H(e^{j\omega})]\} \tag{3.28}$$

then the phase function must include the linear phase term and also accommodate for the sign changes in $B(e^{j\omega})$. Since -1 can be expressed as $e^{\pm j\pi}$, phase jumps of $\pm\pi$ will occur at frequencies where $B(e^{j\omega})$ changes sign. This jump must be added to the phase component that is calculated from the arctangent function. If $B(e^{j\omega}) \geq 0$ for all ω, then no jumps will occur and the phase function is given by $-k\omega$.

Example 3.6. Linear-phase Fourier transform. Let

$$h(0) = \tfrac{1}{2} \qquad h(1) = 1 \qquad h(2) = \tfrac{1}{2}, \qquad h(n) = 0 \text{ otherwise}$$

Then the Fourier transform is equal to

$$H(e^{j\omega}) = \tfrac{1}{2} + e^{-j\omega} + \tfrac{1}{2}e^{-j2\omega}$$

which can be expressed as

$$H(e^{j\omega}) = e^{-j\omega} + \tfrac{1}{2}e^{-j\omega}(e^{j\omega} + e^{-j\omega})$$

$$= (1 + \cos(\omega))e^{-j\omega}$$

This last form is the linear-phase form given above. In this case, the amplitude function is never negative and the phase response is simply $-\omega$.

PHASE JUMPS. From the previous examples, we note that there are two occasions for which the phase function experiences discontinuities, or *jumps*.

1. A jump of $\pm 2\pi$ occurs to maintain the phase function within the principal value range of $[-\pi, \pi]$.
2. *A jump of $\pm \pi$ occurs when $B(e^{j\omega})$ undergoes a change in sign.*

The sign of the phase jump is chosen such that the resulting phase function is odd and, after the jump, lies in the range $[-\pi, \pi]$.

Example 3.7. Magnitude and phase response of the first-order recursive filter. From Example 3.2, we have $H(e^{j\omega}) = (1 - a\,e^{-j\omega})^{-1}$. The magnitude response is most easily obtained by calculating the squared magnitude function, or

$$|H(e^{j\omega})|^2 = H(e^{j\omega})H^*(e^{j\omega})$$

$$= (1 - a\,e^{-j\omega})^{-1}(1 - a\,e^{j\omega})^{-1}$$

$$= (1 - 2a\cos(\omega) + a^2)^{-1}$$

To find the real and imaginary parts, we make the denominator real by multiplying by its complex conjugate:

$$H(e^{j\omega}) = \frac{1}{1 - a\,e^{-j\omega}} \frac{1 - a\,e^{j\omega}}{1 - a\,e^{j\omega}}$$

$$= \frac{1 - a\,e^{j\omega}}{1 - a(e^{j\omega} + e^{-j\omega}) + a^2}$$

$$= \frac{1 - a\cos(\omega) - ja\sin(\omega)}{1 - 2a\cos(\omega) + a^2}$$

Since the denominator is real, we can now separate the numerator terms to determine the real and imaginary parts. The real part is then

$$H_R(e^{j\omega}) = \frac{1 - a\cos(\omega)}{1 - 2a\cos(\omega) + a^2}$$

The real part is an even function of frequency. The imaginary part is equal to

$$H_I(e^{j\omega}) = \frac{-a\sin(\omega)}{1 - 2a\cos(\omega) + a^2}$$

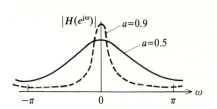

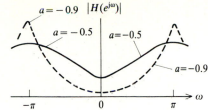

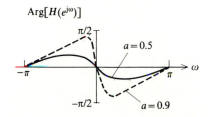

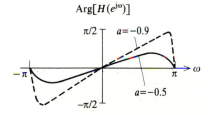

FIGURE 3-4
Magnitude and phase responses of first-order recursive filter for different values of a.

and is an odd function of frequency. The phase response is then equal to

$$\text{Arg}[H(e^{j\omega})] = \arctan[H_I(e^{j\omega})/H_R(e^{j\omega})]$$

$$= \arctan\left[\frac{-a\sin(\omega)}{1 - a\cos(\omega)}\right]$$

These responses are shown in Fig. 3.4 for different values of a. When $0 < a < 1$, a lowpass filter results, and when $-1 < a < 0$, a highpass filter results.

Example 3.8. Magnitude and phase response of the second-order recursive filter. From Example 3.3, we have

$$H(e^{j\omega}) = \frac{1 - r\cos(\omega_0)\,e^{-j\omega}}{1 - 2r\cos(\omega_0)\,e^{-j\omega} + r^2 e^{-j2\omega}}$$

For simplicity of notation, let us write $H(e^{j\omega})$ in the following form

$$H(e^{j\omega}) = \frac{1 + a\,e^{-j\omega}}{1 + \alpha\,e^{-j\omega} + \beta\,e^{-j2\omega}} = \frac{N(e^{j\omega})}{D(e^{j\omega})}$$

where $N(e^{j\omega})$ is the numerator and $D(e^{j\omega})$ is the denominator. The magnitude response is then

$$|H(e^{j\omega})| = \left(\frac{N(e^{j\omega})\,N^*(e^{j\omega})}{D(e^{j\omega})\,D^*(e^{j\omega})}\right)^{1/2}$$

$$= \frac{[1 + 2a\cos(\omega) + a^2]^{1/2}}{[1 + \alpha^2 + \beta^2 + 2\alpha(1 + \beta)\cos(\omega) + 2\beta\cos(2\omega)]^{1/2}}$$

Note from this rather complicated form that ω appears only as an argument of

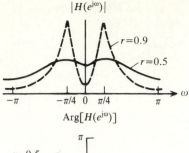

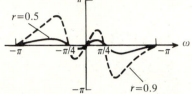

FIGURE 3-5
Magnitude and phase responses of second-order recursive filter for different values of r.

the cosine function, which is an even function of ω. This is reasonable, since $|H(e^{j\omega})|$ is itself an even function of ω.

The real and imaginary parts are obtained by computing

$$H(e^{j\omega}) = \frac{N(e^{j\omega})}{D(e^{j\omega})}\frac{D^*(e^{j\omega})}{D^*(e^{j\omega})}$$

$$= \frac{1 + a\alpha + a\,e^{-j\omega} + (a\beta + \alpha)\,e^{j\omega} + \beta\,e^{j2\omega}}{1 + \alpha^2 + \beta^2 + 2\alpha(1 + \beta)\cos(\omega) + 2\beta\cos(2\omega)}$$

From this expression we find

$$H_R(e^{j\omega}) = \frac{1 + a\alpha + (a + a\beta + \alpha)\cos(\omega) + \beta\cos(2\omega)}{1 + \alpha^2 + \beta^2 + 2\alpha(1 + \beta)\cos(\omega) + 2\beta\cos(2\omega)}$$

and

$$H_I(e^{j\omega}) = \frac{(a\beta + \alpha - a)\sin(\omega) + \beta\sin(2\omega)}{1 + \alpha^2 + \beta^2 + 2\alpha(1 + \beta)\cos(\omega) + 2\beta\cos(2\omega)}$$

The phase response is then given by

$$\text{Arg}[H(e^{j\omega})] = \arctan\left[\frac{(a\beta + \alpha - a)\sin(\omega) + \beta\sin(2\omega)}{1 + a\alpha + (a + a\beta + \alpha)\cos(\omega) + \beta\cos(2\omega)}\right]$$

The magnitude and phase response curves are shown in Fig. 3.5 for $\omega_0 = \pi/4$ and various values of r.

The magnitude response shows a sharp peak close to the frequency $\omega = \pi/4 = \omega_0$, which is called the *resonant frequency*. This peak will be employed later to implement the *passband* features of a digital filter.

Example 3.9. Magnitude and phase response of causal 3-sample averager. Let

$$h(n) = \begin{cases} \frac{1}{3} & \text{for } 0 \le n \le 2 \\ 0 & \text{otherwise} \end{cases}$$

Then

$$H(e^{j\omega}) = \sum_{n=-\infty}^{\infty} h(n)\, e^{-j\omega n} = \frac{1}{3} \sum_{n=0}^{2} e^{-j\omega n}$$

We shall evaluate this sum using three methods: the mathematically tedious approach, the enlightened direct approach and by applying the finite geometric sum formula.

Mathematically tedious method. Evaluating the sum, we get

$$\begin{aligned}
H(e^{j\omega}) &= \tfrac{1}{3}[1 + e^{-j\omega} + e^{-j2\omega}] \\
&= \tfrac{1}{3}[1 + \cos(\omega) - j\sin(\omega) + \cos(2\omega) - j\sin(2\omega)] \\
&= \tfrac{1}{3}[1 + \cos(\omega) + \cos(2\omega)] - \tfrac{1}{3}j[\sin(\omega) + \sin(2\omega)] \\
&= H_R(e^{j\omega}) + jH_I(e^{j\omega})
\end{aligned}$$

$$\begin{aligned}
|H(e^{j\omega})|^2 &= H_R^2(e^{j\omega}) + H_I^2(e^{j\omega}) \\
&= \tfrac{1}{9}[1 + 2(1 + \cos(2\omega)) + 4\cos(\omega)] \\
&= \tfrac{1}{9}[1 + 4\cos^2(\omega) + 4\cos(\omega)] = \tfrac{1}{9}[1 + 2\cos(\omega)]^2
\end{aligned}$$

Taking the square root, we get

$$|H(e^{j\omega})| = \tfrac{1}{3}|1 + 2\cos(\omega)|$$

This result is shown in Fig. 3.6.

Direct method. To allow an Euler identity substitution, we factor out $e^{-j\omega}$:

$$\begin{aligned}
H(e^{j\omega}) &= \tfrac{1}{3}e^{-j\omega}(e^{j\omega} + 1 + e^{-j\omega}) \\
&= \tfrac{1}{3}(1 + 2\cos(\omega))e^{-j\omega} = B(e^{j\omega})e^{-j\omega}
\end{aligned}$$

From which we get

$$|H(e^{j\omega})| = \tfrac{1}{3}|1 + 2\cos(\omega)|$$

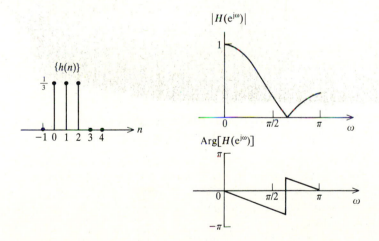

FIGURE 3-6
Magnitude and phase responses of causal 3-sample averager.

and

$$\text{Arg}[H(e^{j\omega})] = \begin{cases} -\omega & \text{for } -2\pi/3 < \omega < 2\pi/3, \quad \text{or when } B(e^{j\omega}) > 0 \\ -\omega \pm \pi & \text{for } -\pi \le \omega < -2\pi/3 \quad \text{or when } B(e^{j\omega}) < 0. \end{cases}$$
$$\text{and } 2\pi/3 < \omega \le \pi,$$

The phase is undefined at frequency points where $|H(e^{j\omega})| = 0$.

Geometric sum method. Applying the finite geometric sum formula with $r = e^{-j\omega}$, we get

$$H(e^{j\omega}) = \frac{1}{3} \frac{1 - e^{-j3\omega}}{1 - e^{-j\omega}}$$

The above form can be simplified by substituting the following identity:

$$1 - e^{-j\alpha} = e^{-j\alpha/2}(e^{j\alpha/2} - e^{-j\alpha/2})$$
$$= 2j\, e^{-j\alpha/2} \sin(\alpha/2)$$

After this substitution, we get

$$H(e^{j\omega}) = \frac{e^{-j3\omega/2}(e^{j3\omega/2} - e^{-j3\omega/2})}{3\, e^{-j\omega/2}(e^{j\omega/2} - e^{-j\omega/2})}$$

$$= \tfrac{1}{3}e^{-j\omega} \frac{\sin(3\omega/2)}{\sin(\omega/2)} = B(e^{j\omega})e^{-j\omega}$$

This last form is the amplitude with linear-phase form described above in Eq. (3.27). From that discussion, we find

$$|H(e^{j\omega})| = \frac{|\sin(3\omega/2)|}{|3 \sin(\omega/2)|}$$

and

$$\text{Arg}[H(e^{j\omega})] = \begin{cases} -\omega & \text{when } B(e^{j\omega}) > 0 \\ -\omega \pm \pi & \text{when } B(e^{j\omega}) < 0 \end{cases}$$

The magnitude response, an even function of ω, is expressed as the ratio of two odd functions of ω. The equivalence of this form with that obtained with the previous two methods is left as an exercise.

3.5 PROGRAM TO EVALUATE THE FOURIER TRANSFORM BY COMPUTER

To provide computer verification of the analysis above, we will describe a simple program to evaluate the Fourier transform of a real-valued sequence at N_P points over the frequency range $0 \le \omega < 2\pi$. In the next chapter, we will present a more efficient but mathematically more complicated algorithm for performing this procedure.

Although not the most efficient method, the evaluation of the real and imaginary parts at a frequency point ω_k can easily be performed directly from the definition. For example, let $h(n)$, for $0 \le n \le N_H - 1$, be a sequence of real-valued elements. Then, the real part of the Fourier transform evaluated at frequency ω_k, denoted by $H_R(e^{j\omega_k})$, is given by

$$H_R(e^{j\omega_k}) = \sum_{n=0}^{N_H-1} h(n)\cos(\omega_k n) \qquad (3.29)$$

The imaginary part, $H_I(e^{j\omega_k})$, is given by

$$H_I(e^{j\omega_k}) = -\sum_{n=0}^{N_H-1} h(n)\sin(\omega_k n) \qquad (3.30)$$

To obtain a discrete-frequency approximation to the continuous-frequency function $H(e^{j\omega})$, we can evaluate the real and imaginary parts at a set of N_P equally-spaced points over $0 \le \omega < 2\pi$, i.e. at $\omega_k = 2k\pi/N_P$, for $0 \le k \le N_P - 1$.

When processing arrays whose indexing starts at one, rather than zero, the above equations for the real and imaginary parts must be modified. If $H(K)$, for $1 \le K \le NH$, is the array of discrete-time values, with $H(1)$ equal to $h(0)$, then the array containing the real part, denoted by $HR(K)$, is given by

$$HR(K) = \sum_{N=1}^{NH} H(N) * \cos(2\pi(K-1)(N-1)/N_P) \qquad \text{for } 1 \le K \le N_P \quad (3.31)$$

The frequency of the Kth cosine sequence is equal to $2\pi(K-1)/N_P$, for $1 \le K \le N_P$. The array containing the imaginary part, $H_I(K)$, is

$$HI(K) = -\sum_{N=1}^{NH} H(N) * \sin(2\pi(K-1)(N-1)/N_P) \qquad \text{for } 1 \le K \le N_P \quad (3.32)$$

Using the programming style described in the previous chapter, the computation of the Fourier transform should be performed with the subprogram, denoted by

```
FT(H,NH,HR,HI,NP)
```

where

H	(real input array) contains the elements of the discrete-time sequence;
NH	(integer input) is the number of elements in H;
HR	(real output array) contains the real part of the Fourier transform;
HI	(real output array) contains the imaginary part of the Fourier transform;
NP	(integer input) is the number of frequency points over $0 \le \omega < 2\pi$.

For displaying the Fourier transform, the magnitude and phase responses are conventionally used. The magnitude and phase responses should be computed using the subprogram denoted by

```
MAGPHS(HR,HI,HMAG,HPHS,NP)
```

where

HR   (input real array) contains the real part of the Fourier transform;

HI   (input real array) contains the imaginary part of the Fourier transform;

HMAG   (output real array) contains the magnitude values;

HPHS   (output real array) contains the phase values; and

NP   (input integer) is the number of points over $0 \leq \omega < 2\pi$.

The array of magnitude values can be computed as

$$\text{HMAG(K)} = \text{SQRT}[\text{HR(K)} * \text{HR(K)} + \text{HI(K)} * \text{HI(K)}] \qquad \text{for } 1 \leq \text{K} \leq \text{NP} \tag{3.33}$$

The array of phase values can be computed as

$$\text{HPHS(K)} = \text{ATAN2(HI(K), HR(K))} \qquad \text{for } 1 \leq \text{K} \leq \text{NP} \tag{3.34}$$

When both HI(K) and HR(K) are zero, the arctangent function should not be used, but rather a value of zero should be assigned for the phase value.

Some programming languages only have the ATAN(Y/X) function, which generates the phase in the first and fourth quadrants, i.e. in the range $[-\pi/2, \pi/2]$. It is left as an exercise for the student to determine the principal value of the phase using the ATAN function and the sign of the real part.

Because of the symmetry properties for real-valued sequences, we need to evaluate the plot the Fourier transform only over the range $0 \leq \omega \leq \pi$. Since both the frequency points $\omega = 0$ and $\omega = \pi$ are included, the number of points to be computed is equal to one more than half the points covering one entire period, or $(N_P/2) + 1$.

3.6 USE OF THE FOURIER TRANSFORM IN SIGNAL PROCESSING

The popularity of the Fourier transform for digital signal processing is due to the fact that the complicated convolutional operation in the time domain is converted to a much simpler multiplicative operation in the frequency domain. To demonstrate this, let $\{y(n)\}$ be the output of a LTI system. Then, its Fourier transform is given by

$$Y(e^{j\omega}) = \sum_{n=-\infty}^{\infty} y(n) e^{-j\omega n} \tag{3.35}$$

Applying the convolutional definition of the output, we have

$$Y(e^{j\omega}) = \sum_{n=-\infty}^{\infty} \left[\sum_{k=-\infty}^{\infty} h(k) x(n-k) \right] e^{-j\omega n} \tag{3.36}$$

Multiplying this last equation by unity, in the form $e^{j\omega k} e^{-j\omega k}$, we get

$$Y(e^{j\omega}) = \sum_{n=-\infty}^{\infty} \sum_{k=-\infty}^{\infty} [h(k) e^{-j\omega k} x(n-k) e^{-j\omega(n-k)}] \qquad (3.37)$$

Since $h(k) e^{-j\omega k}$ is not a function of n, it can be factored out of the summation with respect to n, giving

$$Y(e^{j\omega}) = \sum_{k=-\infty}^{\infty} \left[h(k) e^{-j\omega k} \sum_{n=-\infty}^{\infty} x(n-k) e^{-j\omega(n-k)} \right] \qquad (3.38)$$

In the sum over n, the term $(n-k)$ represents a shift of $\{x(n)\}$ by k samples. Further, note that the value of the exponent follows the index of $\{x(n)\}$. Since the summation is over the entire sequence, this shift is inconsequential in evaluating the sum. By replacing $(n-k)$ with the dummy variable m, we note that the second summation becomes the definition of $X(e^{j\omega})$, the Fourier transform of $\{x(n)\}$. Since $X(e^{j\omega})$ is not a function of k, we can factor it out of the summation over k. The remaining terms in the sum then define the Fourier transform of $\{h(n)\}$. Hence, we have arrived at the desired result

$$Y(e^{j\omega}) = H(e^{j\omega}) X(e^{j\omega}) \qquad (3.39)$$

This last result indicates that the input spectrum $X(e^{j\omega})$ is changed through a multiplicative operation by the filter transfer function $H(e^{j\omega})$ to produce the output spectrum $Y(e^{j\omega})$. Applying the complex vector notation to the output spectrum, we obtain

$$\begin{aligned} Y(e^{j\omega}) &= |Y(e^{j\omega})| \exp\{j \, \text{Arg}[Y(e^{j\omega})]\} \\ &= |H(e^{j\omega})| \exp\{j \, \text{Arg}[H(e^{j\omega})]\} \, |X(e^{j\omega})| \exp\{j \, \text{Arg}[X(e^{j\omega})]\} \quad (3.40) \\ &= |H(e^{j\omega})| \, |X(e^{j\omega})| \exp\{j(\text{Arg}[H(e^{j\omega})] + \text{Arg}[X(e^{j\omega})])\} \end{aligned}$$

Equating the magnitudes and the phases, we find

$$|Y(e^{j\omega})| = |H(e^{j\omega})| \, |X(e^{j\omega})| \qquad (3.41)$$

and

$$\text{Arg}[Y(e^{j\omega})] = \text{Arg}[H(e^{j\omega})] + \text{Arg}[X(e^{j\omega})] \qquad (3.42)$$

In words, the output magnitude spectrum is equal to the product of the input magnitude spectrum and the filter magnitude response. The output phase spectrum is equal to the sum of the input phase spectrum and the filter phase response.

Example 3.10. Signal processing with the Fourier transform. Let the input sequence be given by

$$x(n) = \begin{cases} \frac{1}{3} & \text{for } -1 \le n \le 1 \\ 0 & \text{otherwise} \end{cases}$$

and the filter unit-sample response by

$$h(n) = \begin{cases} a^n & \text{for } n \geq 0 \\ 0 & \text{otherwise} \end{cases}$$

From Examples 3.1 and 3.2, we have

$$X(e^{j\omega}) = \tfrac{1}{3}(1 + 2\cos(\omega))$$

and

$$H(e^{j\omega}) = \frac{1}{1 - a\,e^{-j\omega}}$$

The output spectrum is then equal to

$$Y(e^{j\omega}) = H(e^{j\omega})\,X(e^{j\omega}) = \frac{1 + 2\cos(\omega)}{3(1 - a\,e^{-j\omega})}$$

The output squared-magnitude spectrum is equal to

$$|Y(e^{j\omega})|^2 = \left(\frac{1 + 2\cos(\omega)}{3(1 - a\,e^{-j\omega})}\right)\left(\frac{1 + 2\cos(\omega)}{3(1 - a\,e^{j\omega})}\right)$$

$$= \frac{(1 + 2\cos(\omega))^2}{9(1 - 2a\cos(\omega) + a^2)}$$

The output phase spectrum can be computed in two ways. First, it can be obtained from the real and imaginary components of $Y(e^{j\omega})$, determined from

$$Y(e^{j\omega}) = \left(\frac{1 + 2\cos(\omega)}{3(1 - a\,e^{-j\omega})}\right)\left(\frac{1 - a\,e^{j\omega}}{1 - a\,e^{j\omega}}\right)$$

$$= \frac{(1 - a\cos\omega)(1 + 2\cos(\omega)) - ja\sin(\omega)(1 + 2\cos(\omega))}{3(1 - 2a\cos(\omega) + a^2)}$$

Then

$$\text{Arg}[Y(e^{j\omega})] = \arctan\left(\frac{-a\sin(\omega)(1 + 2\cos(\omega))}{(1 - a\cos(\omega))(1 + 2\cos(\omega))}\right) = \arctan\left(\frac{-a\sin(\omega)}{1 - a\cos(\omega)}\right)$$

The phase can also be determined by adding the phase response of the filter to the phase spectrum of the input:

$$\text{Arg}[Y(e^{j\omega})] = \text{Arg}[H(e^{j\omega})] + \text{Arg}[X(e^{j\omega})]$$

$$= 0 + \arctan\left(\frac{-a\sin(\omega)}{1 - a\cos(\omega)}\right)$$

3.7 FOURIER TRANSFORM OF SPECIAL SEQUENCES

In this section, we consider the Fourier transform of several special sequences that simplify the computations. These results will be helpful for the analysis of digital filters later.

FOURIER TRANSFORM OF A DELAYED SEQUENCE. Let us examine the effect on the Fourier transform when a sequence is delayed by k sample points, as shown in Fig. 3.7. Let us denote the delayed sequence by $\{h_k(n)\} = \{h(n-k)\}$. Then its Fourier transform, denoted by $H_k(e^{j\omega})$, is equal to

$$H_k(e^{j\omega}) = \sum_{n=-\infty}^{\infty} h_k(n)\, e^{-j\omega n}$$

$$= \sum_{n=-\infty}^{\infty} h(n-k)\, e^{-j\omega n} \tag{3.43}$$

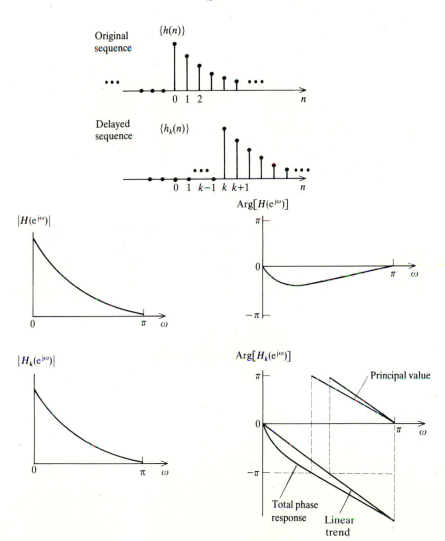

FIGURE 3-7
Comparison of the magnitude and phase spectra for original and delayed sequences.

Multiplying the right side by $e^{-j\omega k} e^{j\omega k}$, we have

$$H_k(e^{j\omega}) = \sum_{n=-\infty}^{\infty} h(n-k) e^{-j\omega(n-k)} e^{-j\omega k}$$

$$= e^{-j\omega k} \sum_{n=-\infty}^{\infty} h(n-k) e^{-j\omega(n-k)} \tag{3.44}$$

By substituting $m = n - k$ in the above summation, we recognize the definition of $H(e^{j\omega})$. Hence,

$$H_k(e^{j\omega}) = H(e^{j\omega}) e^{-j\omega k} \tag{3.45}$$

By expressing this result in complex vector form, we note

$$|H_k(e^{j\omega})| = |H(e^{j\omega})| \tag{3.46}$$

and

$$\text{Arg}[H_k(e^{j\omega})] = \text{Arg}[H(e^{j\omega})] - \omega k \tag{3.47}$$

In words, when we delay a sequence by k samples, where k is any positive or negative integer, the magnitude of the Fourier transform is unchanged. The phase is altered by adding a linear-with-frequency phase term whose slope is equal to $-k$. The utility of this result will become evident in computing the Fourier transform of symmetric sequences in the next section.

FOURIER TRANSFORM OF SEQUENCES WITH SYMMETRIES. In the design of digital filters, we commonly encounter sequences that have a symmetry about some time point, denoted by N_S. We consider sequences that have even symmetry and those that have odd symmetry separately.

Even-symmetric sequences are those whose elements about the point N_S have equal value, or

$$h(N_S + m) = h(N_S - m) \qquad \text{for all } m \tag{3.48}$$

The two cases, depending on whether N_S is an integer, are shown in Fig. 3.8. The special case when $N_S = 0$ defines an *even* sequence, for which

$$h(n) = h(-n) \qquad \text{for all } n \tag{3.49}$$

Antisymmetric sequences are those whose elements about the point N_S are of equal magnitude, but of opposite sign, or

$$h(N_S + n) = -h(N_S - n) \qquad \text{for all } n \tag{3.50}$$

The two cases, depending on whether N_S is an integer, are shown in Fig. 3.8. When N_S is an integer, $h(N_S)$ must equal zero for $\{h(n)\}$ to be an antisymmetric sequence. The special case when $N_S = 0$ defines an *odd* sequence, for which

$$h(n) = -h(-n) \qquad \text{for all } n \tag{3.51}$$

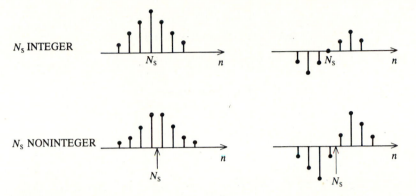

FIGURE 3-8
Symmetric and antisymmetric sequences shown for integer and noninteger values of symmetry point N_S.

Fourier transform of even-symmetric sequences. Let us first consider the Fourier transform of an even sequence, followed by that of a general even-symmetric sequence. Starting with the definition of the Fourier transform, we have

$$H(e^{j\omega}) = \sum_{n=-\infty}^{\infty} h(n) e^{-j\omega n}$$

$$= \sum_{n=-\infty}^{-1} h(n) e^{-j\omega n} + h(0) + \sum_{n=1}^{\infty} h(n) e^{-j\omega n} \qquad (3.52)$$

Letting $m = -n$ in the first sum, we obtain

$$H(e^{j\omega}) = \sum_{m=1}^{\infty} h(-m) e^{j\omega m} + h(0) + \sum_{n=1}^{\infty} h(n) e^{-j\omega n} \qquad (3.53)$$

Noting that $h(n) = h(-n)$, we have

$$H(e^{j\omega}) = h(0) + \sum_{n=1}^{\infty} h(n)(e^{-j\omega n} + e^{j\omega n})$$

$$= h(0) + 2 \sum_{n=1}^{\infty} h(n) \cos(\omega n) \qquad (3.54)$$

Hence, the Fourier transform of an even sequence is a real-valued function of frequency. The magnitude function is then

$$|H(e^{j\omega})| = \begin{cases} H(e^{j\omega}) & \text{when } H(e^{j\omega}) \geq 0 \\ -H(e^{j\omega}) & \text{when } H(e^{j\omega}) < 0 \end{cases} \qquad (3.55)$$

Since its imaginary part is identically zero, the phase function computed from the arctangent function is zero. The phase response then only accommodates the sign change of $H(e^{j\omega})$:

$$\text{Arg}[H(e^{j\omega})] = \begin{cases} 0 & \text{for } \omega \text{ over which } H(e^{j\omega}) > 0 \\ \pm\pi, & \text{for } \omega \text{ over which } H(e^{j\omega}) < 0 \end{cases} \tag{3.56}$$

Recall that the phase is not defined when $H(e^{j\omega}) = 0$.

Example 3.11. **Fourier transform of an even sequence.** Let

$$h(n) = 1 \quad \text{for } -2 \le n \le 2, \quad = 0 \text{ otherwise}$$

Then its Fourier transform is given by

$$\begin{aligned} H(e^{j\omega}) &= \sum_{n=-2}^{2} e^{-j\omega n} \\ &= 1 + e^{j\omega} + e^{-j\omega} + e^{j2\omega} + e^{-j2\omega} \\ &= 1 + 2\cos(\omega) + 2\cos(2\omega) \end{aligned}$$

Any even-symmetric sequence can be obtained from an even sequence by applying a suitable delay. The Fourier transform of a delayed sequence was shown above to have the same magnitude function as the original sequence, but also to contain an additional linear-with-frequency component in the phase function. To illustrate this in a more general fashion, let us consider the even sequence $\{h_E(n)\}$, which is defined for $-M \le n \le M$, as shown in Fig. 3.9. If $\{h_E(n)\}$ is the unit-sample response of a digital filter, then the filter is noncausal because of the nonzero elements for $n < 0$. If a causal filter having the same magnitude response is desired, we need only add sufficient delay so that the first nonzero element occurs at $n = 0$. As shown in Fig. 3.9, we can obtain a causal symmetric sequence $\{h_S(n)\}$ by delaying $\{h_E(n)\}$ by M samples, or

$$h_S(n) = h_E(n - M) \quad \text{for } 0 \le n \le 2M \tag{3.57}$$

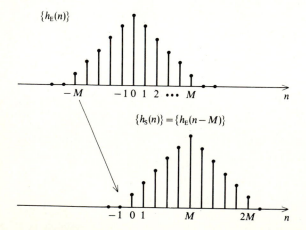

FIGURE 3-9
Delaying an even sequence to produce a causal symmetric sequence.

Then, if $H_S(e^{j\omega})$ is the Fourier transform of $\{h_S(n)\}$ and $H_E(e^{j\omega})$ is that of $\{h_E(e^{j\omega})\}$, we have

$$|H_S(e^{j\omega})| = |H_E(e^{j\omega})| \tag{3.58}$$

and

$$\text{Arg}[H_S(e^{j\omega})] = \text{Arg}[H_E(e^{j\omega})] - M\omega \tag{3.59}$$

Example 3.12. **Fourier transform of an even-symmetric sequence.** Let

$$h(n) = 1 \qquad \text{for } 0 \le n \le 4$$

This is just the delayed version of the sequence given in Example 3.11. We can get the Fourier transform by simply adding the linear phase function -2ω to the result of Example 3.11 to include the delay.

We verify this result by directly computing the Fourier transform of the above sequence, which is given by

$$H(e^{j\omega}) = \sum_{n=0}^{4} e^{-j\omega n}$$

Applying the finite geometric sum formula, we get

$$H(e^{j\omega}) = \frac{1 - e^{-j5\omega}}{1 - e^{-j\omega}}$$

$$= \frac{e^{-j5\omega/2}}{e^{-j\omega/2}} \frac{e^{j5\omega/2} - e^{-j5\omega/2}}{e^{j\omega/2} - e^{-j\omega/2}}$$

$$= \frac{\sin(5\omega/2)}{\sin(\omega/2)} e^{-j2\omega}$$

The phase is a linear function of frequency with slope equal to -2. It is left as an exercise for the student to show that the two amplitude expressions are equivalent.

Recall that an even sequence has zero phase, except for the phase jumps of $\pm\pi$ caused by the sign reversal of the amplitude of the Fourier transform. If these jumps are ignored, we are left with a phase that is linear with frequency, which will be called a *linear-phase function*. This linear-phase property will be important for filter design in the following chapters.

FOURIER TRANSFORM OF ANTISYMMETRIC SEQUENCES. We first consider the Fourier transform of an odd sequence, followed by that for the general antisymmetric sequence. For an odd sequence, $h(0) = 0$, and the Fourier transform can be written as

$$H(e^{j\omega}) = \sum_{n=1}^{\infty} h(n)[e^{-j\omega n} - e^{j\omega n}]$$

$$= -j2 \sum_{n=1}^{\infty} h(n) \sin(\omega n) \tag{3.60}$$

Hence, the Fourier transform of an odd sequence of n is an imaginary function of frequency. Since only sine functions appear in the sum, $H(e^{j\omega})$ is an odd function of ω, i.e. $H(e^{j\omega}) = -H(e^{-j\omega})$. Since the real part is identically zero, and $j = e^{j\pi/2}$, we can write the function in terms of its imaginary part as

$$H(e^{j\omega}) = jH_I(e^{j\omega}) = H_I(e^{j\omega})\, e^{j\pi/2}$$

The magnitude response is then equal to

$$
\begin{aligned}
|H(e^{j\omega})| &= H_I(e^{j\omega}) && \text{for } H_I(e^{j\omega}) > 0 \\
&= -H_I(e^{j\omega}) && \text{for } H_I(e^{j\omega}) < 0
\end{aligned}
\tag{3.61}
$$

and the phase is equal to

$$
\text{Arg}[H(e^{j\omega})] = \begin{cases} \pi/2 & \text{for } \omega \text{ over which } H_I(e^{j\omega}) > 0 \\ \pi/2 \pm \pi & \text{for } \omega \text{ over which } H_I(e^{j\omega}) < 0 \end{cases}
\tag{3.62}
$$

Example 3.13. **Magnitude and phase response of an odd sequence.** Let

$$
h(n) = \begin{cases} -1 & \text{for } n = -2 \\ 1 & \text{for } n = 2 \\ 0 & \text{otherwise} \end{cases}
$$

The Fourier transform is equal to

$$H(e^{j\omega}) = -e^{j2\omega} + e^{-j2\omega} = -2j\sin(2\omega) = 2\sin(2\omega)\, e^{-j\pi/2}$$

The imaginary part is then equal to

$$H_I(e^{j\omega}) = -2\sin(2\omega)$$

The magnitude spectrum is given by

$$|H(e^{j\omega})| = |2\sin(2\omega)|$$

and the phase spectrum is

$$
\text{Arg}[H(e^{j\omega})] = \begin{cases} \pi/2 & \text{for } H_I(e^{j\omega}) > 0, & \text{or for } \pi/2 < \omega < \pi \\ -\pi/2 & \text{for } H_I(e^{j\omega}) < 0, & \text{or for } 0 < \omega < \pi/2 \end{cases}
$$

The magnitude and phase functions are shown in Fig. 3.10.

For odd sequences, a jump of π degrees will always occur at $\omega = 0$. This happens because $H_I(e^{j\omega})$ is an *odd function of* ω, which means that a sign reversal in the amplitude function must occur at $\omega = 0$.

An antisymmetric sequence can be obtained from an odd sequence by applying a delay. Let us consider the odd sequence $\{h_0(n)\}$, which is defined for $-M \le n \le M$. We can obtain a causal antisymmetric sequence $\{h_A(n)\}$ by delaying $\{h_0(n)\}$ by M samples:

$$h_A(n) = h_0(n - M) \qquad \text{for } 0 \le n \le 2M \tag{3.63}$$

Then, we have

$$|H_A(e^{j\omega})| = |H_0(e^{j\omega})| \tag{3.64}$$

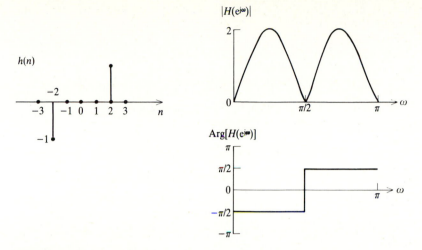

FIGURE 3-10
Magnitude and phase spectra for an odd sequence.

and

$$\text{Arg}[H_A(e^{j\omega})] = \text{Arg}[H_0(e^{j\omega})] - M\omega \qquad (3.65)$$

Ignoring the phase jumps due to sign reversals in the amplitude function, including the one at $\omega = 0$, we are also left with a linear-phase function for antisymmetric sequences. This result is true even when M is not an integer, although a strictly odd sequence does not exist in this case. This is illustrated in the next example.

Example 3.14. Fourier transform of an antisymmetric sequence. In this example, we will consider the case when N_S is not an integer. As shown in Fig. 3.11,

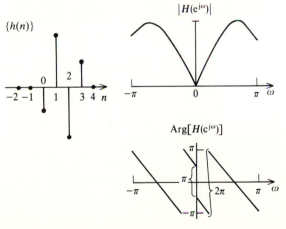

FIGURE 3-11
Magnitude and phase spectra for an antisymmetric sequence.

let $h(0) = -\frac{1}{2}$, $h(1) = 1$, $h(2) = -1$, $h(3) = \frac{1}{2}$ and $h(n) = 0$, otherwise. Then, the Fourier transform is given by

$$
\begin{aligned}
H(e^{j\omega}) &= -\tfrac{1}{2} + e^{-j\omega} - e^{-j2\omega} + \tfrac{1}{2}e^{-j3\omega} \\
&= -\tfrac{1}{2}(1 - e^{-j3\omega}) + (e^{-j\omega} - e^{-j2\omega}) \\
&= -e^{-j3\omega/2}[\tfrac{1}{2}(e^{j3\omega/2} - e^{-j3\omega/2}) + e^{j\omega/2} - e^{-j\omega/2}] \\
&= -je^{-j3\omega/2}(\sin(3\omega/2) + 2\sin(\omega/2))
\end{aligned}
$$

Substituting $-j = e^{-j\pi/2}$, we finally get

$$
H(e^{j\omega}) = (\sin(3\omega/2) + 2\sin(\omega/2))e^{-j(3\omega + \pi)/2}
$$

When N_S is not an integer, the phase spectrum still has a linear slope equal to $-N_S\omega$. In terms of Eq. (3.65), $M = \frac{3}{2}$.

3.8 THE INVERSE FOURIER TRANSFORM

The inverse Fourier transform allows us to determine the unit-sample response of a filter for which we know only the transfer function. Similarly, it also helps us determine the output discrete-time sequence of a filter from the output spectrum.

The Fourier transform $H(e^{j\omega})$ is a periodic function of the continuous-valued variable ω, with period 2π. Hence, it can be expanded as a Fourier series. We show that the coefficients of the series terms are equal to the values of the discrete-time sequence $\{h(n)\}$. We first state the *Fourier series argument*:

Let $s(t)$ be a periodic function of the continuous-valued variable t with period T. Then $s(t)$ can be expressed in the following Fourier series expansion:

$$
s(t) = \sum_{n=-\infty}^{\infty} a_n e^{-j\Omega_n t} \tag{3.66}
$$

where $\Omega_n = 2\pi n/T$ is the nth harmonic frequency, and the coefficients are given by

$$
a_n = \frac{1}{T} \int_{-T/2}^{T/2} s(t) e^{j\Omega_n t} dt \tag{3.67}
$$

In exactly the same fashion, $H(e^{j\omega})$ can be expressed as the following Fourier series expansion

$$
H(e^{j\omega}) = \sum_{n=-\infty}^{\infty} c_n e^{-j\theta_n \omega} \tag{3.68}
$$

where $\theta_n = 2\pi n/2\pi = n$, can be thought of as a harmonic frequency, and

$$
c_n = \frac{1}{2\pi} \int_{-\pi}^{\pi} H(e^{j\omega}) e^{j\omega n} d\omega \tag{3.69}
$$

Note the similarity of the definition of the Fourier transform in Eq. (3.4) and the Fourier series expansion of $H(e^{j\omega})$ given in Eq. (3.68). This similarity allows us to define the *inverse Fourier transform* as

$$h(n) = \frac{1}{2\pi} \int_{-\pi}^{\pi} H(e^{j\omega}) e^{j\omega n}\, d\omega \tag{3.70}$$

This integral solution for the inverse Fourier transform is useful for analytic purposes, as shown later, but it is usually very difficult to evaluate for typical functional forms of $H(e^{j\omega})$. An alternate, and occasionally more helpful, method of determining the values of $h(n)$ follows directly from the definition of the Fourier transform, equal to

$$H(e^{j\omega}) = \sum_{n=-\infty}^{\infty} h(n)\, e^{-j\omega n}$$

If $H(e^{j\omega})$ can be expressed as a series of complex exponentials, then $h(n)$ is simply the coefficient of $e^{-j\omega n}$. This is illustrated in the following example.

Example 3.15.   Inverse Fourier transform of the 3-sample averager. Let

$$H(e^{j\omega}) = \tfrac{1}{3}(1 + 2\cos(\omega))$$

Employing the Euler identity for the cosine we obtain

$$H(e^{j\omega}) = \tfrac{1}{3}(1 + e^{j\omega} + e^{-j\omega}).$$

Noting that $h(n)$ is the coefficient of $e^{-j\omega n}$, we have by inspection

$$h(0) = h(1) = h(-1) = \tfrac{1}{3},$$
$$h(n) = 0 \quad \text{for } n < -1 \text{ and } n > 1$$

Alternately, we could perform the integration. The value of $h(n)$ is then

$$h(n) = \frac{1}{2\pi} \int_{-\pi}^{\pi} \tfrac{1}{3}(1 + e^{j\omega} + e^{-j\omega}) e^{j\omega n}\, d\omega$$

$$= \frac{1}{6\pi} \int_{-\pi}^{\pi} [e^{j\omega n} + e^{j\omega(n+1)} + e^{j\omega(n-1)}]\, d\omega$$

$$= \frac{1}{6\pi} \left(\frac{e^{j\pi n} - e^{-j\pi n}}{jn} + \frac{e^{j\pi(n+1)} - e^{-j\pi(n+1)}}{j(n+1)} + \frac{e^{j\pi(n-1)} - e^{-j\pi(n-1)}}{j(n-1)} \right)$$

$$= \frac{\sin(n\pi)}{3n\pi} + \frac{\sin[(n+1)\pi]}{3(n+1)\pi} + \frac{\sin[(n-1)\pi]}{3(n-1)\pi}$$

Recall that $\sin(n\pi) = 0$ for integer values of n. When the denominator is also zero, an indeterminate form results. Applying L'Hospital's argument, we find that

$$\lim_{\varepsilon \to 0} \sin(\varepsilon)/\varepsilon = 1$$

Applying this approach to each of the three terms above, we obtain

$$h(n) = \begin{cases} \frac{1}{3} & \text{for } n = -1, 0, 1 \\ 0 & \text{otherwise} \end{cases}$$

If $H(e^{j\omega})$ is a rational function of $e^{-j\omega n}$, then a series expansion can be obtained either by performing a division or by computing the Taylor series expansion, as illustrated in Example 3.16.

Example 3.16.   Inverse Fourier transform for the first-order recursive filter. Let

$$H(e^{j\omega}) = (1 - a\,e^{-j\omega})^{-1}$$

Computing the Taylor series expansion, we have

$$H(e^{j\omega}) = 1 + a\,e^{-j\omega} + a^2\,e^{-j2\omega} + a^3\,e^{-j3\omega} + \cdots + a^k\,e^{-j\omega k} + \cdots$$

Equating the coefficient of $e^{-j\omega k}$ with $h(k)$, we have

$$h(n) = \begin{cases} a^n & \text{for } n \ge 0 \\ 0, & \text{otherwise} \end{cases}$$

Example 3.17 illustrates the application of frequency domain analysis to signal processing, in which the output sequence of a filter is determined from the output spectrum.

Example 3.17. Determining the output sequence from the output spectrum. The output spectrum of Example 3.10 is given by

$$Y(e^{j\omega}) = \frac{1}{3} \frac{e^{j\omega} + 1 + e^{-j\omega}}{1 - a\,e^{-j\omega}}$$

$$= \frac{1}{3} \left(\frac{e^{j\omega}}{1 - a\,e^{-j\omega}} + \frac{1}{1 - a\,e^{-j\omega}} + \frac{e^{-j\omega}}{1 - a\,e^{-j\omega}} \right)$$

Expanding each term in the parenthesis in a Taylor series expansion, we get

$$Y(e^{j\omega}) = \tfrac{1}{3}[e^{j\omega} + a + a^2\,e^{-j\omega} + a^3\,e^{-j2\omega} + a^4\,e^{-j3\omega} + \cdots$$
$$+ 1 + a\ e^{-j\omega} + a^2\,e^{-j2\omega} + a^3\,e^{-j3\omega} + \cdots$$
$$+ \quad e^{-j\omega} + a\ e^{-j2\omega} + a^2\,e^{-j3\omega} + \cdots]$$

Combining all the terms having the same complex exponential factor and equating $h(k)$ with the coefficient of $e^{-j\omega k}$, the output sequence is found to be

$$y(n) = \tfrac{1}{3}[a^{n+1}u(n+1) + a^n u(n) + a^{n-1}u(n-1)]$$

3.9   FOURIER TRANSFORM OF THE PRODUCT OF TWO DISCRETE-TIME SEQUENCES

In the design of nonrecursive filters, we commonly encounter the product of two sequences $\{w(n)\}$ and $\{h(n)\}$. Let the sequence of term-by-term products be denoted by $\{v(n)\}$, where

$$v(n) = w(n)\,h(n) \qquad \text{for all } n \tag{3.71}$$

Then its Fourier transform is given by

$$V(e^{j\omega}) = \sum_{n=-\infty}^{\infty} w(n)\,h(n)\,e^{-j\omega n} \tag{3.72}$$

To determine a more useful form for $V(e^{j\omega})$, we first express $h(n)$ as an inverse Fourier transform given by

$$h(n) = \frac{1}{2\pi} \int_{-\pi}^{\pi} H(e^{j\alpha})\,e^{j\alpha n}\,d\alpha \tag{3.73}$$

We thus get

$$V(e^{j\omega}) = \sum_{n=-\infty}^{\infty} w(n)\left(\frac{1}{2\pi} \int_{-\pi}^{\pi} H(e^{j\alpha})\,e^{j\alpha n}\,d\alpha\right) e^{-j\omega n} \tag{3.74}$$

Interchanging the order of summation and integration, we have

$$V(e^{j\omega}) = \frac{1}{2\pi} \int_{-\pi}^{\pi} \left(\sum_{n=-\infty}^{\infty} w(n)\,e^{-j(\omega-\alpha)n}\right) H(e^{j\alpha})\,d\alpha \tag{3.75}$$

The term in the parentheses is similar to the definition of the Fourier transform of $\{w(n)\}$, but at a frequency argument of $\omega - \alpha$, or $W(e^{j(\omega-\alpha)})$. Including this Fourier transform into the integral, we get

$$V(e^{j\omega}) = \frac{1}{2\pi} \int_{-\pi}^{\pi} H(e^{j\alpha})\,W(e^{j(\omega-\alpha)})\,d\alpha$$

$$= H(e^{j\omega}) \circledast W(e^{j\omega}) \tag{3.76}$$

This form is that of a convolution operation involving two functions of the continuous-valued variable ω. Because both these functions are periodic, the integral is evaluated over one period. We denoted this operation as a *circular convolution*. Hence, we have shown that multiplication in the time domain corresponds to convolution in the frequency domain. We will use this result in the next section to determine the Fourier transform of sampled continuous-time data.

> **Example 3.18.   Fourier transform of the product of two sequences.** The most common application of this result occurs in truncating infinite-duration sequences, as illustrated here.
>   Let us consider the infinite duration sequence given by
>
> $$h(n) = \frac{\sin(\pi n/4)}{\pi n/4} \qquad \text{for all } n$$
>
> The corresponding $H(e^{j\omega})$ is equal to the ideal lowpass filter characteristic shown in Fig. 3.12. To truncate this sequence to $-3 \le n \le 3$, we multiply $h(n)$ by the

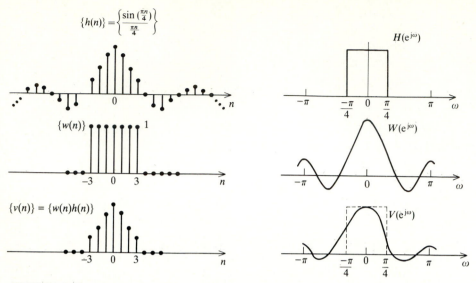

FIGURE 3-12
Multiplication of two discrete-time sequences with the corresponding spectral convolution operation.

following sequence, called a *window function,*

$$w(n) = \begin{cases} 1 & \text{for } -3 \le n \le 3 \\ 0 & \text{otherwise} \end{cases}$$

The windowed sequence is then

$$v(n) = w(n)\,h(n) = \sin(\pi n/4)/(\pi n/4) \qquad \text{for } -3 \le n \le 3$$

and is shown in Fig. 3.12. To find $V(e^{j\omega})$, we must first compute $W(e^{j\omega})$ and perform the convolution. The Fourier transform of $\{w(n)\}$ is equal to

$$W(e^{j\omega}) = \sum_{n=-3}^{3} e^{-j\omega n} = e^{j3\omega} \sum_{n=0}^{6} e^{-j\omega n}$$

$$= e^{j3\omega} \frac{1 - e^{-j7\omega}}{1 - e^{-j\omega}} = \frac{\sin(3\omega/2)}{\sin(\omega/2)}$$

The result of the convolution is shown in Fig. 3.12. The main effect of the convolution is that the sharp transitions in $H(e^{j\omega})$ have been made smoother. We will explore this multiplication with a window sequence in more detail in the design of finite-impulse response filters in a later chapter.

3.10 SAMPLING A CONTINUOUS FUNCTION TO GENERATE A SEQUENCE

In practice, discrete-time sequences are often obtained by sampling continuous-time signals. Typically, this occurs when signals are to be processed by computer; situations include speech analysis and extracting information from sensors for bioengineering and robotic applications. We will apply these results to sampling the Fourier transform in the next chapter. An application of sampling continuous-time signals will also occur in a later chapter in the design of digital filters. In this section we relate the spectrum of the continuous-time signal to that of the discrete-time sequence that is obtained by sampling.

We proceed by first stating the relationship between the continuous-time and discrete-time spectra and then provide the rather involved mathematical proof. Let us define the continuous-time, or *analog*, signal by $c_A(t)$. Its spectrum is then given by $C_A(j\Omega)$, where Ω is the continuous-time frequency, having units *radians per second*. For continuous-time signals, the Fourier transform is defined by

$$C_A(j\Omega) = \int_{-\infty}^{\infty} c_A(t)\, e^{-j\Omega t}\, dt \tag{3.77}$$

which is similar to the definition of the Fourier transform of discrete-time signals, but with the sum replaced by an integral. When $c_A(t)$ is sampled every T_s seconds, we obtain the discrete-time sequence $\{c_A(nT_s)\}$, whose values are equal to

$$c_A(nT_s) = c_A(t)|_{t=nT_s}. \tag{3.78}$$

T_s is commonly called the *sampling period* and has units *seconds per sample interval*. The discrete-time Fourier transform for such sampled data sequences is defined by

$$C(e^{j\omega}) = \sum_{n=-\infty}^{\infty} c_A(nT_s)\, e^{-jn\omega} \tag{3.79}$$

where ω has units *radians per sample interval*. This discrete-time spectrum is related to the continuous-time spectrum $C_A(j\Omega)$ by

$$C(e^{j\omega}) = \frac{1}{T_s} \sum_{k=-\infty}^{\infty} C_A(j\omega/T_s + j2k\pi/T_s) \tag{3.80}$$

In words, the spectrum of the discrete-time sequence, derived by sampling a continuous-time function, is equal to the sum of an infinite set of shifted continuous-time spectra. A graphical interpretation of this identity is shown in Fig. 3.13.

Proof. To show this relationship, we take the following approach. First, we treat sampling as a *multiplicative* operation involving the continuous-time signal $c_A(t)$ and a train of Dirac delta functions occurring every T_s seconds. This multiplication produces the intermediate function of continuous-time $c_S(t)$. The continuous-time spectrum of $c_S(t)$, denoted by $C_S(j\Omega)$, is shown to have the same analytic form as $C_S(e^{j\omega})$. Finally, we determine $C_S(j\Omega)$ in terms of the original analog spectrum $C_A(j\Omega)$.

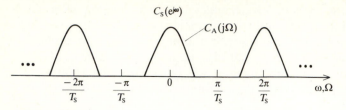

FIGURE 3-13
Spectrum of a discrete-time sequence obtained by sampling a continuous-time signal.

We first define the Dirac delta function and present its properties that are important for this application. By definition, the Dirac delta function, denoted by $\delta(t)$, has infinite height and zero width, but its area, or *weight*, is equal to unity. To illustrate how we can approach a Dirac delta function, consider the rectangular function $r_\varepsilon(t)$, whose width is ε and whose height is $1/\varepsilon$, so its area is equal to one, as shown in Fig. 3.14. We can define $\delta(t)$ as

$$\delta(t) = \lim_{\varepsilon \to 0} r_\varepsilon(t) \tag{3.81}$$

The following three properties of the Dirac delta function are useful here.

Property 1. The product of a function and a delta function. In multiplying any continuous-time function $f(t)$, the weight of the delta function becomes equal to the value of the function at the instant of the delta function, or

$$f(t)\,\delta(t - t_0) = f(t_0)\,\delta(t - t_0) \tag{3.82}$$

Property 2. The integral of the product (The sifting property). The integral of the product of any function $f(t)$ and a delta function is a constant, equal to the value of the function at the instant of the delta function. To show this, we integrate the product to give

$$\int_{-\infty}^{\infty} f(t)\,\delta(t - t_0)\,dt = \int_{-\infty}^{\infty} f(t_0)\,\delta(t - t_0)\,dt \tag{3.83}$$

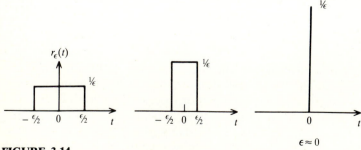

FIGURE 3-14
Limiting procedure to obtain a Dirac delta function.

by Property 1. Since $f(t_0)$ is not a function of t, we can take it out of the integral to give

$$f(t_0) \int_{-\infty}^{\infty} \delta(t - t_0) \, dt = f(t_0).$$

This last equality follows because the area of $\delta(t)$ is unity. The Dirac delta function is used to select, or sift, the value of the function at $t = t_0$.

Property 3. Convolving a function with a delta function (The shifting property). When the function $f(t)$ is convolved with $\delta(t - t_0)$, the origin of $f(t)$ is shift to $t = t_0$, or

$$\delta(t - t_0) * f(t) = f(t - t_0)$$

To show this, we write out the continuous-time convolution equation,

$$\int_{-\infty}^{\infty} \delta(\lambda - t_0) f(t - \lambda) \, d\lambda \tag{3.84}$$

and define the variable $\alpha = \lambda - t$. We then have

$$\int_{-\infty}^{\infty} f(-\alpha) \, \delta(\alpha + t - t_0) \, d\alpha = f(t - t_0) \tag{3.85}$$

This last result follows by Property 2.

An *infinite train* of Dirac delta functions occurring every T_s seconds is denoted by $\delta_{T_s}(t)$ and can be expressed as

$$\delta_{T_s}(t) = \sum_{n=-\infty}^{\infty} \delta(t - nT_s) \tag{3.86}$$

We can now use these properties of the Dirac delta function to determine the relationship between the spectra. The sampling process is modeled by the multiplication operation given by

$$c_S(t) = c_A(t) \cdot \delta_{T_s}(t) \qquad \text{for all } t$$

$$= c_A(t) \cdot \sum_{n=-\infty}^{\infty} \delta(t - nT_s) \qquad \text{for all } t \tag{3.87}$$

as shown in Fig. 3.15. Since $c_S(t)$ is a continuous-time function, its spectrum $C_S(j\Omega)$ is given by

$$C_S(j\Omega) = \int_{-\infty}^{\infty} c_S(t) \, e^{-j\Omega t} \, dt \tag{3.88}$$

$$= \int_{-\infty}^{\infty} \left[c_A(t) \sum_{n=-\infty}^{\infty} \delta(t - nT_s) \right] e^{-j\Omega t} \, dt \tag{3.89}$$

Changing the order of integration and summation, and applying the sifting

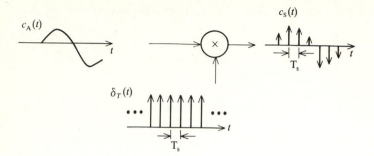

FIGURE 3-15
Sampling a continuous-time signal can be treated as a multiplicative process with a train of Dirac impulses.

property of the Dirac delta function, we have

$$C_S(j\Omega) = \sum_{n=-\infty}^{\infty} \left[\int_{-\infty}^{\infty} \delta(t - nT_s) \, c_A(t) \, e^{-j\Omega t} \, dt \right]$$

$$= \sum_{n=-\infty}^{\infty} c_A(nT_s) \, e^{-jn\Omega T_s} \tag{3.90}$$

We have arrived at our first intermediate result. By comparing Eq. (3.90) with (3.79), we observe

$$C(e^{j\omega}) = C_S(j\Omega)\big|_{\Omega=\omega/T_s} \tag{3.91}$$

That is, the form of $C(e^{j\omega})$ can be obtained by determining $C_S(j\Omega)$ and replacing Ω by ω/T_s. The division of ω by T_s scales the values of ω to be the same as Ω. For sampled-data applications, it is conventional to express ω in the range $[-\pi/T_s, \pi/T_s)$, to have the same numerical values as Ω. The units of ω/T_s are radians per second, just as for the continuous-time frequency.

To complete the proof, we must determine $C_S(j\Omega)$ in terms of the continuous-time spectrum $C_A(j\Omega)$. Since $\delta_{T_s}(t)$ is a periodic function, it can be expressed in the following Fourier series expansion given by

$$\delta_{T_s}(t) = \sum_{n=-\infty}^{\infty} a_n \, e^{-j\Omega_n t} \tag{3.92}$$

where $\Omega_n = 2\pi n/T_s$, and

$$a_n = \frac{1}{T_s} \int_{-T_s/2}^{T_s/2} \delta_{T_s}(t) \, e^{j\Omega_n t} \, dt \tag{3.93}$$

Over the interval $-T_s/2 \le t < T_s/2$, the impulse train $\delta_{T_s}(t)$ is represented by only the single Dirac impulse situated at $t = 0$, or $\delta(t)$. Making this substitution and applying the sifting property, we get

$$a_n = \frac{1}{T_s} \int_{-T_s/2}^{T_s/2} \delta(t) \, e^{j\Omega_n t} \, dt = \frac{1}{T_s} e^{j0} = \frac{1}{T_s} \qquad \text{for all } n \tag{3.94}$$

Substituting this Fourier series expansion into the sampling process, we get

$$c_S(t) = c_A(t) \cdot \delta_{T_s}(t)$$

$$= c_A(t) \cdot \frac{1}{T_s} \sum_{n=-\infty}^{\infty} e^{-j\Omega_n t}$$

$$= \frac{1}{T_s} \sum_{n=-\infty}^{\infty} c_A(t) e^{-j\Omega_n t} \tag{3.95}$$

The continuous-time spectrum of $c_S(t)$ is then equal to

$$C_S(j\Omega) = \int_{-\infty}^{\infty} c_S(t) e^{-j\Omega t} \, dt$$

$$= \frac{1}{T_s} \int_{-\infty}^{\infty} \left[\sum_{n=-\infty}^{\infty} c_A(t) e^{-j\Omega_n t} \right] e^{-j\Omega t} \, dt \tag{3.96}$$

Interchanging the order of summation and integration, we obtain

$$C_S(j\Omega) = \frac{1}{T_s} \sum_{n=-\infty}^{\infty} \left[\int_{-\infty}^{\infty} c_A(t) e^{-j(\Omega + \Omega_n)t} \, dt \right] \tag{3.97}$$

Evaluating the kth term of the sum, denoted by $C_{S,k}(j\Omega)$, we find

$$C_{S,k}(j\Omega) = \int_{-\infty}^{\infty} c_A(t) e^{-j(\Omega + \Omega_k)t} \, dt$$

$$= C_A(j\Omega + j\Omega_k) \tag{3.98}$$

This result is simply the original analog spectrum that has been shifted from $\Omega = 0$ to $\Omega = -\Omega_k$, as shown in Fig. 3.16. The total spectrum is then the sum of these shifted versions and is given by

$$C_S(j\Omega) = \frac{1}{T_s} \sum_{k=-\infty}^{\infty} C_{S,k}(j\Omega) = \frac{1}{T_s} \sum_{k=-\infty}^{\infty} C_A(j\Omega + j2k\pi/T_s) \tag{3.99}$$

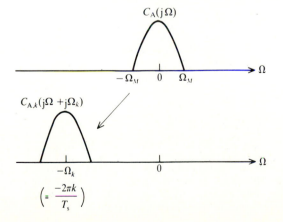

FIGURE 3-16
Shifted version of the analog spectrum.

$C_S(j\Omega)$ is a periodic function of Ω, with period $2\pi/T_s$. More concisely, $C_S(j\Omega)$ is called the *periodic extension* of $C_A(j\Omega)$.

This last identity is the second desired intermediate result. Combining this result with that given by (3.91), we arrive at the final desired identity given in (3.80). (Q.E.D.)

NYQUIST CRITERION. When sampling a continuous-time signal $c_A(t)$ to produce the sequence $\{c_A(nT_s)\}$, we want to ensure that all the information in the original signal is retained in the samples. There will be no information loss if we can exactly recover the continuous-time signal from the samples. To determine the condition under which there is no information loss, let us consider $c_A(t)$ to have a bandlimited spectrum, or one for which

$$C_A(j\Omega) = 0 \qquad \text{for } |\Omega| > \Omega_M \tag{3.100}$$

as shown in Fig. 3.17(a). As demonstrated above, when $c_A(t)$ is sampled with sampling period T_s, then the spectrum of the sampled signal $C_S(j\Omega)$ is the periodic extension of $C_A(j\Omega)$ with period $2\pi/T_s$, as shown in Fig. 3.17(b). The

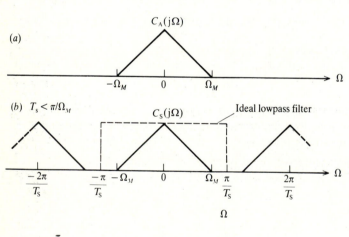

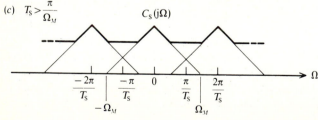

FIGURE 3-17
Relationship between continuous-time spectrum of a signal and the spectrum of the discrete-time sequence obtained by sampling the signal with sampling period T_s. (a) Original spectrum of continuous-time signal; (b) spectrum of sampled sequence when $\Omega_M < \pi/T_s$; (c) spectrum of sampled sequence when $\Omega_M > \pi/T_s$. The latter case illustrates the aliasing error.

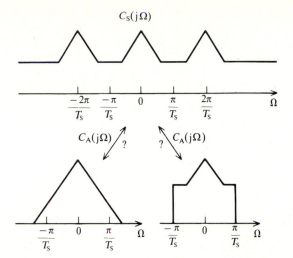

FIGURE 3-18

Two candidates for the continuous-time spectrum when aliasing occurs.

form of $C_S(j\Omega)$ in the frequency range $[-\pi/T_s, \pi/T]$ is identical to $C_A(j\Omega)$ if

$$\pi/T_s > \Omega_M \qquad \text{or} \qquad T_s < \pi/\Omega_M \qquad (3.101)$$

In this case, there is no overlap in the adjacent spectral components and we can then extract $C_A(j\Omega)$ *exactly* from $C_S(j\Omega)$ by employing the ideal lowpass filter magnitude response shown in Fig. 3.17(*b*).

If T_s is chosen to be greater than π/Ω_M, spectral overlap occurs in the periodic extension, given by Eq. (3.99), and the form of $C_S(j\Omega)$ in the range $-\pi/T_s \le \Omega < \pi/T_s$ is then no longer similar to $C_A(j\Omega)$ as shown in Fig. 3.17(*c*). This overlap, caused by sampling at too low a rate, produces an irretrievable error in the spectral values, called *aliasing*. The true spectral shape is irretrievable since many different $C_A(j\Omega)$ functions can produce the same $C_S(j\Omega)$. Two possible candidates are shown in Fig. 3.18.

A useful sampling measure is the *sampling rate*, $f_s = 1/T_s$. Recalling that $\Omega = 2\pi f$, and defining f_M as the highest frequency component in the signal, then no spectral overlap will occur if

$$f_s > 2f_M \qquad (3.102)$$

To prevent aliasing error, more than two samples are required per period of the highest frequency sinusoidal component present in the signal. The smallest sampling rate before aliasing occurs for a particular continuous-time signal is called the *Nyquist rate*. The effect of aliasing is most easily demonstrated by considering a sinusoidal signal in Example 3.19.

Example 3.19. Demonstration of aliasing. Let $c_A(t) = \cos(\Omega_0 t)$, for all t, with $\Omega_0 = 100\pi$. This function is sampled to produce the sequence

$$c_S(nT_s) = \cos(n\Omega_0 T_s) \qquad \text{for all } n$$

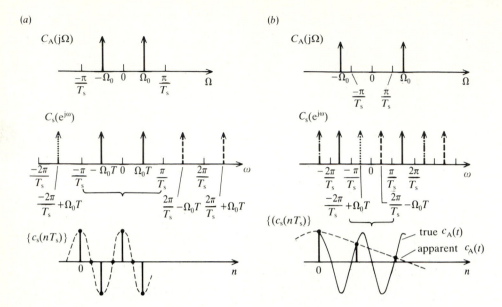

FIGURE 3-19
Demonstration of aliasing caused by choosing T_s too large. (a) Unaliased version, when $T_s < \pi/\Omega_0$; (b) aliasing caused when $T_s > \pi/\Omega_0$. In this latter case a low-frequency cosine function is apparent when the sequence is interpolated.

The continuous-time spectrum of $c_A(t)$ is equal to two Dirac delta functions

$$C_A(j\Omega) = \frac{\pi}{2}\delta(\Omega - \Omega_0) + \frac{\pi}{2}\delta(\Omega + \Omega_0)$$

as shown in Fig. 3.19. Sampling $c_A(t)$ causes the spectrum to be extended periodically:

$$C_S(j\Omega) = \frac{\pi}{2T_s}\sum_{n=-\infty}^{\infty} [\delta(\Omega - \Omega_0 - 2n\pi/T_s) + \delta(\Omega + \Omega_0 - 2n\pi/T_s)]$$

To demonstrate aliasing we note the position of the impulses situated within $-\pi/T_s \leq \Omega < \pi/T_s$, as we vary the sampling period T_s. The sampling period dictated by the sampling rate is $T_s < \pi/\Omega_0 (=10^{-2}\,\text{s})$.

For $T_s < \pi/\Omega_0$, say $T_s = 10^{-3}\,\text{s}$, the delta functions falling within $-\pi/T_s \leq \Omega < \pi/T_s$, or $-1000\pi \leq \Omega < 1000\pi$, are only those for the $n = 0$ term above. These occur at $\Omega = \pm\Omega_0$. In the discrete-time domain, these impulses will produce the desired sequence $\{\cos(n\Omega_0 T_s)\}$.

For $T_s > \pi/\Omega_0$, say $T_s = 1.1 \times 10^{-2}\,\text{s}$, the impulse in the range $0 < \Omega < \pi/T_s$, or $0 < \Omega < 90\pi$, is at

$$\Omega = (2\pi/T_s) - \Omega_0 = 180\pi - 100\pi = 80\pi \text{ rad s}^{-1}$$

from the term $\delta(\Omega + \Omega_0 - 2\pi/T_s)$, or the $k = 1$ term. The impulse located in the range $-\pi/T_s < \Omega \leq 0$ is at

$$\Omega = \Omega_0 - 2\pi/T_s = 100\pi - 180\pi = -80\pi \text{ rad s}^{-1}$$

from $\delta(\Omega - \Omega_0 + 2\pi//T_s)$, or the $k = -1$ term. In this case, the sampled sequence appears to be derived from a continuous-time function at a lower frequency $2\pi/T_s - \Omega_0 = 80\pi$ rad s^{-1}. This is the aliasing phenomenon.

This effect of aliasing errors is occasionally observed in films of moving cars in which the wheels appear to be turning in the direction opposite to that expected from the motion of the car.

ANTI-ALIASING FILTER. In practice, the frequency range containing significant power in the original continuous-time signal may be larger than that containing the desired information. This commonly occurs when a low-frequency signal is corrupted by high-frequency noise or in modulated signals, such as broadcast radio. If sampled at the rate dictated by the Nyquist criterion for the desired analog signal, unwanted high frequency signals would cause aliasing errors to occur.

To prevent aliasing errors caused by these undesired high-frequency signals, an analog lowpass filter, called an *anti-aliasing filter,* must be used. This filter is applied to the analog signal prior to sampling and reduces the power in the analog signal for the frequency range beyond $\Omega = \pi/T_s$. In practice, the spectral magnitude level for $\Omega \geq \pi/T_s$ should be less than 1% (-40 dB) of the important spectral features in the desired signal spectrum to prevent significant aliasing.

For example, let the observed analog signal contain power in an arbitrarily large frequency range, but have *relevant* information only in the frequency range $-\Omega_R \leq \Omega \leq \Omega_R$. We wish to determine the appropriate sampling period T_s. The Nyquist criterion tells us that T_s must be less than π/Ω_R. If sampled at this rate, the higher frequency components in the analog signal will be aliased into the relevant discrete-time signal. The anti-aliasing filter must satisfy two conditions:

Condition 1. The analog signal components with frequencies less than Ω_R must be negligibly attenuated by the filter.

Condition 2. To prevent aliasing in the relevant analog frequency range, the components for $|\Omega| \geq 2\pi/T_s - \Omega_R$ must be attenuated. After sampling, the components in this frequency range fall into the range $-\Omega_R \leq \Omega \leq \Omega_R$ when the periodic extension is formed.

We consider how these conditions can be met in Chapter 8, where the design of analog lowpass filters is discussed.

3.11 RECONSTRUCTION OF CONTINUOUS-TIME SIGNALS FROM DISCRETE-TIME SEQUENCES

In many practical applications, discrete-time sequences are required to be transformed into continuous-time signals. One example is the digital audio

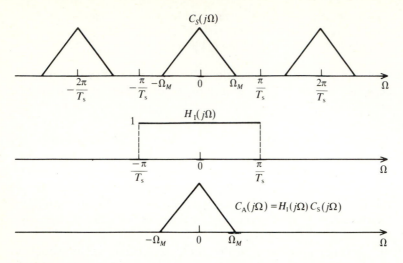

FIGURE 3-20

The spectrum of an analog signal can be obtained by multiplying the spectrum of the sample values with the transfer function of an ideal analog lowpass filter. In practice, $H_I(j\Omega)$ must be approximated.

compact disk system, in which the audio signal is stored as digital samples on the disk and must ultimately be converted into the analog signal that drive the speakers. Another is the digital controller in a robot system, that uses the digital computer to process the sensor data, but requires an analog signal to control the actuator motors. In these cases, we are faced with the converse problem from that of sampling $c_A(t)$ to obtain $\{c_A(nT_s)\}$. The relevant question now becomes how to recover the analog signal $c_A(t)$ from the sampled sequence $\{c_S(nT_s)\}$.

We begin by considering the unaliased spectrum of $C_S(j\Omega)$, shown in Fig. 3.20. $C_A(j\Omega)$ then has the same form as $C_S(j\Omega)$ over $-\pi/T_s \le \Omega < \pi/T_s$. Let us consider an analog lowpass filter with the following ideal frequency response

$$H_I(j\Omega) = \begin{cases} 1 & \text{for } -\pi/T_s \le \Omega < \pi/T_s \\ 0 & \text{otherwise} \end{cases} \tag{3.103}$$

We could retrieve $C_A(j\Omega)$ from $C_S(j\Omega)$ by filtering with an ideal lowpass filter, or

$$C_A(j\Omega) = H_I(j\Omega)C_S(j\Omega), \qquad \text{for all } \Omega \tag{3.104}$$

To interpret this result in the time domain, let us examine the corresponding convolution equation given by

$$c_A(t) = \int_{-\infty}^{\infty} c_S(\alpha)h_I(t - \alpha) \, d\alpha \tag{3.105}$$

The function $c_S(t)$ is the continuous-time sampled signal composed of a series

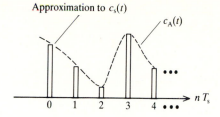

Approximation to $c_s(t)$

$c_A(t)$

$n T_s$

0 1 2 3 4 •••

FIGURE 3-21
Approximating the sequence of weighted Dirac
delta functions with a sequence of narrow pulses
having different amplitudes.

of weighted Dirac delta functions. In practice, this continuous-time function
could be approximated by narrow pulses whose amplitudes are proportional to
the corresponding sample values $c_s(nT_s)$, as shown in Fig. 3.21. The function
$h_I(t)$ is the continuous-time impulse response of the ideal lowpass filter, found
by the inverse Fourier transform given by

$$h_I(t) = \frac{1}{2\pi} \int_{-\infty}^{\infty} H_I(j\Omega) \, e^{j\Omega t} \, d\Omega$$

$$= \frac{1}{2\pi} \int_{-\pi/T_s}^{\pi/T_s} e^{j\Omega t} \, d\Omega$$

$$= \frac{e^{j\pi t/T_s} - e^{-j\pi t/T_s}}{j2\pi t}$$

$$= \frac{1}{T_s} \frac{\sin(\pi t/T_s)}{\pi t/T_s} = \frac{1}{T_s} \text{Sinc}(t/T_s) \tag{3.106}$$

The $\text{Sinc}(x)$ function is shown in Fig. 3.22. The function is equal to zero for
integral values of the sampling time, i.e. $t = nT_s$, where n is a non-zero-valued
integer. For $t = 0$, we find

$$h_I(0) = \lim_{t \to 0} \frac{1}{T_s} \frac{\sin(\pi t/T_s)}{\pi t/T_s}$$

$$= \frac{1}{T_s} \lim_{t \to 0} \frac{(\pi t/T_s) - (1/3!)(\pi t/T_s)^3 + \cdots}{\pi t/T_s} = 1/T_s \tag{3.107}$$

Because $c_s(t)$ is composed of a series of weighted Dirac delta functions,
the convolution produces a sum of shifted and scaled versions of $h_I(t)$, as
demonstrated by Property 3 of the Dirac delta function. As shown in Fig. 3.23,

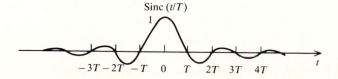

Sinc (t/T)

1

$-3T -2T -T \quad 0 \quad T \quad 2T \quad 3T \quad 4T$

t

FIGURE 3-22
The $\text{Sinc}(t/T)$ function, used
for interpolation, is the impulse response of the ideal
analog lowpass filter.

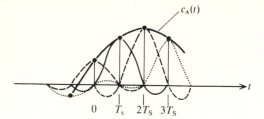

FIGURE 3-23
In the time domain, the interpolation process is treated as the convolution of weighted and delayed Sinc(t/T) functions. The values of the sequence $\{c_S(nT_s)\}$ are shown as dots. The continuous-time function is shown in heavy solid line.

the original analog signal can be reconstructed from the samples by

$$c_A(t) = \sum_{n=-\infty}^{\infty} c_S(nT_s)\, h_I(t - nT_s)$$

$$= \frac{1}{T_s} \sum_{n=-\infty}^{\infty} c_S(nT_s) \frac{\sin(\pi[t - nT_s]/T_s)}{\pi[t - nT_s]/T_s} \tag{3.108}$$

The action of $h_I(t)$ in this convolution is to fill in, or *interpolate,* the values of the continuous-time function between the sample points. Such functions are called *interpolation functions.* Because of the infinite limits of the Sinc(x) function, it is awkward to apply practically. In practice, the ideal interpolating filter must be approximated by a causal lowpass analog filter.

PRACTICAL INTERPOLATION FILTERS. The problem with the ideal interpolation function $h_I(t)$ is that it extends in time from minus infinity to plus infinity. Hence, it incorporates *the entire sequence,* all the future as well as all the past samples, into the computation of the interpolated function at any time point. Especially in control situations, we cannot usually wait to observe the entire sequence before an interpolation is performed. In this section we consider some of the properties that practical interpolators should have and provide examples of two commonly employed interpolators.

Let us consider the impulse response of a practical continuous-time interpolating filter to be given by $\phi(t)$. In practice $\phi(t)$ should have the following three properties:

(1) $$\phi(0) = 1$$

(2) $$\phi(nT_s) = 0 \qquad \text{for } n \neq 0 \tag{3.109}$$

(3) $$\int_{-\infty}^{\infty} |\phi(t)|\, dt < \infty$$

The first two properties permit the interpolated function to equal the sample values at the sampling instants. Property (3) guarantees that the interpolated function remains finite between the sampling instants. Example 3.20 considers two commonly employed interpolation functions that have these properties.

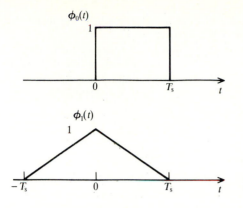

FIGURE 3-24
Two commonly used interpolation functions.

Example 3.20. Zero-order hold and linear interpolators. The impulse response of the zero-order hold interpolator $\phi_0(t)$ is given by

$$\phi_0(t) = \begin{cases} 1 & \text{for } 0 \le t < T_s \\ 0 & \text{otherwise} \end{cases}$$

That of the linear interpolator $\phi_1(t)$ is given by

$$\phi_1(t) = \begin{cases} 1 - |t|/T, & \text{for } -T \le t \le T \\ 0 & \text{otherwise} \end{cases}$$

These interpolating functions are shown in Fig. 3.24. Notice that both these functions have the three desirable properties.

Examples of interpolated functions are shown in Fig. 3.25. The zero-order hold is very easy to implement by quickly charging and discharging a capacitor at the sampling instants, but the interpolated function $c_{A0}(t)$ exhibits discontinuities

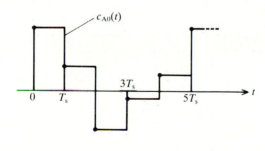

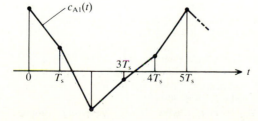

FIGURE 3-25
Results of applying the two interpolation functions to the same sequence.

at the sampling times. These discontinuities may be filtered by the mechanical inertia of the system that is driven with the signal.

The linear interpolator produces a smoother function $c_{A1}(t)$. But this interpolation is noncausal, since it must look ahead to determine the slope value. To make it causal, a delay can be introduced so that the interpolated function lags behind the samples by one sampling period. In control systems, this delay is crucial and may restrict the use of the linear interpolator.

3.12   SUMMARY

In this chapter, we presented the frequency-domain procedures for analyzing digital filters. The Fourier transform of the unit-sample response provided the frequency filtering characteristics of the digital filter, while the Fourier transform of a data sequence indicated its frequency content. The Fourier transform was shown to be a periodic function of ω, with period 2π, and for real-valued sequences to have a magnitude function that is an even function of frequency and a phase function that is odd. This allows the entire function to be determined from the values for $0 \le \omega \le \pi$. Even or odd real-valued sequences were shown to have purely real or imaginary Fourier transforms, while sequences with symmetries have linear phase functions. The operations of convolution and multiplication were shown to be converted into the other when the Fourier transform was computed. The inverse Fourier transform was defined through a Fourier series argument, but found to be complicated to apply directly. An alternate method involving a Taylor expansion was found to be simpler, but did not provide an analytic form for the time series. The relationship between continuous-time and discrete-time frequency functions was discussed and so were procedures to convert a continuous-time function into a discrete-time sequence. This chapter ended with a short discussion on reconstructing the analog signal from the sampled sequence by using an interpolating filter. In the next chapter, we will investigate the signal processing implications of evaluating the Fourier transform at a finite number of points by computer.

FURTHER READING

Antoniou, Andreas: *Digital Filters: Analysis and Design,* McGraw-Hill, New York, 1979.

Bracewell, R. N.: *The Fourier Transform and Its Applications,* 2d ed., McGraw-Hill, New York, 1978.

Childers, D. G. (ed.): *Modern Spectral Analysis,* IEEE Press, New York, 1978. [This is a book of reprints which presents a historical perspective of the evolution of spectral analysis.]

Gold, Bernard and Charles M. Rader: *Digital Processing of Signals,* McGraw-Hill, New York, 1969.

Hamming, R. W.: *Digital Filters,* 2d ed., Prentice-Hall, Englewood Cliffs, NJ, 1983.

Jong, M. T.: *Methods of Discrete Signal and System Analysis,* McGraw-Hill, New York, 1982.

Oetkin, G., T. W. Parks, and H. W. Schussler: "New results in the design of digital interpolators", *IEEE Trans. ASSP* **ASSP-23**: 301–309 (1975).

Oppenheim, Alan, V., and Ronald W. Schafer: *Digital Signal Processing,* Prentice-Hall, Englewood Cliffs, NJ, 1975.

Papoulis, A.: *The Fourier Integral and its Applications*, McGraw-Hill, New York, 1962.

Papoulis, A.: *Signal Analysis*, McGraw-Hill, New York, 1977. [This book contains the continuous-time results referred to in this chapter.]

Peled, A. and B. Liu: *Digital Signal Processing: Theory, Design and Implementation*. Wiley, New York, 1976.

Phillips, C. L. and H. T. Nagle, Jr.: *Digital Control Systems Analysis and Design*, Prentice-Hall, Englewood Cliffs, NJ, 1984.

Rabiner, Lawrence R. and Bernard Gold: *Theory and Application of Digital Signal Processing*, Prentice-Hall, Englewood Cliffs, NJ, 1975.

Schwartz, Mischa; *Information, Transmission, Modulation and Noise*, McGraw-Hill, New York, 1980. [This book contains the continuous-time results referred to in this chapter.]

Schwartz, Mischa, and Leonard Shaw: *Signal Processing: Discrete Spectral Analysis, Detection and Estimation*, McGraw-Hill, New York, 1975.

PROBLEMS

3.1. Given the following unit-sample response sequences compute the Fourier transform and sketch the magnitude and phase functions.
(a) $h(-1) = 1$, $h(1) = 1$, $h(n) = 0$, otherwise.
(b) $h(0) = 1$, $h(1) = -1$, $h(n) = 0$, otherwise.
(c) $h(-1) = 1$, $h(0) = -2$, $h(1) = 1$, $h(n) = 0$, otherwise.

3.2. Determine whether the following sequences have Fourier transforms and compute the Fourier transform when it exists.
(a) $h(n) = 1$, for $n \geq 0$, $= 0$, otherwise.
(b) $h(n) = 1$, $0 \leq n \leq N$, $= 0$, otherwise.
(c) $h(n) = 1$, for all n.
(d) $h(n) = 1$, for $-N \leq n \leq N$, $= 0$, otherwise.

3.3. An important theorem of signal processing is *Parseval's theorem*, that can be stated as

$$\sum_{n=-\infty}^{\infty} x(n) x^*(n) = \frac{1}{2\pi} \int_{-\pi}^{\pi} X(e^{j\omega}) X^*(e^{j\omega}) \, d\omega$$

Prove this theorem by substituting the definition of $X(e^{j\omega})$ into the integral and evaluating.

3.4. Show that a causal filter that has two or more nonzero elements in its unit-sample response cannot have a phase response that is zero for all frequencies.

3.5. Consider the sequence $\{g(n)\}$ that is the time-reversed version of $\{h(n)\}$, or

$$g(n) = h(-n) \qquad \text{for all } n$$

(a) Let the $H(e^{j\omega})$ be the transfer function of the filter whose unit-sample response is $\{h(n)\}$. Compute the transfer function $G(e^{j\omega})$ in terms of $H(e^{j\omega})$.
(b) Consider taking the output of the filter described by $H(e^{j\omega})$ and letting it be the input to a second filter having the transfer function $G(e^{j\omega})$. The total transfer function is then $G(e^{j\omega}) H(e^{j\omega})$. What are the magnitude and phase responses of this cascade combination?

3.6. For the following difference equations, determine the transfer function of the filter and sketch the magnitude and phase responses.
(a) $y(n) = x(n) - x(n - N)$. Let $N = 4$.
(b) $y(n) = 2r \cos(\theta) y(n - 1) + r^2 y(n - 2) + x(n)$. Let $\theta = \pi/4$ and $r = 0.9$.

3.7. For the following magnitude and phase functions, determine and sketch the unit-sample response, the digital filter implementation and the difference equation. Assume $H_R(e^{j\omega}) > 0$ at $\omega = 0$.

(a) $|H(e^{j\omega})| = [20 + 16 \cos(2\omega)]^{1/2}$,

$$\text{Arg}[H(e^{j\omega})] = \arctan\left(\frac{-\sin(2\omega)}{2 + \cos(2\omega)}\right)$$

(b) $|H(e^{j\omega})| = [17 - 8 \cos(2\omega)]^{1/2}$,

$$\text{Arg}[H(e^{j\omega})] = \arctan\left(\frac{4 \sin(2\omega)}{1 - 4 \cos(2\omega)}\right)$$

(c) $|H(e^{j\omega})| = [1.81 - 1.8 \cos(\omega)]^{1/2}$,

$$\text{Arg}[H(e^{j\omega})] = -\omega + \arctan\left(\frac{0.9 \sin(\omega)}{1 - 0.9 \cos(\omega)}\right)$$

3.8. Show that the two different analytic forms for the magnitude response in Example 3.9 are equivalent.

3.9. Some computer languages provide only the ATAN(Y/X) function, which produces the radian angles in the range $-\pi/2 \le \omega \le \pi/2$. Describe a procedure to determine the principal value of the arctangent over the range $-\pi \le \omega < \pi$, from the result of the ATAN(Y/X) function and the values of X and Y.

3.10. If we wanted to generate the sequence

$$x(n) = \cos(\pi n/10), \qquad \text{for all } n$$

by sampling the continuous-time signal

$$x(t) = \cos(\Omega_0 t), \qquad \text{for all } t$$

where $\omega_0 = \Omega_0 T_s$. What is the value of the sampling period T_s?

3.11. Exploring the aliasing phenomenon. The following continuous-time Fourier transform pairs for the sinusoids of infinite duration are given by:

if $\quad u(t) = \cos(\Omega_0 t) \quad$ then $\quad U(j\Omega) = (\pi/2)[\delta(\Omega - \Omega_0) + \delta(\Omega + \Omega_0)]$

and

if $\quad v(t) = \sin(\Omega_0 t) \quad$ then $\quad V(j\Omega) = (j\pi/2)[\delta(\Omega - \Omega_0) - \delta(\Omega + \Omega_0)]$

Determine and sketch the real and imaginary parts of the spectra $U_S(j\Omega)$ and $V_S(j\Omega)$ for the continuous-time *sampled* functions for the following three sampling periods:

(i) $T_s < \pi/\Omega_0$

(ii) $T_s > \pi/\Omega_0$

(iii) $T_s = \pi/\Omega_0$

3.12. Sampling a bandlimited signal. Let $c_A(t)$ have a spectrum $C_A(j\Omega)$ that is nonzero only over the frequency range $\Omega_1 \le |\Omega| \le \Omega_2$, as shown in Fig. P3.12. It can be shown that $c_A(t)$ can be sampled at a rate that has to be greater than twice the bandwidth, $\Omega_2 - \Omega_1$, rather than twice the highest frequency Ω_2, and yet have no information loss. This can be demonstrated by doing the following steps.

(a) Let $T_s = \pi/(\Omega_2 - \Omega_1)$. Determine and sketch the periodically extended spectrum of the time series.

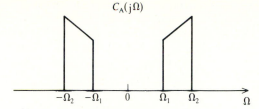

$C_A(j\Omega)$

FIGURE P3-12
Bandpass spectrum of continuous-time signal.

(b) Sketch the magnitude response of the bandpass filter that retrieves the desired signal spectrum.

(c) The impulse response response of this filter acts as the interpolation function. Assume the magnitude response of part (b) and a zero phase response to find the impulse response. Sketch this continuous-time function. Consider whether this waveform is reasonable by considering its values at the sampling instants.

COMPUTER PROJECTS

3.1. High frequency sequences
Object. Observe the shape of sinusoidal sequences for frequencies in the range $\pi/2 < \omega < \pi$, since they can often appear to be confusing.

Computer results. Use TEST0 to obtain plots of either the sine or cosine sequence for the following values of frequency: $\omega = \pi/2$, $3\pi/4$, $7\pi/8$.

Analytic results. Determine and sketch the corresponding continuous-time sinusoid that connects the dots of the computer plot.

3.2. Computer evaluation of the Fourier transform
Object. Write and verify the program that evaluates the Fourier transform of a real-valued sequence at a specified number of frequency points in the range $0 \le \omega < 2\pi$.

Analytic results. Determine and sketch the magnitude and phase spectra of the sequences given below and compare them with the computer results. In all cases, evaluate the sum in the Fourier transform.

Computer results. Write the subprogram

    FT(H, NH, HR, HI, NFT)

where H is the input array containing the real-valued discrete-time sequence of length NH, HR and HI are the output arrays of size NFT containing the real and imaginary values of the Fourier transform, with the Kth element corresponding to the frequency point $\omega_K = 2\pi(K-1)/\text{NFT}$, for $1 \le K \le \text{NFT}$. For this project, let NFT = 128, which will allow comparison with the fast Fourier transform in the next chapter.

A clever way of obtaining the value of π in the program is by using the arctangent function: PI = 4. * ATAN(1.).

To compute the magnitude and phase spectral sequences, write

    MAGPHS(HR, HI, HMAG, HPHS, NFT)

which converts the input real HR and imaginary HI arrays of the FT of size NFT into the output magnitude HMAG array and phase HPHS array. Use the FORTRAN function ATAN2(XI, XR), where XI is the imaginary component and XR is the real component, to calculate the principal value of the phase, i.e., in the range $[-\pi, \pi)$. Undefined values of the phase, when HR(K) = HI(K) = 0, should be set to zero.

To plot the magnitude and phase sequences, use the PLOTMR and PLOTPR subroutines given in Appendix A. These plot the 65 elements of HMAG and HPHS that correspond to the range $0 \le \omega \le \pi$.

Compute the spectra for the following sequences.
(a) $h(0) = \frac{1}{3}$, $h(1) = \frac{2}{3}$, $h(2) = 1$, $h(3) = \frac{2}{3}$, $h(4) = \frac{1}{3}$ and $h(n) = 0$, otherwise.
(b) The unit-sample response of first-order recursive filter for the following values of a: 0.8, 0.95, -0.8, and -0.95 (compute first 128 elements).

3.3. Becoming familiar with the phase spectrum
Object. Illustrate the jumps in the phase response curves.

Analytic results. Determine and sketch the phase spectra of the sequences given below and compare them with the computer results. In all cases, evaluate the sum in the Fourier transform. Indicate the phase jump location and type on the sketches and computer plots.

Computer project. Use the programs FT and MAGPHS written in Project 3.2 to compute the phase and plot with the subroutine PLOTPR given in Appendix A. Use NFT = 128.
(a) Jumps of 2π in the phase spectrum. Compute the phase spectrum of the delayed unit-sample sequence $\{d(n - k)\}$ for $k = 2$. Repeat for $k = 3$ and $k = 10$. Sketch the linear phase functions on the plots.
(b) Jumps of π in the phase spectrum. Compute the phase spectrum of the following sequence: $h(0) = h(1) = h(2) = 1$, and $h(n) = 0$, otherwise.
(c) Both jumps of π and 2π. Compute the phase spectrum for the following sequence: $h(n) = 1$ for $1 \le n \le 5$, and $h(n) = 0$, otherwise. Indicate the causes of the jumps.

3.4. Demonstration of aliasing
Object. Demonstrate the aliasing phenomenon both analytically and numerically.

Analytic results. Do Problem 3.11 and compare the results with those produced by the computer.

Computer results. For this project use the TEST0.FOR program in Appendix A to generate both the sine and cosine sequences. Rather than varying the sampling period with respect to the frequency of the sinusoid, we will keep the sampling period constant, equal to 1, and vary the frequency. Demonstrate the aliasing effect by plotting the time-domain sequences which result when the frequency Ω_0 is varied to correspond to the three cases in Problem 3.11.

3.5. Frequency domain interpretation of interpolating sequences
Object. Explore the frequency-domain behavior of two discrete-time interpolation functions.

Consider the sequence
$$v(n) = \cos(\omega_0 n) \qquad \text{for } 0 \le n \le NV - 1$$

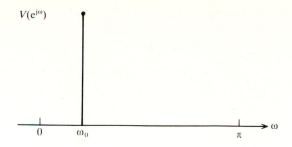

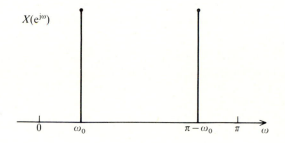

FIGURE Proj 3-5
Spectra for Project 3.5.

with $\omega_0 = \pi/16$ and $NV = 128$ and $x(n)$ such that

$$x(n) = v(n) \qquad \text{for } n \text{ even}, \quad 0 \le n \le NV - 1$$
$$x(n) = 0 \qquad \text{for } n \text{ odd}, \quad 0 \le n \le NV - 1.$$

The idea is to determine the values of $\{x(n)\}$ that were zeroed, by employing interpolation functions as was done in Project 2.5.

The spectra of the two sequences, denoted by $V(e^{j\omega})$ and $X(e^{j\omega})$ are shown in Fig. Proj. 3.5. They are related by

$$X(e^{j\omega}) = \sum_{k=-\infty}^{\infty} V(e^{j(\omega + k\pi)})$$

In words, $X(e^{j\omega})$ is the periodic extension of $V(e^{j\omega})$, but with period π.

Analytic results. We will compare the performance of two discrete-time interpolators:

Zero-order interpolator, where

$$h_0(n) = \begin{cases} 1 & \text{for } 0 \le n \le 1 \\ 0 & \text{otherwise} \end{cases}$$

First-order interpolator, where

$$h_1(n) = \begin{cases} 0.5 & \text{for } n = -1 \\ 1.0 & \text{for } n = 0 \\ 0.5 & \text{for } n = 1 \\ 0 & \text{otherwise} \end{cases}$$

(a) Determine the magnitude responses $|H_0(e^{j\omega})|$ and $|H_1(e^{j\omega})|$. Sketch both responses on the same graph for $0 \le \omega \le \pi$.

(b) Let $\{y_0(n)\}$ and $\{y_1(n)\}$ be the outputs of the zero-order and first-order interpolators, respectively, when $\{x(n)\}$ is the input. Compute the magnitude spectra $|Y_0(e^{j\omega})|$ and $|Y_1(e^{j\omega})|$. Sketch both spectra on the same graph.

Computer results. The interpolation can be implemented as a nonrecursive filter:

```
SUBROUTINE NONREC(X, NX, BCOEF, NB, Y)
```

where X is the input array of size NX, BCOEF is the coefficient array of size NB, and Y is the output array of size NX + NCOEF − 1. For the nonrecursive filter recall that the coefficient sequence is equal to the unit-sample response. Since NX = 128, dimension the Y array sufficiently large to contain the output sequence.

(a) Generate $\{v(n)\}$ and $\{x(n)\}$ for $\omega_0 = \pi/16$. Plot the first 55 samples. Implement the two interpolators and plot 55 samples of the output sequences $\{y_0(n)\}$ and $\{y_1(n)\}$. The first-order interpolator corresponds to a noncausal filter. Repeat the above for $\omega_0 = \pi/4$.

(b) Compute and plot the magnitude spectra of $\{v(n)\}$ and $\{x(n)\}$. Use NFT = 128. Do the same for $y_0(n)$ and $y_1(n)$, for $0 \le n \le 127$.

(c) Explain why there is a difference in the interpolator performance for the two values of ω_0.

3.6. Detection filters for the FSK system

Object. Implement and investigate the filters that differentiate between the two sequences generated by the FSK subroutine written in project in Chapter 2.

In the FSK subroutine, 32 samples of two different frequencies were used to code the binary values 0 and 1 to be transmitted over communication lines. In this project, the samples of the two sinusoids will serve a second purpose, other than the code. They will correspond to the unit-sample response of the two receiver filters, but in time-reversed order. This is, if $\{h_0(n)\}$ is the unit-sample response of the "zero-frequency (K_0) filter", then

$$h_0(n) = \cos[2\pi K_0(31 - n)/32] \quad \text{for } 0 \le n \le 31$$

The unit-sample response of the "one-frequency (K_1) filter" is given by

$$h_1(n) = \cos[2\pi K_1(31 - n)/32] \quad \text{for } 0 \le n \le 31$$

When the unit-sample response is matched in this manner to the desired input signal, the filter is called a *matched filter*.

Analytic results

(a) Sketch the block diagram of the filter as a nonrecursive structure, i.e. containing only feed-forward coefficients.

(b) Determine the magnitude response of each filter. (Hint: use the finite geometric sum formula.)

Computer results

(a) Compute and plot the magnitude spectra of the two 32-sample sequences (use NFT = 128). Use PLOTMR to plot only the first 65 points. Indicate the location of the two FSK frequency values on each magnitude spectrum plot.

(b) Implement the two filters using the NONREC subprogram. Compute and plot the magnitude response of the two filters (use NFT = 128). Compare these to part a.

(c) Generate and plot the four discrete-time sequences that are obtained by passing each of the 32-bit sequences into each of the two filters.

(d) Generate and plot the discrete-time output sequence for each filter when the input is the FSK sequence corresponding to bit pattern 010. Based on these output patterns, describe a simple procedure to determine which binary value was transmitted.

CHAPTER
4

THE DISCRETE FOURIER TRANSFORM

4.1 INTRODUCTION

The discrete Fourier transform (DFT) sequence, denoted by $\{H(k)\}$, allows us to evaluate the Fourier transform $H(e^{j\omega})$ on a digital computer. This complex-valued sequence is obtained by sampling one period of the Fourier transform at a finite number of frequency points. The DFT is important for two reasons. First, it allows us to determine the frequency content of a signal, that is, to perform spectral analysis. The second application of the DFT is to perform filtering operations in the frequency domain. After defining the DFT and the inverse discrete Fourier transform (IDFT), we examine some of its properties that are relevant for signal processing. The DFT can be applied directly to causal, finite-duration sequences. Difficulties arise, however, when we try to apply it to noncausal or infinite-duration sequences. These difficulties are examined and solutions are suggested. The fast Fourier transform (FFT) algorithm is presented as an efficient computational procedure for evaluating the DFT. Its efficiency is the main reason much current signal processing is performed in the frequency domain. The FFT algorithm is described in both an intuitive and analytical form using the decimation-in-time approach.

4.2 THE DEFINITION OF THE DISCRETE FOURIER TRANSFORM (DFT)

The DFT is the finite-duration discrete-frequency sequence that is obtained by sampling one period of the Fourier transform. This sampling is conventionally performed at N equally spaced points over the period extending over $0 \leq \omega < 2\pi$, or at

$$\omega_k = 2\pi k/N \qquad \text{for } 0 \leq k \leq N-1 \tag{4.1}$$

If $\{h(n)\}$ is a discrete-time sequence with the Fourier transform $H(e^{j\omega})$, then the DFT, denoted by $\{H(k)\}$, is defined as

$$H(k) = H(e^{j\omega})\big|_{\omega = \omega_k = 2\pi k/N} \qquad \text{for } 0 \leq k \leq N-1 \tag{4.2}$$

The DFT sequence starts at $k = 0$, corresponding to $\omega = 0$, but does not include $k = N$, corresponding to $\omega = 2\pi$.

Let us consider the consequences of sampling $H(e^{j\omega})$. In the previous chapter, it was shown that the Fourier transform $H(e^{j\omega})$ is periodic in ω, with period 2π. Its Fourier series expansion was then computed and the coefficients were found to be equal to the discrete-time sequence $\{h(n)\}$. It was also shown that when a function of continuous-valued time is sampled with sampling period T_s, then the spectrum of the resulting discrete-time sequence becomes a periodic function of frequency with period $2\pi/T_s$. A similar result can be shown to be true here. As shown in Fig. 4.1, when $H(e^{j\omega})$ is sampled with sampling period $\omega_s = 2\pi/N$, the corresponding discrete-time sequence, denoted by $\{\bar{h}(n)\}$, becomes periodic in time with period $2\pi/\omega_s = N$. This

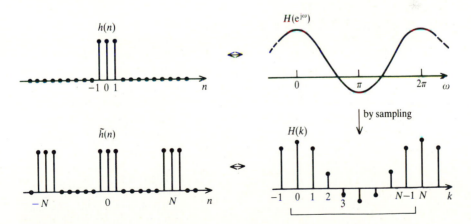

FIGURE 4-1
A periodic extension of a time sequence results when a spectrum is sampled. Case for $N = 8$ samples of spectrum over $0 \leq \omega < 2\pi$. Bracket indicates N-point DFT sequence.

periodic discrete-time sequence can be expressed in terms of $\{h(n)\}$ as

$$\bar{h}(n) = \sum_{m=-\infty}^{\infty} h(n + mN) \tag{4.3}$$

The sequence $\{\bar{h}(n)\}$ is called the *periodic extension* of $\{h(n)\}$. The number of sample points in one period of the spectrum N is also equal to the period of $\{\bar{h}(n)\}$.

The importance of this result is apparent when we employ the DFT for signal processing and filtering in the frequency domain, for it is $\{\bar{h}(n)\}$ that is the corresponding sequence that is being used in the time domain. This chapter describes the implications and pitfalls that accompany this result.

We are free to choose the value of N, the number of samples of $H(e^{j\omega})$ over $0 \le \omega < 2\pi$. Since N is also the period of $\{\bar{h}(n)\}$, we must be careful not to choose N too small. We now illustrate the considerations involved in choosing the value of N. Let $\{h(n)\}$ be a causal finite-duration sequence containing M samples. Infinite-duration and noncausal sequences are discussed later. As shown in Fig. 4.2, if $M \le N$, then

$$\bar{h}(n) = \begin{cases} h(n) & \text{for } 0 \le n \le M - 1 \\ 0 & \text{for } M \le n \le N - 1 \end{cases} \tag{4.4}$$

Two points are noteworthy about this result when $M \le N$. The first is that the finite-duration sequence $h(n)$, for $0 \le n \le M - 1$, can be recovered uniquely from the first period of $\{\bar{h}(n)\}$. The second point is that the excess number of points in one period of $\{\bar{h}(n)\}$, or those for $(k - 1)N + M \le n \le kN - 1$ for any integer k, are equal to zero.

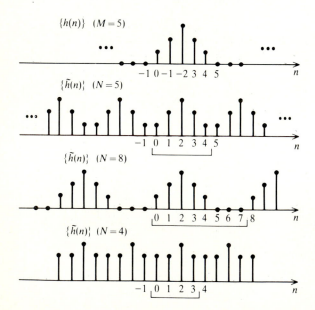

FIGURE 4-2
Relationship of duration of sequence M and number of sample points in spectrum N. When $N < M$, a time-aliasing effect occurs. Underbar indicates sequence produced by inverse DFT.

If $M > N$, however, the duration of $\{h(n)\}$ is longer than the period of $\{\bar{h}(n)\}$. An overlap then occurs when the periodic extension is formed, as shown in Fig. 4.2. This overlap results in an error when we try to retrieve $\{h(n)\}$ from the first period of $\{\bar{h}(n)\}$. In this case, *time-aliasing* is said to occur. To prevent time-aliasing, we must sample $H(e^{j\omega})$ over $0 \leq \omega < 2\pi$ at least as many times as there are elements in $\{h(n)\}$, or $N \geq M$ times. Clearly, this poses a difficulty when trying to apply this result to infinite-duration sequences. These problems will be considered below.

4.3 COMPUTING THE DISCRETE FOURIER TRANSFORM FROM THE DISCRETE-TIME SEQUENCE

We now derive the relationship of the discrete Fourier transform to the discrete-time sequence. First, we note that since $\{\bar{h}(n)\}$ is a periodic sequence, it can be expressed as the following Fourier series expansion

$$\bar{h}(n) = \frac{1}{N} \sum_{k=-\infty}^{\infty} \bar{H}(k)\, e^{j2\pi kn/N} \tag{4.5}$$

where $\{\bar{H}(k)\}$ are a set of complex-valued coefficients to be determined, and the $1/N$ factor has been accepted as the convention. Because the discrete-time complex exponential sequence $\{e^{j2\pi kn/N}\}$ is itself periodic in k, with period N, there are only a finite number of distinct sequences, equal to $e^{j2\pi kn/N}$, for $0 \leq k \leq N-1$. To show this, let $0 \leq k \leq N-1$ and consider a frequency outside this range, say $e^{j2\pi(k+N)n/N}$. Then

$$e^{j2\pi(k+N)n/N} = e^{j2\pi kn/N}\, e^{j2\pi n} = e^{j2\pi kn/N} \tag{4.6}$$

which is identical to the sequence $e^{j2\pi kn/N}$. Hence, only a finite set of frequencies need to be included in the above summation, and $\{\bar{h}(n)\}$ can be expressed in terms of the following N frequency components

$$\bar{h}(n) = \frac{1}{N} \sum_{k=0}^{N-1} \bar{H}(k)\, e^{j2\pi nk/N} \tag{4.7}$$

To determine the values of $\bar{H}(k)$, for $0 \leq k \leq N-1$, we multiply both sides of Eq. (4.7) by $e^{-j2\pi nr/N}$ and sum over $0 \leq n \leq N-1$, to get

$$\sum_{n=0}^{N-1} \bar{h}(n)\, e^{-j2\pi nr/N} = \frac{1}{N} \sum_{n=0}^{N-1} \left[\sum_{k=0}^{N-1} \bar{H}(k)\, e^{j2\pi nk/N} \right] e^{-j2\pi nr/N} \tag{4.8}$$

Changing the order of summations and noting that $\bar{H}(k)$ is not a function of n, we get

$$\sum_{n=0}^{N-1} \bar{h}(n)\, e^{-j2\pi nr/N} = \frac{1}{N} \sum_{k=0}^{N-1} \bar{H}(k) \left[\sum_{n=0}^{N-1} e^{j2\pi n(k-r)/N} \right] \tag{4.9}$$

The sum in the square brackets can be evaluated by applying the finite geometric sum formula, to obtain

$$\sum_{n=0}^{N-1} e^{j2\pi n(k-r)/N} = \frac{1 - e^{j2\pi(k-r)}}{1 - e^{j2\pi(k-r)/N}} \tag{4.10}$$

Since $(k - r)$ takes on only integer values, the numerator term is zero for all values of k and r. However, for $k = r + mN$, where m is an integer, the denominator is also zero. In this case, each term in the sum is equal to 1, and the sum evaluates to N, or

$$\sum_{n=0}^{N-1} e^{j2\pi n(k-r)/N} = \begin{cases} N & \text{for } k = r + mN \\ 0 & \text{otherwise} \end{cases} \tag{4.11}$$

Since k lies in the range from zero to $N - 1$ in Eq. (4.9), we can apply this result for $k = r$, or for $m = 0$. This results in the desired solution for $\bar{H}(r)$ given by

$$\bar{H}(r) = \sum_{n=0}^{N-1} \bar{h}(n) e^{-j2\pi nr/N} \qquad \text{for } 0 \le r \le N - 1 \tag{4.12}$$

Replacing r with k gives the final desired solution for $\bar{H}(k)$.

We now relate this result to the Fourier transform of the original sequence $\{h(n)\}$. Let $\{h(n)\}$ be a causal, finite-duration sequence containing M samples. Then its Fourier transform is given by

$$H(e^{j\omega}) = \sum_{n=0}^{M-1} h(n) e^{-j\omega n} \tag{4.13}$$

We can increase the duration of $\{h(n)\}$ from M to N samples by appending the appropriate number of zeros. Since zero-valued elements contribute nothing to the sum, the Fourier transform is equal to

$$H(e^{j\omega}) = \sum_{n=0}^{N-1} h(n) e^{-j\omega n} \tag{4.14}$$

Sampling $H(e^{j\omega})$ at N equally spaced points over $0 \le \omega < 2\pi$ produces the sequence

$$H(k) = H(e^{j\omega})\big|_{\omega = 2\pi k/N} = \sum_{n=0}^{N-1} h(n) e^{-j2\pi kn/N} \qquad 0 \le k \le N - 1 \tag{4.15}$$

Since the number of sample points N is equal to the number of elements in the original discrete-time sequence, the first N samples of $\{\bar{h}(n)\}$ are equal to $\{h(n)\}$, or $h(n) = \bar{h}(n)$, for $0 \le n \le N - 1$. Only the first N values of $\{\bar{h}(n)\}$ are needed in the sum in Eq. (4.12). Comparing (4.15) with (4.12), we note that $H(k) = \bar{H}(k)$, for $0 \le k \le N - 1$. Applying this result to (4.7), we get

$$\bar{h}(n) = \frac{1}{N} \sum_{k=0}^{N-1} H(k) e^{j2\pi nk/N} \qquad \text{for } 0 \le n \le N - 1 \tag{4.16}$$

Summarizing these results, we arrive at the desired DFT relationships:

$$H(k) = \sum_{n=0}^{N-1} h(n)\, e^{-j2\pi nk/N} \qquad \text{for } 0 \le k \le N-1 \qquad (4.17)$$

which is called an *N-point DFT*. The inverse discrete Fourier transform, or *IDFT*, by which we compute the discrete time sequence from the DFT sequence, is given by

$$h(n) = \frac{1}{N} \sum_{k=0}^{N-1} H(k)\, e^{j2\pi nk/N} \qquad \text{for } 0 \le n \le N-1 \qquad (4.18)$$

Equations (4.17) and (4.18) form the basis for the computer algorithms that evaluate the DFT, and are directly applicable to *causal, finite-duration* sequences. The number N of frequency points in one period of the DFT is equal to the number of elements in the discrete-time sequence. To obtain more points in the DFT sequence, we can always increase the duration of $\{h(n)\}$ by adding additional zero-valued elements. The procedure is called *padding with zeros*. These zero-valued elements contribute nothing to the sum in Eq. (4.17), but act to decrease the frequency spacing $e^{j2\pi k/N}$. We will encounter this padding operation in the computation of the convolution below.

Example 4.1.   DFT of the causal 3-sample averager. Let

$$h(n) = \begin{cases} \frac{1}{3} & \text{for } 0 \le n \le 2 \\ 0 & \text{otherwise} \end{cases}$$

To compute the N-point DFT, where $N \ge 3$, the kth complex coefficient $H(k)$, for $0 \le k \le N-1$, is given by

$$
\begin{aligned}
H(k) &= \sum_{n=0}^{N-1} h(n)\, e^{-j2\pi nk/N} = \sum_{n=0}^{2} \tfrac{1}{3}\, e^{-j2\pi nk/N} \\
&= \tfrac{1}{3}\bigl(1 + e^{-j2\pi k/N} + e^{-j4\pi k/N}\bigr) \\
&= \tfrac{1}{3}\, e^{-j2\pi k/N}\bigl(e^{j2\pi k/N} + 1 + e^{-j2\pi k/N}\bigr) \\
&= e^{-j2\pi k/N}\, \frac{1 + 2\cos(2\pi k/N)}{3}
\end{aligned}
$$

Comparing this result with Example 3.9, we see that the magnitude and phase values of $\{H(k)\}$ are samples of $H(e^{j\omega})$ at N points equally spaced over $0 \le \omega < 2\pi$. The DFT sequences for different values of N are shown in Fig. 4.3.

The DFT is specified over the range $0 \le \omega < 2\pi$, rather than $-\pi \le \omega < \pi$, the range used for the Fourier transform in the previous chapter. This is the accepted convention for all the DFT algorithms. It is convenient for interpreting the results of computer processing, since the dc value, i.e. $\omega = 0$, then corresponds to the first element of the DFT sequence.

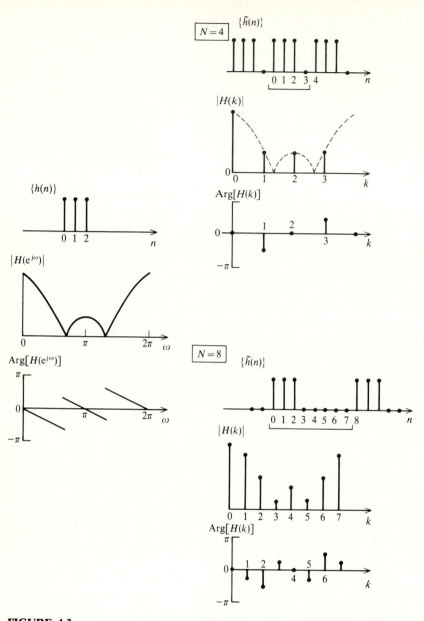

FIGURE 4-3
Relationship of Fourier transform and discrete Fourier transform for a causal sequence. Cases for $N = 4$ and $N = 8$ are shown.

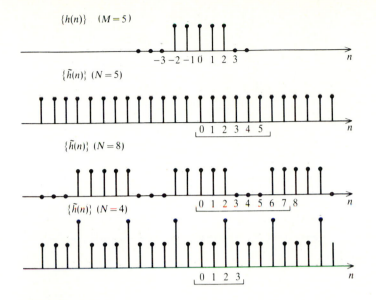

FIGURE 4-4
Relationship of duration of noncasual sequence M and number of sample points in spectrum N. The discrete-time sequence is conventionally taken as the first N samples of the inverse DFT. When $N < M$, a time-aliasing effect occurs.

NONCAUSAL SEQUENCES. Let $\{h_{NC}(n)\}$ be a finite-duration noncausal sequence containing M samples. The period extension is given by

$$\bar{h}(n) = \sum_{m=-\infty}^{\infty} h_{NC}(n + mN) \tag{4.19}$$

as shown in Fig. 4.4. However, the conventional definition of the IDFT given by Eq. (4.18) indicates that the first N samples of the periodic extension are obtained, or $\bar{h}(n)$, for $0 \le n \le N - 1$, when the IDFT is evaluated. The noncausal part of $\{h_{NC}(n)\}$ will then appear *at the end* of the resulting sequence, being the noncausal part of the second period of the periodic extension, as shown in Fig. 4.4 for $N = 8$.

Example 4.2.   The DFT of noncausal 3-sample averager. Let

$$h(n) = \begin{cases} \frac{1}{3} & \text{for } -1 \le n \le 1 \\ 0 & \text{otherwise} \end{cases}$$

In the definition of the DFT pair, the time sequence used in the computation is equal to $\bar{h}(n)$, for $0 \le n \le N - 1$. Hence, as shown in Fig. 4.5, the sequence that must be used for evaluating an N-point DFT is

$$h(n) = \begin{cases} \frac{1}{3} & \text{for } n = 0, 1, N - 1 \\ 0 & \text{otherwise} \end{cases}$$

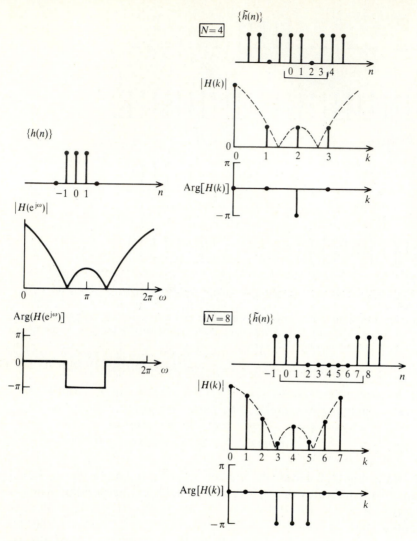

FIGURE 4-5
Relationship of Fourier transform and discrete Fourier transform for a noncausal sequence.
Cases for $N = 4$ and $N = 8$ are shown.

Then

$$H(k) = \tfrac{1}{3}(1 + e^{-j2\pi k/N} + e^{-j2(N-1)\pi k/N})$$

Since adding 2π to the exponent of the complex exponential does not change its
value, we find

$$e^{-j2\pi(N-1)\pi k/N} = e^{-j[2(N-1)\pi k/N + 2\pi N/N]} = e^{j2\pi k/N}$$

and

$$H(k) = \tfrac{1}{3}(1 + e^{-j2\pi k/N} + e^{j2\pi k/N})$$
$$= \frac{1 + 2\cos(2\pi k/N)}{3}$$

These are samples of the Fourier transform of the 3-sample averager, as shown in Fig. 4.5.

We would have obtained the same result if we allowed the DFT sum to range over negative index values, to include the $h(-1)$ term. But, since the standard computer algorithms that evaluate the DFT are defined in terms of non-negative index values, these equivalent terms from the next period of $\{\bar{h}(n)\}$ are included in the sum. This convention is often a source of confusion when interpreting the results of the DFT and IDFT computations.

INFINITE-DURATION SEQUENCES. For infinite-duration sequences, there is always some amount of time-aliasing which results when the DFT relationships are applied. In practice, the value of N is usually chosen to be large enough to make this error negligible for a particular application. We will consider the effects of choosing the proper value of N in the chapter on IIR filter design.

> **Example 4.3. DFT of an infinite-duration sequence.** Let us consider the familiar infinite-duration unit-sample response of the first-order recursive filter, given by
>
> $$g(n) = a^n u(n) \qquad \text{for all } n$$
>
> As shown in Example 3.2, the Fourier transform is equal to
>
> $$H(e^{j\omega}) = [1 - a\, e^{-j\omega}]^{-1}$$
>
> Sampling this function produces the DFT sequence given by
>
> $$H(k) = H(e^{j\omega})\big|_{\omega = 2\pi k/N} = [1 - a\, e^{-j2\pi k/N}]^{-1} \qquad \text{for } 0 \le k \le N-1$$
>
> Applying the IDFT formula to determine the corresponding discrete-time sequence, we have
>
> $$h(n) = \frac{1}{N} \sum_{k=0}^{N-1} \frac{e^{j2\pi kn/N}}{1 - a\, e^{-j2\pi k/N}} \qquad \text{for } 0 \le n \le N-1$$
>
> Expanding the denominator in a Taylor series, we get
>
> $$h(n) = \frac{1}{N} \sum_{k=0}^{N-1} e^{j2\pi kn/N} \left[\sum_{r=0}^{\infty} a^r e^{-j2\pi kr/N} \right] \qquad \text{for } 0 \le n \le N-1$$
>
> $$= \frac{1}{N} \sum_{r=0}^{\infty} a^r \left[\sum_{k=0}^{N-1} e^{j2\pi k(n-r)/N} \right] \qquad \text{for } 0 \le n \le N-1$$
>
> From (4.11), the sum in the brackets is zero, except when $r = n + mN$, where m is

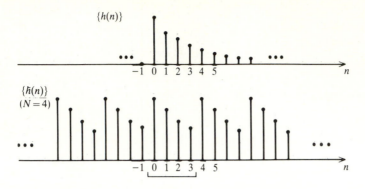

FIGURE 4-6
Time-aliasing effect for an infinite-duration sequence. Underbar indicates sequence produced by inverse DFT when $N = 4$.

an integer. The sum then evaluates to N. Hence, we have

$$h(n) = \sum_{\substack{r=0 \\ r=n+mN}}^{\infty} a^r \quad \text{for } 0 \le n \le N-1$$

For r to be greater than zero, m must also be greater than or equal to 0. Summing over the infinite number of values of m, we have

$$h(n) = \sum_{m=0}^{\infty} a^{(n+mN)} = a^n \sum_{m=0}^{\infty} (a^N)^m = \frac{a^n}{1-a^N} \quad \text{for } 0 \le n \le N-1$$

This result is *not* equal to $\{g(n)\}$, although it approaches $\{g(n)\}$ as N becomes infinite, as shown in Fig. 4.6. Since the unit-sample response of a stable filter approaches zero with time, a sufficiently large value of N can be found to make the aliasing effects negligible.

Alternatively, the periodic extension is given by

$$\bar{h}(n) = \sum_{m=-\infty}^{\infty} a^{n+mN} u(n + mN) \quad \text{for } 0 \le n \le N-1$$

$$= \sum_{m=0}^{\infty} a^{n+mN} = a^n \sum_{m=0}^{\infty} (a^N)^m = \frac{a^n}{1-a^N} \quad \text{for } 0 \le n \le N-1$$

This is the same as the result above.

4.4 PROPERTIES OF THE DFT

Since the DFT sequence consists of samples of the Fourier transform, the linearity, periodicity and symmetry properties of the Fourier transform are also true for the DFT. These properties are listed in Table 4.1 and discussed below.

Because of the periodicity and symmetry of the Fourier transform for real-valued sequences, we need to plot only the values of $H(k)$, for $0 \le k \le (N/2) + 1$, which corresponds to $H(e^{j\omega})$, for $0 \le \omega \le \pi$.

TABLE 4.1
Properties of the discrete Fourier transform

1. Definition: For a finite-duration sequence $h(n)$, for $0 \le n \le N-1$,

$$H(k) = \sum_{n=0}^{N-1} h(n) e^{-j2\pi kn/N} \qquad \text{for } 0 \le k \le N-1 \qquad \text{(DFT)}$$

and

$$h(n) = \frac{1}{N} \sum_{k=0}^{N-1} H(k) e^{j2\pi kn/N} \qquad \text{for } 0 \le n \le N-1 \qquad \text{(IDFT)}$$

2. Periodic extension:

$$\bar{h}(n) = \sum_{m=-\infty}^{\infty} h(n+mN)$$

3. Linearity: If $\{x(n)\} = \{h_1(n)\} + \{h_2(n)\}$, then $X(k) = H_1(k) + H_2(k)$.
4. Periodicity: $H(k) = H(k+N)$.
5. Magnitude and phase functions: If $H(k) = H_R(k) + jH_I(k)$, then

$$|H(k)|^2 = H_R^2(k) + H_I^2(k)$$
$$\text{Arg}[H(k)] = \arctan[H_I(k)/H_R(k)]$$

6. Fourier transform of delayed sequence:

$$\text{if} \quad \{y(n)\} = \{x(n-n_0)\} \quad \text{then} \quad Y(k) = X(k) e^{-j2\pi kn_0/N}$$

7. Fourier transform of the convolution of two sequences:

$$\text{if} \quad \{y(n)\} = \{h(n)\} * \{x(n)\} \quad \text{then } Y(k) = H(k)X(k)$$

A linear convolution of $\{h(n)\}$ and $\{x(n)\}$ results when N_y-point DFT sequences are computed for $\{h(n)\}$ and $\{x(n)\}$, where N_y is the duration of $\{y(n)\}$. Otherwise, a circular convolution of $\{h(n)\}$ and $\{x(n)\}$ results.

8. Fourier transform of the product of two sequences:

$$\text{if} \quad \{y(n)\} = \{h(n)\,x(n)\} \quad \text{then} \quad Y(k) = H(k) \circledast X(k)$$

$$\text{(circular convolution)}$$

9. For the *real-valued sequence* $h(n)$, for $0 \le n \le N-1$:

$$H_R(k) = \sum_{n=0}^{N-1} h(n) \cos(2\pi kn/N), \qquad \text{for } 0 \le k \le N-1$$

$$H_I(k) = -\sum_{n=0}^{N-1} h(n) \sin(2\pi kn/N) \qquad \text{for } 0 \le k \le N-1$$

(a) Complex-conjugate symmetry: $\qquad H(k) = H^*(N-k)$
(b) Real component is even function: $\qquad H_R(k) = H_R(N-k)$
(c) Imaginary component odd function: $\qquad H_I(k) = -H_I(N-k)$
(d) Magnitude function is even function: $\qquad |H(k)| = |H(N-k)|$
(e) Phase function is odd function: $\qquad \text{Arg}[H(k)] = -\text{Arg}[H(N-k)]$
(f) If $\{h(n)\} = \{h(-n)\}$ (even sequence), then $H(k)$ is purely real.
(g) If $\{h(n)\} = \{-h(-n)\}$ (odd sequence), then $H(k)$ is purely imaginary.

4.5 CIRCULAR CONVOLUTION

Given $h(n)$, for $0 \leq n \leq N - 1$, and $\{x(n)\}$, their convolution is equal to

$$y(n) = \sum_{k=0}^{N-1} h(k) x(n - k) \qquad \text{for all } n \qquad (4.20)$$

To differentiate it from the operation performed with the DFT, described below, we denote Eq. (4.20) as being the *linear convolution*. In the previous chapter, the Fourier transform was found to be useful for converting a convolutional operation into a multiplicative one. The values of $\{y(n)\}$ are then computed from the inverse Fourier transform $Y(e^{j\omega}) = H(e^{j\omega}) X(e^{j\omega})$.

Here, we would like to determine $\{y(n)\}$ from the inverse DFT of $\{Y(k)\}$, where $Y(k) = H(k) X(k)$, for $0 \leq k \leq N - 1$. But caution must be exercised in applying the DFT relations because the sequences in the DFT relations are *periodic* with period N, being the periodic extension of the original sequences. The convolution that results is then called a *circular convolution*, defined by

$$\bar{y}(n) = \sum_{k=0}^{N-1} \bar{h}(k) \bar{x}(n - k) \qquad \text{for } 0 \leq n \leq N - 1 \qquad (4.21)$$

or, more concisely,

$$\{\bar{y}(n)\} = \{\bar{h}(n)\} \circledast \{\bar{x}(n)\} \qquad (4.22)$$

where $\circledast$ denotes the circular convolution operation. The steps involved in computing this convolution are identical to those for the linear convolution, except that the sum is taken over only one period. The resulting $\{\bar{y}(n)\}$ is also periodic, with period N.

The main consequence of the periodic nature of the sequences in the convolution is that the shift produced by the index $(n - k)$ actually represents a *rotation*. The periodicity makes the sequence values appear to *wrap around* from the beginning of the sequence to the end. The circular convolution is illustrated in Example 4.4.

Example 4.4. Circular convolution with the DFT relations. Let

$$x(n) = \begin{cases} 1 & \text{for } n = 0 \\ 0.5 & \text{for } n = 1 \\ 0 & \text{otherwise} \end{cases}$$

and

$$h(n) = \begin{cases} 0.5 & \text{for } n = 0 \\ 1 & \text{for } n = 1 \\ 0 & \text{otherwise} \end{cases}$$

We want to determine the linear convolution of the two sequences

$$\{y(n)\} = \{x(n)\} * \{h(n)\}$$

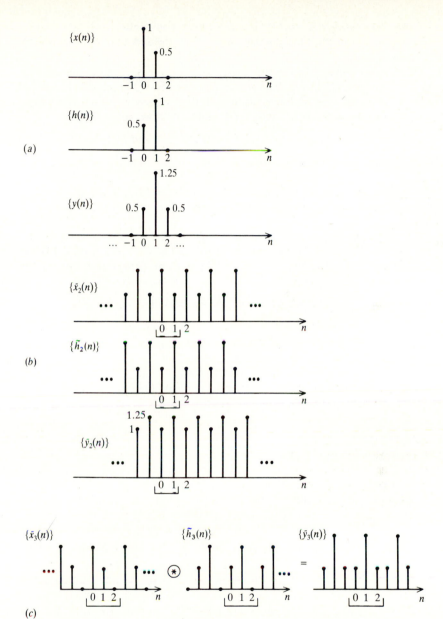

FIGURE 4-7

(a) Linear convolution of two finite-duration sequences. (b) Circular convolution result when 2-point DFT sequences are used for the sequences shown in (a). The minimum size of the DFT in this example is 3. (c) Circular convolution result when 3-point DFT sequences are used for the sequences shown in (a).

Linear convolution. By performing the linear convolution with the finite-duration sequences in the time domain, we get the desired $\{y(n)\}$ as shown in Fig. 4.7(*a*).

Circular convolution. Since both $\{x(n)\}$ and $\{h(n)\}$ have durations of two samples, it may be reasonable (*but wrong!*) to take two 2-point DFTs. Let us consider the consequences of doing this in both the frequency and time domains.

Frequency domain. Performing the 2-point DFTs we get

$$X(0) = 1.5 \text{ and } X(1) = 0.5 \quad \text{and} \quad H(0) = 1.5 \text{ and } H(1) = -0.5$$

Multiplying $\{X(k)\}$ by $\{H(k)\}$ to compute $\{Y(k)\}$ and evaluating the product for $k = 0, 1$, we get $Y(0) = 2.25$ and $Y(1) = -0.25$.

Applying the inverse DFT, we get the sequence $y(0) = 1$ and $y(1) = 1.25$, which is not the same as the desired linear convolution result. This sequence is shown in Fig. 4.7(*b*).

Time domain. Since we computed 2-point DFTs, the corresponding periodic convolution was between periodic sequences $\{\bar{x}_2(n)\}$ and $\{\bar{h}_2(n)\}$ having a period of two samples, which were obtained from $\{x(n)\}$ and $\{h(n)\}$ by periodic extension. Performing the periodic convolution, we get

$$\bar{y}_2(0) = \bar{x}_2(0)\,\bar{h}_2(0) + \bar{x}_2(1)\,\bar{h}_2(-1) = 1$$
$$\bar{y}_2(1) = \bar{x}_2(0)\,\bar{h}_2(1) + \bar{x}_2(1)\,\bar{h}_2(0) = 1.25$$
$$\bar{y}_2(2) = \bar{x}_2(0)\,\bar{h}_2(2) + \bar{x}_2(1)\,\bar{h}_2(1) = 1$$

$$\cdots$$

The sequence $\{\bar{y}_2(n)\}$ is periodic with period 2 and the first period has the same values as those produced by the 2-point IDFT above.

4.6 PERFORMING A LINEAR CONVOLUTION WITH THE DFT

To perform a linear convolution with the DFT, it is necessary to evaluate the DFT at a sufficiently large number of points so that time-aliasing effects do not occur. To demonstrate this, let $\{x(n)\}$ and $\{h(n)\}$ be finite-duration sequences of length N_X and N_H, respectively, and $\{y(n)\} = \{x(n)\} * \{h(n)\}$. After performing the linear convolution of $\{x(n)\}$ and $\{h(n)\}$, the sequence $\{y(n)\}$ has duration $N_Y = N_X + N_H - 1$.

To determine the linear convolution result in the frequency domain, the DFT of $\{y(n)\}$ must be evaluated at N_Y points to retain the information that is present in $\{Y(e^{j\omega})\}$, i.e. $\{Y(k)\}$ must contain N_Y samples of $Y(e^{j\omega})$ over $0 \le \omega < 2\pi$. Then, to compute the product $Y(k) = X(k)\,H(k)$, $0 \le k \le N_Y - 1$, both $\{X(k)\}$ and $\{H(k)\}$ must also be evaluated at N_Y points. Example 4.5 illustrates this point.

Example 4.5. Performing a linear convolution with the DFT. Let

$$x(n) = \begin{cases} 1 & \text{for } n = 0 \\ 0.5 & \text{for } n = 1 \\ 0 & \text{otherwise} \end{cases}$$

and

$$h(n) = \begin{cases} 0.5 & \text{for } n = 0 \\ 1 & \text{for } n = 1 \\ 0 & \text{otherwise} \end{cases}$$

Frequency domain. The linear convolution will produce a 3-sample sequence. To avoid time-aliasing we convert the 2-sample input sequences into 3-sample sequences by padding with zeros. The 3-point DFT sequences are given by

$$X(0) = 1.5 \qquad X(1) = 1 + 0.5e^{-j2\pi/3} \qquad \text{and} \qquad X(2) = 1 + 0.5\,e^{-j4\pi/3}$$

and

$$H(0) = 1.5 \qquad H(1) = 0.5 + e^{-j2\pi/3} \qquad \text{and} \qquad H(2) = 0.5 + e^{-j4\pi/3}$$

Computing the product

$$Y(k) = H(k)\, X(k) \qquad \text{for } 0 \le k \le 2$$

we get

$$Y(0) = 2.25$$

$$Y(1) = 0.5 + 1.25\,e^{-j2\pi/3} + 0.5\,e^{-j4\pi/3}$$

$$Y(2) = 0.5 + 1.25\,e^{-j4\pi/3} + 0.5\,e^{-j8\pi/3}$$

Computing the IDFT, we get

$$y(0) = 0.5 \qquad y(1) = 1.25 \qquad \text{and} \qquad y(2) = 0.5$$

These are the same values as the linear convolution result obtained in the previous example.

Time domain. Since we have calculated 3-point DFTs, the result should be the same as the periodic convolution between $\{\bar{x}_3(n)\}$ and $\{\bar{h}_3(n)\}$. The values of the result for different values of n are equal to

$$\bar{y}_3(0) = \bar{x}_3(0)\,\bar{h}_3(0) + \bar{x}_3(1)\,\bar{h}_3(-1) + \bar{x}_3(2)\,\bar{h}_3(-2) = 0.5$$

$$\bar{y}_3(1) = \bar{x}_3(0)\,\bar{h}_3(1) + \bar{x}_3(1)\,\bar{h}_3(0) + \bar{x}_3(2)\,\bar{h}_3(-1) = 1.25$$

$$\bar{y}_3(2) = \bar{x}_3(0)\,\bar{h}_3(2) + \bar{x}_3(1)\,\bar{h}_3(1) + \bar{x}_3(2)\,\bar{h}_3(0) = 0.5$$

. . .

The sequence $\{\bar{y}_3(n)\}$ is periodic with period 3 and the first period has the same values as those produced by the 3-point IDFT, as shown in Fig. 4.7(c).

In Chapter 9, we will indicate how this procedure can be extended to evaluate a linear convolution when one of the sequences has an infinite duration. This situation commonly occurs when processing long data records, such as those of digitized music.

4.7 COMPUTATIONS FOR EVALUATING THE DFT

In this section, we describe the computations required for the direct evaluation of the DFT of a complex-valued sequence. In the previous chapter, we presented a simple relationship for evaluating the Fourier transform of a real-valued sequence at a collection of frequency points. By considering complex-valued sequences here, we find that both the transform and its inverse can be evaluated with the same algorithm. Even though the time sequence may be real-valued, the DFT sequence is complex-valued. Hence, it is necessary to consider complex-valued sequences.

Let us consider the complex-valued sequence given by

$$h(n) = h_R(n) + jh_I(n) \qquad \text{for } 0 \le n \le N-1 \tag{4.23}$$

Including this complex form in the definition of the DFT, we obtain

$$
\begin{aligned}
H(k) &= \sum_{n=0}^{N-1} h(n)\, e^{-j2\pi nk/N} \qquad \text{for } 0 \le k \le N-1 \\
&= \sum_{n=0}^{N-1} [h_R(n) \cos(2\pi kn/N) + h_I(n) \sin(2\pi kn/N)] \\
&\quad -j \sum_{n=0}^{N-1} [h_R(n) \sin(2\pi kn/N) - h_I(n) \cos(2\pi kn/N)] \tag{4.24}
\end{aligned}
$$

The DFT sequence can also be expressed explicitly in terms of its real and imaginary parts as

$$H(k) = H_R(k) + jH_I(k) \qquad \text{for } 0 \le k \le N-1 \tag{4.25}$$

The IDFT relationship is then given by

$$
\begin{aligned}
h(n) &= \frac{1}{N} \sum_{k=0}^{N-1} H(k)\, e^{j2\pi nk/N} \qquad \text{for } 0 \le n \le N-1 \\
&= \frac{1}{N} \sum_{k=0}^{N-1} [H_R(k) \cos(2\pi kn/N) - H_I(k) \sin(2\pi kn/N)] \\
&\quad + \frac{j}{N} \sum_{k=0}^{N-1} [H_R(k) \sin(2\pi kn/N) + H_I(k) \cos(2\pi kn/N)] \tag{4.26}
\end{aligned}
$$

Except for the difference in the signs of the exponents and the $1/N$ scaling factor in the inverse, the same operations are performed in both transforms. Both the DFT and IDFT computations require $4N$ real-valued multiplications per point, and $4N^2$ such operations are necessary to compute the N-point transform. To obtain a simple, but still reasonably accurate, estimate of the operation count, we ignore the observation that the first real and imaginary terms are multiplications by 1 and 0 respectively, and hence, not true multiplications. We then find that for large values of N, the number $4N^2$, is a good approximation for the required number of real-valued multiplications to compute an N-point discrete Fourier transform or its inverse. For complex-valued sequences, the symmetries of the real and imaginary components, which are valid for real-valued sequences, no longer hold.

In the next section we describe the implementation of a program to directly evaluate the DFT from (4.24) and the IDFT from (4.26). Because of the similarity in the operations for computing the DFT and IDFT, if an algorithm were found to perform the evaluation in an efficient manner, the benefits would be applicable to both the transform and its inverse. Later in this chapter, we will consider an alternate algorithm to the direct evaluation, called the *fast Fourier transform* (*FFT*). This algorithm requires N to be a power of 2 and takes into account the periodicity of the DFT to reduce the number of multiplications from $4N^2$ to $4N \log_2 N$.

4.8 PROGRAMMING THE DISCRETE FOURIER TRANSFORM

The program to evaluate the Fourier transform of a finite-duration sequence $h(n)$, for $0 \le n \le N_H - 1$, at a set of discrete-frequency points, was described in Chapter 3. This same program can also be used to compute the discrete Fourier transform after making two modifications. First, the number of frequency points in the DFT, denoted by NDFT, should be equal to the number of elements in the discrete-time sequence N_H. To be consistent with the conventional algorithms and the popular fast Fourier transform (FFT), we take NDFT to be a power of 2, or

$$\text{NDFT} = 2^{\text{NLOG2}}$$

where NLOG2 is an integer. The reason for this is explained when we explore the mathematically involved FFT algorithm later.

The second modification involves reducing the number of arrays used in the computation to conserve memory requirements. The standard DFT algorithms conventionally employ the same arrays for both the input and output, as described below.

Incorporating these two modifications, the subprogram is given by

```
DFT(NLOG2, HR, HI, NTYPE)
```

where

NLOG2	(integer input) is the log to the base 2 of the size of the arrays
HR	(real input and output array) is the array of real values
HI	(real input and output array) is the array of imaginary values
NTYPE	(integer input) is the operation to be performed: NTYPE = 1 causes the DFT to be computed, NTYPE = −1 produces the IDFT

To compute the DFT, the HR and HI arrays contain the real and imaginary parts of the discrete-time sequence, padded with zeros if necessary. If the sequence is real, all the elements of HI should be set to zero. After the routine has been accomplished, HR and HI contain the real and imaginary parts of the DFT sequence. To compute the IDFT, the HR and HI arrays contain the real and imaginary parts of the DFT before the call. After execution, HR and HI contain the real and imaginary parts of the discrete-time sequence. If the DFT sequence corresponds to that for a real-valued sequence, the elements in the HI array evaluate to values that are approximately equal to zero, allowing for round-off errors in the computations.

To allow the same arrays to be used in the calling argument, it is helpful to use temporary arrays, TR and TI, of size NDFT, for the computation of the real and imaginary parts of the transform within the subprogram. The FFT algorithm computes the DFT without needing this temporary storage, as described later.

As in the previous program, a problem arises with the indexing of the arrays, which usually starts at one, rather than zero. For the Kth element of the DFT sequence, $H(k)$, for $1 \leq K \leq NDFT$, the corresponding frequency value, denoted by OMK, is then equal to

$$OMK = 2\pi(K - 1)/NDFT \qquad \text{for } 1 \leq K \leq NDFT \qquad (4.27)$$

The evaluation is straightforward, being the direct implementation of (4.24) for the DFT and (4.26) for the IDFT. The real part of the Kth element of the DFT computation then becomes

$$TR(K) = \sum_{M=1}^{NDFT} [HR(M) \cos((M - 1)OMK) + HI(M) \sin((M - 1)OMK)]$$

$$\text{for } 1 \leq K \leq NDFT \quad (4.28)$$

A similar equation can be written for the array TI containing the imaginary part. For evaluating the DFT, the HR and HI arrays contain the real and imaginary discrete-time sequence values. After the TR and TI arrays are computed, these values are transferred to HR and HI, respectively, thus completing the transform. For evaluating the IDFT, the HR and HI arrays contain the real and imaginary parts of the DFT sequence.

Evaluating the DFT and IDFT with the above algorithm is a simple procedure. In many cases, this DFT routine is sufficient for occasional use for determining the spectrum of a discrete-time sequence. However, if spectral analysis is to be employed routinely as part of a repetitive signal processing task, this DFT routine is too time-consuming, especially when the number of points is large, say NDFT = 1024. Recall that the number of multiplications alone is $4NDFT^2$. In the next section, we investigate procedures to increase the speed of evaluating the required $4NDFT^2$ computations. In the following section, we describe the revolutionary fast Fourier transform, which reduces the multiplication count to $4NDFT \log_2 NDFT$.

4.9 INCREASING THE COMPUTATIONAL SPEED OF THE DFT

If the DFT subroutine is to be used often for spectral analysis and filtering operations, the N^2 operations required to evaluate the N-point DFT become tediously time-consuming. In the next section, we describe the fast Fourier transform (FFT) algorithm that produces exactly the same results in $N \log_2 N$ operations. For certain applications, however, the FFT may not be the best way to proceed. For example, some applications require only a small number of frequency points, say at the M frequencies $\omega_1, \omega_2, \ldots,$ and ω_M, where $M \ll N$, rather than at the full set of N frequency values $\omega_k = 2\pi k/N$, for $0 \leq k \leq N-1$. A second example is when the M frequency points are not harmonically related, but rather an arbitrary collection of *interesting frequencies*. In this case the FFT will, in general, not evaluate the spectrum at these exact frequency values, but only at the harmonic frequencies that approximate these.

In this section, we discuss programming and implementation procedures to improve the speed of the N^2 computations that are required for the N-point DFT. These programming techniques save computation time by storing the sine and cosine coefficients in a table and accessing the discrete-time array values efficiently. The discussion below illustrates the general principle of computer processing of a *memory/speed tradeoff*, that is, a given computation can be done more quickly if more memory is allowed.

To illustrate the various points below, let us consider computing the DFT of a real-valued sequence $x(n)$, for $0 \leq n \leq N-1$. The ideas discussed below can be extended to complex-valued sequences in a straightforward way. The kth value of the DFT sequence is given by

$$X(k) = \sum_{n=0}^{N-1} x(n)\, e^{-j2\pi nk/N}$$

$$= \sum_{n=0}^{N-1} x(n) \cos(2\pi kn/N) - j \sum_{n=0}^{N-1} x(n) \sin(2\pi kn/N) \qquad \text{for } 0 \leq k \leq N-1$$

$$(4.29)$$

COMPUTATION OF THE TRIGONOMETRIC COEFFICIENT VALUES. This DFT can be programmed directly, as was done in the DFT routine in the previous section. However, note that for each value of the time sequence, the sine and cosine coefficients must be evaluated. In a computer, this evaluation is usually performed by computing a series expansion of the trigonometric identity to the accuracy of the number representation, a time-consuming operation. For example, the Taylor expansion of the cosine is equal to

$$\cos(\alpha) = 1 - \alpha^2/2! + \alpha^4/4! - \alpha^6/6! + \cdots$$

One way to reduce the computation time is to note that successive values of the complex-exponential sequence differ by a constant factor $e^{-j2\pi k/N}$, or

$$e^{-j2\pi k(n+1)/N} = e^{-j2\pi kn/N}\, e^{-j2\pi k/N} \qquad (4.30)$$

Hence, for each frequency point k, successive values of the sine and cosine coefficients can be determined by evaluating the constant factor

$$e^{-j2\pi k/N} = \cos(2\pi k/N) - j \sin(2\pi k/N) \qquad (4.31)$$

The two coefficients can then be calculated recursively as

$$\cos[2\pi k(n+1)/N] = \cos[2\pi kn/N]\cos[2\pi k/N] - \sin[2\pi kn/N]\sin[2\pi k/N]$$

$$(4.32)$$

and

$$\sin[2\pi k(n+1)/N] = \cos[2\pi kn/N]\sin[2\pi k/N] + \sin[2\pi kn/N]\cos[2\pi k/N]$$

The recursion starts with $\cos(0) = 1$ and $\sin(0) = 0$.

The undesirable feature of this procedure is that the constant factor in Eq. (4.31) is computed to the accuracy of the machine. The resulting error in the coefficients, even though small for small values of n, accumulates as n increases. This effect is most noticeable for large N-point DFT computations. A method that does not have this error, at the expense of additional memory requirements, is to generate and store the coefficient values in an array, as described next.

STORING DFT COEFFICIENTS IN A TABLE. If the DFT is to be performed repetitively, as in spectral analysis of a set of data records, the computation speed can be increased by computing the necessary coefficients once and storing these values in a table consisting of an array of cosine coefficients, CTAB, and an array of sine coefficients, STAB. These arrays are then used in the DFT computations.

The memory required for storing CTAB and STAB can be reduced by noting two points. First, because the frequencies in the DFT computations are harmonically related, or $\omega_k = 2\pi k/N$, all the needed coefficient values exist in the set of N coefficients for the fundamental frequency, or $\omega = \omega_1$. These are shown in Fig. 4.8 for $N = 32$. Let us consider a pointer variable, NPNTR, that points to the desired coefficient value in the array. To obtain the coefficients for the spectral value at ω_1, the pointer is incremented by 1 each time a new coefficient is needed. For the spectral value at ω_2, the desired coefficient set is equal to every second value in the table, starting from the first value. The pointer is then incremented by 2. The coefficients for the kth spectral value are then obtained by incrementing the pointer by k. When the end of the table is encountered, or when NPNTR $> N$, the pointer is brought back into the appropriate place in the table by subtracting N, or NPNTR = NPNTR $- N$.

As shown in Fig. 4.8, the sine and cosine coefficients take on the same values, with the elements of CTAB only shifted with respect to STAB. Hence, both sets of coefficients can be obtained from the CTAB table. But now there are two pointers, NCPNTR that points to the cosine values and NSPNTR that points to the sine values. These pointers are initialized to the values given by NCPNTR = 1 and NSPNTR = $(3N/4) + 1$, to produce the correct sequence of coefficient values. When either pointer value exceeds N, the size of the array, it is set back to the correct position by substracting N, as before.

CTAB VALUES

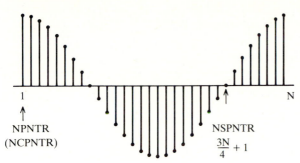

NPNTR
(NCPNTR)

NSPNTR

$\dfrac{3N}{4}+1$

STAB VALUES

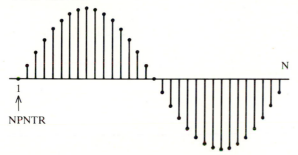

NPNTR

FIGURE 4-8
Cosine (CTAB) and sine (STAB) coefficent values for evaluating N-point DFT. Case for $N = 32$ is shown. Pointers NPNTR, NCPNTR and NSPNTR indicate the coefficient values that are needed for a particular computation step.

ARRAY INDEX COMPUTATION OVERHEAD. In computing the DFT relationship, the discrete-time sequence values are processed sequentially. In FORTRAN, the array index is stored and accessed from memory every time it is needed. This index overhead also contributes to slowing the computation. In some programming languages, like C, for example, there is an array *auto-increment* feature that eliminates the need for an array index variable. With this feature, the next value of the array can be accessed automatically. This feature can be used to increase the computation speed.

GOERTZEL ALGORITHM FOR COMPUTING THE DFT. Let us consider a second-order filter whose unit-sample response is the cosine sequence given by

$$h(n) = \cos(2\pi kn/N) \qquad \text{for } n \geq 0 \tag{4.33}$$

This filter is not stable by our definition, but we will use it for only a finite number of time points, to insure that the output does not become infinite. Such a filter can be obtained from Example 2.13 by letting $r = 1$. The filter structure is shown in Fig. 4.9(a). If the input to this filter is the real-valued sequence $x(n)$, for $0 \le n \le N-1$, the output value at time $n = N$ can be determined by the convolution equation and is given by

$$y(N) = \sum_{m=0}^{N-1} x(m)\, h(N-m)$$

$$= \sum_{m=0}^{N-1} x(m) \cos[2\pi k(N-m)/N] \tag{4.34}$$

(a)

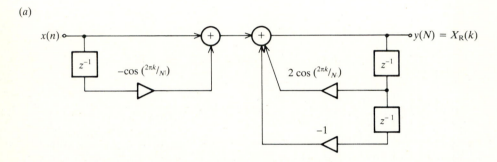

(b)

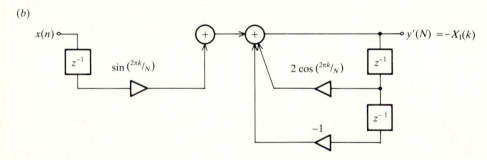

(c)

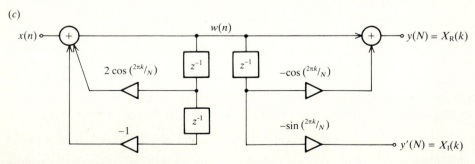

FIGURE 4-9
Filter structures for evaluating the DFT of a sequence $x(n)$, for $0 \le n \le N-1$, with the Goertzel algorithm. a) $y(N)$ equals the real part of $X(k)$. b) $y'(N)$ equals the negative imaginary part of $X(k)$. c) Real and imaginary parts computed simultaneously at $n = N$.

Since

$$\cos[2\pi k(N-m)/N] = \cos[-2\pi km/N + 2\pi k] = \cos[2\pi km/N]$$

we can write

$$y(N) = \sum_{m=0}^{N-1} x(m)\cos(2\pi km/N) \tag{4.35}$$

Comparing this equation with the definition of the DFT, we find that by putting the finite-duration sequence $x(n)$, for $0 \le n \le N-1$, through the filter, the output at time $n = N$ is equal to the real part of the DFT element $X(k)$. This result is true when the filter has a zero initial condition. In a similar manner, consider a second filter whose unit-sample response is given by

$$h'(n) = \sin(2\pi kn/N) \qquad \text{for } n \ge 0 \tag{4.36}$$

This filter is shown in Fig. 4.9(b) and discussed in Example 5.4. If $\{x(n)\}$ is applied to this filter, the output at time N is given by

$$y'(N) = \sum_{m=0}^{N-1} x(m)\sin(2\pi km/N) \tag{4.37}$$

Comparing this equation with the definition of the DFT, we find that by putting the finite-duration sequence $x(n)$, for $0 \le n \le N-1$, through the filter, the output at time $n = N$ is equal to the negative imaginary part of the DFT element, or $-X_I(k)$.

These two filters are combined in Fig. 4.9(c). By changing the recursive and nonrecursive parts and including the sequence $\{w(n)\}$ that is common to both filters, we find that the kth DFT element is equal to

$$X(k) = w(N) - \cos(2\pi k/N)\, w(N-1) - j\sin(2\pi k/N)\, w(N-1) \tag{4.38}$$

Doing a multiplication count on this filter, we find a multiplication by $2\cos(2\pi k/N)$ for each of N time points. The multiplication by -1 is not counted as a multiplication. At time $n = N$, the two additional multiplications given in Eq. (4.38) are performed, for a total of $N + 2$ for each element in the DFT sequence. This is approximately one-half of that required for the direct evaluation of the DFT for a real-valued sequence. Computing the DFT with this procedure is called the *Goertzel algorithm*. With this approach the entire set of coefficient values is not needed, since these values are computed in the filter as the unit-sample response. The same problem exists of expressing the recursive coefficient $2\cos(2\pi k/N)$ sufficiently accurately, as described above in the recursive calculation of the coefficient value set, given in (4.32).

To compute the entire DFT sequence, a separate filter is included for each spectral value. One desirable feature of this algorithm is that, if computation time is to be minimized, these filters can be implemented and operate on the input sequence in parallel. Hence, the entire DFT sequence could be computed shortly after observing the Nth sample of the input sequence. For N frequency points, however, the number of multiplications is still proportional to N^2. We next describe the fast Fourier transform, an algorithm that produces the same result with a multiplication count that is proportional to $N \log_2 N$.

4.10 INTUITIVE EXPLANATION FOR THE DECIMATION-IN-TIME FFT ALGORITHM

We describe the FFT algorithm that exploits the symmetry properties of the discrete-time complex exponential to reduce the multiplication count significantly for the computation of the DFT. For evaluating an N-point DFT, the FFT algorithm described here achieves its efficiency when N is a power of 2, or $N = 2^{\text{NLOG2}}$, for some integer NLOG2. This restriction presents no practical problem, since the length of $\{h(n)\}$ can always be increased to the next power of 2 by padding the sequence with an appropriate number of zeros.

To provide some understanding for the steps in the FFT algorithm, let us consider the input discrete-time sequence $h(n)$, for $0 \le n \le N - 1$, to be a real-valued sequence. The same basic ideas are also applicable to complex-valued sequences and to the inverse Fourier transform computation.

For real-valued discrete-time sequences, the definition of the DFT reduces to

$$H(k) = \sum_{n=0}^{N-1} h(n) \cos(2\pi nk/N) - j \sum_{n=0}^{N-1} h(n) \sin(2\pi nk/N) \qquad \text{for } 0 \le k \le N - 1$$

$$(4.39)$$

To compute this N-point DFT sequence requires $2N^2$ real multiplications. The set of values for the sine and cosine coefficients required for evaluating each $H(k)$ term are shown in Fig. 4.10 for $N = 8$. The coefficients for computing the terms k and $(N/2) + k$, for $0 \le k \le (N/2) - 1$, exhibit a curious similarity. The coefficients for even-valued n are the same, while those for odd-valued indices have the same magnitude, but are of opposite sign. To exploit this similarity, we break up the original sequence $\{h(n)\}$ into two subsequences: $\{h_E(n)\}$, containing the elements indexed by the even-values of n; and $\{h_0(n)\}$, containing the elements indexed by the odd-values. These sets of coefficients are shown in Fig. 4.11.

Two features of the even-indexed coefficient set are noteworthy. First, the set of coefficients for frequency points k and $(N/2) + k$, for $0 \le k \le (N/2) - 1$, are identical. Hence, their contribution to the terms $H(k)$ and $H(N/2 + k)$ will be the same. Only one of these terms need be computed, the other being obtained for free. The second feature is that this set of coefficient values is exactly that required for computing the $N/2$-point DFT of $\{h_E(n)\}$. To compute this $N/2$-point DFT, we need to perform $2(N/2)^2 = N^2/2$ real multiplications. A similar result, except for a sign inversion in the coefficients, occurs for the odd-indexed sequence. To compute the $N/2$-point transforms for the even and odd-indexed sequences requires N^2 multiplications, which is smaller than the original $2N^2$. Hence, splitting, or *decimating,* the sequence

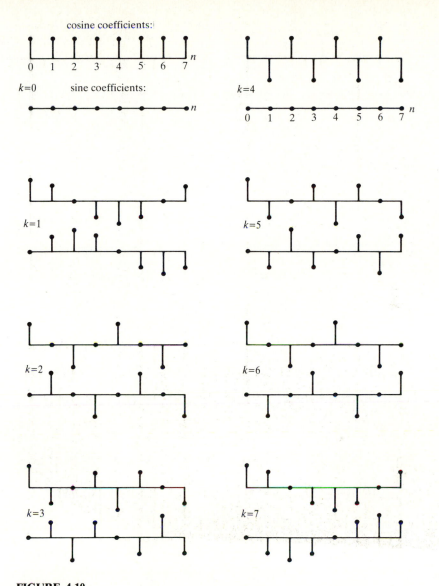

FIGURE 4-10
The set of cosine and sine coefficient values for evaluating the kth element of an 8-point DFT sequence.

into even and odd-indexed subsequences leads to a reduction in the number of computations to evaluate the DFT. We indicate below how these two $N/2$-point DFT transforms are combined to form the N-point DFT of the original sequence.

To reduce further the number of multiplications, we now consider

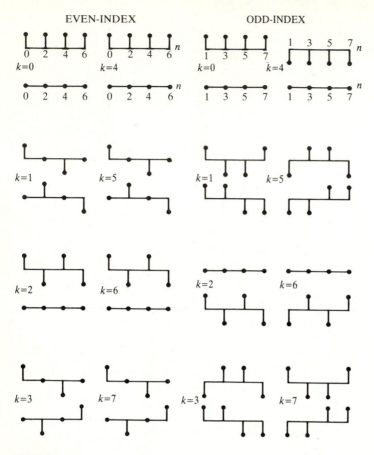

FIGURE 4-11
The set of cosine and sine coefficient values for evaluating the kth element of an 8-point DFT sequence broken up into the even-valued indexed and odd-valued indexed sub-sequences.

$\{h_E(n)\}$ to be the original sequence to be transformed into an $N/2$-point DFT. We again decimate $\{h_E(n)\}$ into its even and odd-indexed subsequences. The corresponding set of coefficients is shown in Fig. 4.12. Again, note the same two features of this set as was apparent above: the set of coefficients for frequency points k and $(N/2) + k$, for $0 \le k \le (N/4) - 1$ are identical, and these are the coefficient values necessary to compute an $N/4$-point DFT. At this stage in the decimation, four subsequences of length $N/4$ are obtained, each requiring $2(N/4)^2 = N^2/8$ multiplications to compute their $N/4$-point DFT, or a total of $N^2/2$. Hence, another reduction by 2 is obtained in the multiplication count.

The decimation operations can continue until only pairs of the original sequence occur. Each of these require a 2-point DFT computation to be

EVEN-INDEX   ODD-INDEX

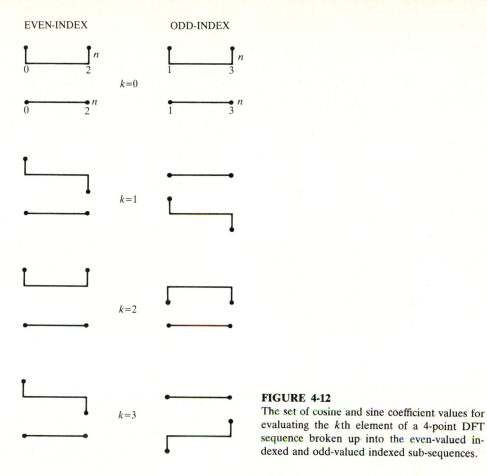

FIGURE 4-12
The set of cosine and sine coefficient values for evaluating the kth element of a 4-point DFT sequence broken up into the even-valued indexed and odd-valued indexed sub-sequences.

performed. At this level, however, all the coefficients are either 0 or ± 1, and do not require a multiplication. This can be observed from Eq. (4.39) by letting $N = 2$ and setting $k = 0$ or 1. Hence, by merely adding and subtracting elements at this lowest possible level of decimation, we can compute 2-point DFTs. Since the original sequence had $N = 2^{\text{NLOG2}}$ elements, it could be decimated into subsequences $\text{NLOG2} - 1$ times, and there are $2^{\text{NLOG2}-1}$ 2-point DFTs. This procedure is called the *decimation-in-time FFT algorithm*, since the discrete-time samples are continuously split into two groups. There is also a decimation-in-frequency FFT algorithm. It will not be covered here, but can be found in the references.

Having described the decimation operation and indicating why there is a net reduction in the number of computations, we now describe the analysis to indicate the procedure of combining the $N/2$-point DFTs to form an N-point DFT and determine the operation count.

4.11 ANALYTIC DERIVATION OF THE DECIMATION-IN-TIME FFT ALGORITHM

In this section, we provide the analysis behind the intuition presented in the previous section. Let $h(n)$, for $0 \le n \le N-1$, where $N = 2^{\text{NLOG2}}$, for some integer NLOG2, be a complex-valued sequence. We split up the DFT defining equation into two summations consisting of the even and odd-indexed sequences given by

$$H(k) = \sum_{n=0}^{N-1} h(n)\, e^{-j2\pi nk/N} \qquad \text{for } 0 \le k \le N-1$$

$$= \sum_{n \text{ even}} h(n)\, e^{-j2\pi nk/N} + \sum_{n \text{ odd}} h(n)\, e^{-j2\pi nk/N} \qquad (4.40)$$

Letting $n = 2r$ in the first sum, and $n = 2r+1$ in the second, we have

$$H(k) = \sum_{r=0}^{(N/2)-1} h(2r)\, e^{-j4\pi rk/N} + \sum_{r=0}^{(N/2)-1} h(2r+1)\, e^{-j2\pi(2r+1)k/N}$$

$$= \sum_{r=0}^{(N/2)-1} h(2r)\, e^{-j2\pi rk/(N/2)} + e^{-j2\pi k/N} \sum_{r=0}^{(N/2)-1} h(2r+1)\, e^{-j2\pi rk/(N/2)} \qquad (4.41)$$

Both sums appear in the form of an $N/2$-point DFT. Let us define $\{h_E(n)\}$ as the even-valued index sequence and $\{h_O(n)\}$ as the odd-valued and $\{H_E(k)\}$ and $\{H_O(k)\}$ as being, respectively, the DFT of $\{h_E(n)\}$ and $\{h_O(n)\}$. Since $\{H_E(k)\}$ and $\{H_O(k)\}$ are the DFTs of $N/2$-point sequences, they are periodic in k with period $N/2$. These two will be combined to form the N-point DFT $\{H(k)\}$ by

$$H(k) = H_E(k) + e^{-j2\pi k/N} H_O(k) \qquad \text{for } 0 \le k \le (N/2) - 1$$

$$= H_E(k - N/2) + e^{-j2\pi k/N} H_O(k - N/2) \qquad \text{for } N/2 \le k \le N-1 \quad (4.42)$$

Since $4N^2$ real-valued multiplications are required to compute the N-point DFT of a complex valued sequence, to calculate $\{H_E(k)\}$ or $\{H_O(k)\}$ requires $4(N/2)^2 = N^2$ real multiplications, or $2N^2$ for both. To combine $\{H_E(k)\}$ and $\{H_O(k)\}$ to form $\{H(k)\}$ requires four real multiplications per point for $0 \le k \le (N/2) - 1$, or $2N$ for the combination procedure. The points for $N/2 \le k \le N-1$ contain the same products, but with different signs. Hence, to compute an N-point DFT by decimating the sequence once requires $2(N^2 + N)$ multiplications, which is less than $4N^2$ for $N > 2$.

If decimating the sequence once reduces the multiplication count, additional decimations should reduce it even further. To show this, the even- and odd-indexed sequences can now be considered as original sequences and also decimated into their own even and odd parts. The DFT of the even-numbered elements of the original sequence can then be written as

$$H_E(k) = H_{E,E}(k) + e^{-j2\pi k/(N/2)} H_{E,O}(k) \quad \text{for } 0 \le k \le (N/4) - 1$$

$$= H_{E,E}(k - N/4) + e^{-j2\pi k/(N/2)} H_{E,O}(k - N/4) \quad \text{for } N/4 \le k \le (N/2) - 1$$

$$(4.43)$$

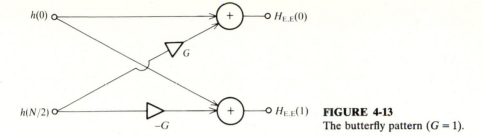

FIGURE 4-13
The butterfly pattern ($G = 1$).

where $\{H_{E,E}(k)\}$ is the DFT of the even-indexed sub-sequence of $\{h_E(n)\}$ and $\{H_{E,O}(k)\}$ is that of the odd-indexed sub-sequence of $\{h_E(n)\}$. A similar equation can be written for $\{H_O(k)\}$.

Since $N = 2^{NLOG2}$, the original sequence can be decimated into even- and odd-indexed sequences $NLOG2 - 1$ times, until $2^{NLOG2-1}$ 2-element sub-sequences are obtained. The first even-indexed two-point sequence is composed of $h(0)$ and $h(N/2)$. This 2-point sequence has the 2-point DFT given by

$$H_{E,E}(k) = h(0) + h(N/2)\, e^{-j\pi k} \qquad \text{for } k = 0, 1 \qquad (4.44)$$

Evaluating this expression, we find

$$H_{E,E}(0) = h(0) + h(N/2) \qquad \text{and} \qquad H_{E,E}(1) = h(0) - h(N/2)$$

This last operation is shown graphically in Fig. 4.13, and is called a *butterfly pattern*. The steps for combining the DFTs of the decimated subsequences are shown as a sequence of butterflies in Fig. 4.14 for $N = 8$.

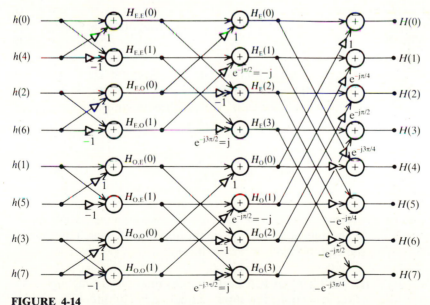

FIGURE 4-14
Decomposition of an N-point DFT into butterfly patterns. Case for $N = 8$ is shown.

TABLE 4.2
Comparison of real multiplication counts for N-point DFT ($N = 2^K$)

NLOG2	N	DFT ($4N^2$)	FFT ($4N \log_2 N$)
5	32	4 096	640
6	64	16 384	1 536
7	128	65 536	3 584
8	256	262 144	8 192
9	512	1 048 576	18 432
10	1 024	4 194 304	40 960

The computations are performed column by column, by operating on all the pairs first, and then proceeding to the right to perform the computations on the results of the first set of butterflies. Notice that as the computation along a particular vertical set of values is completed, these values are no longer needed in future computations. Hence, the memory locations that were used to store these intermediate results are free to be used for storing new values, and ultimately the values of the DFT itself. This *in-place* computation is memory-efficient and a second advantage of the FFT for computers with limited memory.

To obtain an approximate multiplication count, note that there are N complex multiplications, or $4N$ real multiplications, required after each decimation and that there are $\log_2 N - 1$ splits. Hence, $4N(\log_2 N - 1)$ real-valued multiplications are required. What is important in the multiplication count is not the exact number, but how it is related to the number of points N. The evaluation of the DFT using the straightforward evaluation requires on the order of N^2 complex multiplications, while evaluating it with the FFT algorithm requires $N \log_2 N$. These two numbers are compared in Table 4.2. Notice that reductions from 6-fold for a 32-point DFT, to over a hundred-fold for a 1024-point DFT can be achieved. Since this reduction is also directly related to computation time, the overwhelming popularity of the FFT is understandable.

4.12 SOME GENERAL OBSERVATIONS ON THE FFT

In this section, we discuss some observations on the implementation of the FFT. These include the unusual ordering of the input sequence, the memory requirements and the computation of the coefficients.

RE-ORDERING THE INPUT SEQUENCE. In the decimation-in-time algorithm described in the previous section, the input sequence is continuously divided into subsequences having odd- and even-valued indices, until subsequences

TABLE 4.3

Comparison of row location and index number of input sequence

Row number		Sequence index	
Decimal	Binary	Decimal	Decimal
0	(000)	0	(000)
1	(001)	4	(100)
2	(010)	2	(010)
3	(011)	6	(110)
4	(100)	1	(100)
5	(101)	5	(101)
6	(110)	3	(011)
7	(111)	7	(111)

of length 2 are obtained. The index values on the left side of the computation graph shown in Fig. 4.14 indicate an unusual ordering of the input sequence. This ordering has an interesting interpretation, that can be extended to arrays of any duration, when the index value is expressed as a binary number. In Table 4.3, we list the number of the row in the computation diagram and the index of the input sequence that occupies that row. Comparing the binary representation of the row and index values, we note that the index value appears as the *reversed-bit order* of the row number. This reversed-bit transformation of the index values is valid for any size N-point DFT, where N is a power of 2. This is commonly observed in decimation-in-time FFT subroutines, in which the first step involves placing the input sequence values into the appropriate row location. An algorithm to perform this bit reversal by using the decimal index values is left as a problem for the student.

MEMORY REQUIREMENTS. In the direct computation of the DFT using

$$X(k) = \sum_{n=0}^{N-1} x(n)\, e^{-j2\pi nk/N} \qquad \text{for } 0 \le k \le N-1 \qquad (4.45)$$

all the values of $x(n)$ are needed for each value of $X(k)$. If the computation is performed by sequentially computing the $X(k)$ values, as in the DFT routine described in Section 4.8, these spectral values must be stored in a temporary buffer array. After the computations, the contents of this buffer are transferred to the arrays passed in the argument of the subroutine. From the computation diagram for the FFT shown in Fig. 4.14, the data flow is consistently directed from left to right. This indicates that after the values of a particular column are used, they are no longer needed in the computations. Hence, those memory locations can be used for storing the results of future computations. If the computations are performed column by column, the results can be stored at the same memory locations, and no significant buffer memory is required. This memory-efficient procedure is called *in-place* computation and is increasingly important as the size of the DFT is increased.

COEFFICIENT VALUES. As discussed previously for the implementation of the DFT, a *memory/speed tradeoff* also exists for the FFT. As noted from the computation diagram, there are only a finite number of coefficients needed to evaluate the FFT. These coefficient values can be computed as they are needed, directly from the function or recursively, as described previously, or they can be pre-computed and stored in a look-up table. In this case the table look-up procedure is more complicated than for the DFT, since the input sequence values are in reversed-bit order. Many FFT decimation-in-time subroutines exist in standard scientific subroutine packages. As a general rule, the length of the program, in lines of code, and its memory requirements are inversely related to the execution speed. That is, the longer programs operate more quickly and the simple programs require more time. But *any* FFT subroutine is much faster than the direct method of evaluating the DFT.

4.13 OTHER FAST REALIZATIONS OF THE DFT

Although the most popular, the FFT is not the only fast realization to evaluate the DFT. Other procedures offer desirable features of better accuracy or simpler implementation. These procedures fall into the category of *number theoretic transforms*. Since they tend to be esoteric, they are not suitable for discussion in a beginning text such as this one. These procedures are discussed in the reference by Elliot and Rao.

4.14 SUMMARY

In this chapter we have defined the discrete Fourier transform (DFT) as the sequence obtained by sampling one period of the Fourier transform over $0 \le \omega < 2\pi$ at N frequency points, where N is an integer. The DFT was shown to be the Fourier series expansion of the periodic extension of a discrete-time sequence, obtained by repeating the original sequence with a period of N samples. After defining the DFT and its inverse, the IDFT, we examined some of its properties relevant to signal processing. The DFT relations could be used without modification to achieve the desired results when applied to causal, finite-duration sequences. For noncausal sequences, the inverse DFT produces the causal part, followed by the noncausal part. For infinite-duration sequences, an inevitable time-aliasing error occurs, whose effect decreases as N is increased. When attempting to perform the convolution operation in the frequency domain with the DFT, the equivalent time-domain operation is the circular convolution. The desired linear convolution result can be obtained by computing the DFT at a sufficiently large number of frequency points. The fast Fourier transform (FFT) algorithm was presented as an efficient computational procedure for evaluating the DFT. The FFT algorithm was described in both an intuitive and analytical form using the decimation-in-time approach. Some general remarks were made regarding the implementation of the FFT algorithm.

FURTHER READING

Bergland, G. D.: "A guided tour of the fast Fourier transform," *IEEE Spectrum*, **6**: 41–52 (1969)

Blahut, R. E.: *Fast Algorithms for Digital Signal Processing*, Addison-Wesley, Reading, MA, 1984.

Bracewell, R. N.: *The Fourier Transform and Its Applications*, McGraw-Hill, New York, 1978.

Burrus, C. S., and T. W. Parks: *DFT/FFT and Convolution Algorithms*, Wiley-Interscience, New York, 1985. [This book provides a description of the various algorithms to implement a discrete Fourier transform. Program listings are given in FORTRAN and assembly language for the Texas Instruments TMS32010 signal processing chip.]

Childers, D. G., (ed.): *Modern Spectral Analysis*, IEEE Press, New York, 1978. [This is a book of reprints which presents a historical perspective of the evolution of spectral analysis.]

Cooley, J. W., and J. W. Tukey.: "An algorithm for the machine computation of complex Fourier series," *Math. Comp.*, **19**: 297–301 (1965).

Elliott, D. F., and K. R. Rao.: *Fast Transformations: Algorithms, Analysis and Applications*, Academic Press, New York, 1982.

Oppenheim, Alan V., and Ronald W. Schafer: *Digital Signal Processing*, Prentice-Hall, Englewood Cliffs, NJ, 1975.

Peled, A., and B. Liu: *Digital Sgnal processing: Theory, Design and Implementation*. Wiley, New York, 1976.

Rabiner, Lawrence R., and Bernard Gold: *Theory and Application of Digital Signal Processing*, Prentice-Hall, Englewood Cliffs, NJ, 1975.

Schwartz, Mischa and Leonard Shaw: *Signal Processing: Discrete Spectral Analysis, Detection and Estimation*, McGraw-Hill, New York, 1975.

PROBLEMS

4.1. Evaluate and compare the 8-point DFT for the following sequences.

(a) $x(n) = \begin{cases} 1 & \text{for } -3 \le n \le 3 \\ 0 & \text{otherwise} \end{cases}$

(b) $x(n) = \begin{cases} 1 & \text{for } 0 \le n \le 6 \\ 0 & \text{otherwise} \end{cases}$

4.2. Verify the properties of the DFT relationship given in Table 4.1.

4.3. When discussing the Fourier transform, it was shown that an even discrete-time sequence, or one for which $h(n) = h(-n)$, has a Fourier transform that is real-valued. Since the DFT sequence as produced by the computer algorithm can have only positive index values, indicate the symmetry property of a discrete-time sequence whose DFT sequence is real.

4.4. Let us consider the finite-duration sequence $\{x(n)\}$. It is common to process sequences that are defined over index values that do not correspond to the range $0 \le n \le N - 1$ required for the DFT computations. In these cases, the $\{h(n)\}$ must be periodically extended to form $\{\bar{h}(n)\}$, and the sequence $\bar{h}(n)$, for $0 \le n \le N - 1$ is processed to produce the desired results. Determine and sketch the discrete-time sequence to be processed to find the 8-point DFT, when the following sequences are given.

(a) $x(n) = \begin{cases} 1 & \text{for } n = -1 \\ 0 & \text{otherwise} \end{cases}$

(b) $x(n) = \begin{cases} 3 - |n| & \text{for } -3 \le n \le 3 \\ 0 & \text{otherwise} \end{cases}$

(c) $x(n) = \begin{cases} n & \text{for } -3 \le n \le 3 \\ 0 & \text{otherwise} \end{cases}$

4.5. Consider the N-point sequence $\{x(n)\}$ and its N-point DFT $\{X(k)\}$ where N is an even number.

(a) Show that if $\{x(n)\}$ is a symmetric sequence, then $X(N/2) = 0$.

(b) Show that if $\{x(n)\}$ is an antisymmetric sequence, then $X(0) = 0$.

4.6. Compute the N-point DFT of each of the following sequences. Let $N = 8$ and $M = 5$.

(a) $x(n) = \begin{cases} 1 & \text{for } 0 \le n \le N - 1 \\ 0 & \text{otherwise} \end{cases}$

(b) $x(n) = \begin{cases} 1 & \text{for } 0 \le n \le M - 1, \\ 0 & \text{otherwise} \end{cases}$

(c) $x(n) = \begin{cases} 1 & \text{for } n = 0 \\ -1 & \text{for } n = N - 1 \end{cases}$

(d) $x(n) = \begin{cases} 1 & \text{for } n = 0 \\ -1 & \text{for } n = M, \end{cases}$

4.7. The DFT sequence $\{H(k)\}$ represents samples of the Fourier transform $H(e^{j\omega})$ at equally-spaced frequency intervals, or

$$H(k) = H(e^{j\omega})|_{\omega = 2\pi k/N} \qquad \text{for } 0 \le k \le N - 1$$

To obtain a denser sampling, the finite-duration time sequence $h(n)$, containing M samples, where $M < N$, can be increased to any finite duration N by padding with $N - M$ zeros. The DFT relations can then be applied to obtain N samples for any desired N. For causal sequences, these zeros are appended to the end of the given sequence. For noncausal sequences, a more careful padding is necessary, as illustrated in this problem. For the sequences given below, sketch the 8-point sequence that is used for computing the 8-point DFT.

(a) $h(n) = \begin{cases} n & \text{for } 0 \le n \le 3 \\ 0 & \text{otherwise} \end{cases}$

(b) $h(n) = \begin{cases} |n| & \text{for } -1 \le n \le 1 \\ 0 & \text{otherwise} \end{cases}$

4.8. Perform the linear convolution of the two sequences below in the time domain. Compare with the result obtained by using the DFT relations.

$$h(n) = \begin{cases} n & \text{for } 0 \le n \le 3 \\ 0 & \text{otherwise} \end{cases}$$

and

$$x(n) = \begin{cases} |n| & \text{for } -1 \le n \le 1 \\ 0 & \text{otherwise} \end{cases}$$

4.9. Show the symmetry properties of the DFT sequence by computing the 8-point DFT of the real-valued sequence given by

$$x(n) = \begin{cases} 2 - |n| & \text{for } -1 \le n \le 1 \\ 0 & \text{otherwise} \end{cases}$$

4.10. **Computing the DFT of two real-valued sequences with one execution of the FFT algorithm**. The FFT algorithm can operate on a complex-valued sequence. When transforming a real-valued sequence, the imaginary part is set to zero. We can transform two real sequences simultaneously by letting the real part of the complex-valued input be set equal to one real sequence, and the imaginary part equal to the second real sequence. The complex-valued output of the algorithm must then be manipulated to achieve the desired result, consisting of the two DFT sequences of the two real-valued input sequences.

Let $x_1(n)$, for $0 \le n \le N-1$ be a real-valued sequence having the DFT sequence whose real and imaginary parts are given by $R_1(k)$ and $I_1(k)$, for $0 \le k \le N-1$, or

$$R_1(k) = \sum_{n=0}^{N-1} x_1(n) \cos(2\pi kn/N) \qquad \text{for } 0 \le k \le N-1$$

and

$$I_1(k) = - \sum_{n=0}^{N-1} x_1(n) \sin(2\pi kn/N) \qquad \text{for } 0 \le k \le N-1$$

Let $x_2(n)$, for $0 \le n \le N-1$ be a real-valued sequence having the DFT sequence given by $R_2(k)$ and $I_2(k)$, for $0 \le k \le N-1$.

Since the discrete Fourier transform is a linear transformation, it obeys the superposition principle. Define the discrete-time sequence $\{y(n)\}$, is

$$y(n) = x_1(n) + jx_2(n) \qquad \text{for } 0 \le n \le N-1$$

(a) Find the real and imaginary parts of the DFT of $\{y(n)\}$, denoted by $R_y(k)$ and $I_y(k)$ in terms of $R_1(k)$, $R_2(k)$, $I_1(k)$ and $I_2(k)$.

(b) From the above relationships, find the inverse relationship to obtain $R_1(k)$, $R_2(k)$, $I_1(k)$ and $I_2(k)$, to produce the desired DFT sequences.

4.11. Determine the reversed-bit order of the input sequence to compute a 16-point DFT with the FFT algorithm.

4.12. Develop the reversed-bit reordering algorithm based on the decimal number of the sequence index. Demonstrate your algorithm for 8-point DFT and 16-point DFT calculations.

COMPUTER PROJECTS

4.1. **Program to evaluate the DFT**
Object. Write a subroutine that evaluates the discrete Fourier transform and compare the results with the Fast Fourier transform (FFT) algorithm given in Appendix A. It is not necessary to be familiar with the FFT algorithm to perform this project. We consider the FFT to be just like any other packaged program and simply want to verify that it works correctly.

Write the subroutine

    DFT(NLOG2, HR, HI, NTYPE)

as described in the text. The call to the FFT has the identical arguments as above, given by

    FFT(NLOG2, HR, HI, NTYPE).

For the problems below, let NLOG2 = 7 (NDFT = 128).

Analytic results. Determine and sketch the magnitude and phase of the Fourier transform for the sequences below and compare with the computer results.

Computer results

(1) Test the DFT subroutine by computing the DFT of the sequences below and plotting the entire (NDFT) magnitude array and the phase array for the following sequences. Use the PLOT subroutine.

Compare these plots with those obtained with the subroutines PLOTMR for the **m**agnitude **r**esponse and PLOTPR for the **p**hase **r**esponse given in Appendix A. These plotting routines show only (NDFT/2) + 1 points, or 65 points when NDFT = 128.

(a) *Causal sequence.* Let $h(n) = 1$, for $0 \le n \le 3$, and $=0$, otherwise.

(b) *Noncausal sequence.* Let $h(-1) = 0.5$, $h(0) = 1$ and $h(1) = 0.5$.

(2) Compute the DFT sequences of the two discrete-time sequences given above with the FFT algorithm. Compare the PLOTMR and PLOTPR plots with those produced above.

(3) Consider the sequence

$$h(n) = 1 \qquad \text{for } 0 \le n \le \text{NDFT} - 1.$$

Use the DFT subroutine with NDFT = 64,128,256 and 512 to determine the time required for each array size. Do *not* plot or print your results. Use the TIME function on the computer. Your dimension statements should allow for 512-element arrays.

Construct a table of the programming times and compare with those required by the FFT subroutine to compute the same transforms.

Remark. The results produced by DFT and FFT should be identical, but the speed of the FFT is much greater. Hence, we will use only the FFT subroutine for the remainder of the projects in the course.

4.2. DFT of finite-duration sinusoidal sequences

Object. Investigate the effect on the DFT when varying the period of a discrete-time sinusoidal sequence and the length of the sequence.

Computer results. For the problems below, let NLOG2 = 7 (NDFT = 128) and plot the first 65 points of the magnitude sequence corresponding to $0 \le \omega \le \pi$ with PLOTMR. Generate the discrete-time sequence

$$x(n) = \begin{cases} \cos(\omega_0 n) & \text{for } 0 \le n \le K - 1 \\ 0 & \text{for } K \le n \le \text{NDFT} - 1, \end{cases}$$

for the following conditions:

(a) ω_0 on a sampling point of the DFT: let $\omega_0 = \pi/4$, and $K = 64$, 32 and 16.

(b) ω_0 not on a sampling point of DFT: let $\omega_0 = 15.5\pi/64$ and $K = 64$, 32 and 16.

4.3. Performing convolutions with the DFT

Object. Perform a linear convolution with the DFT and the observe the effects of a circular convolution.

We want to compute the linear convolution given by

$$\{y(n)\} = \{h(n)\} * \{x(n)\}$$

where

$$x(n) = \begin{cases} 1 & \text{for } 0 \le n \le 5 \\ 0 & \text{otherwise} \end{cases}$$

and

$$h(n) = \begin{cases} 1 & \text{for } 0 \le n \le 5 \\ 0 & \text{otherwise} \end{cases}$$

Analytic results. Determine $\{y(n)\}$ in closed analytic form and sketch.

Computer results
(1) **Performing the *linear convolution* with the DFT.** List the steps in performing the linear convolution by using the smallest possible number of points NDFT in the DFT, that is still a power of 2. Plot the following by using PLOT:
(a) the original time sequence $\{x(n)\}$ (plot NDFT points),
(b) the magnitude spectrum of (a),
(c) the magnitude spectrum of the product of the two DFT sequences,
(d) the inverse DFT sequence and compare with $\{y(n)\}$.
(2) **Performing the *circular convolution* with the DFT.** Demonstrate the time-aliasing phenomenon produced by choosing NDFT too small. Produce the plots listed in part (1). Explain the results on the plots.

4.4. Interpolating the DFT by padding the time-domain sequence with zeros

Object. Compute the DFT at a denser set of points in the frequency domain by padding the time-domain sequence with zeros.

Computer results Let

$$x(n) = r^n \cos(\omega_0 n) \qquad \text{for } 0 \le n \le 63$$

with $r = 0.9$ and $\omega_0 = 9\pi/64$. Compute the 64-point DFT and plot the 33 points of the magnitude spectrum corresponding to $0 \le \omega \le \pi$. Use the PLOT subroutine.

Generate the 128-point sequence $\{y(n)\}$ by padding $\{x(n)\}$ with zeros:

$$y(n) = \begin{cases} x(n) & \text{for } 0 \le n \le 63 \\ 0 & \text{for } 64 \le n \le 127 \end{cases}$$

Compute the 128-point DFT and plot the 65 points of the magnitude spectrum corresponding to $0 \le \omega \le \pi$. Compare the two plots.

4.5. DFT of the product of two sequences

Object. Verify that the product of two time sequences results in the convolution of their Fourier transforms.

Consider the following sequence: $y(n) = w(n) x(n)$, for all n, where

$$x(n) = \begin{cases} \cos(\pi n/8) & \text{for } 0 \le n \le 127 \\ 0 & \text{otherwise} \end{cases}$$

and

$$w(n) = \begin{cases} 1 & \text{for } 0 \le n \le 15 \text{ and } 113 \le n \le 127 \\ 0 & \text{otherwise} \end{cases}$$

Analytic results. Determine and sketch the discrete Fourier transforms $\{Y(k)\}$, $\{X(k)\}$ and $\{W(k)\}$, and compare with the computer results. (Hint: $\{X(k)\}$ has only two nonzero points: use the finite geometric sum formula for $\{W(k)\}$, and a convolution with an isolated point produces only a shift and a scale change.)

Computer results. Compute and plot the magnitudes of the 128-point DFTs for the three sequences, using PLOTMR.

4.6. Interpolating time-domain sequences by padding the DFT with zeros
Object. Illustrate how a time-domain sequence can be interpolated by modifying the frequency-domain function.

Computer results. Let $x(n) = \cos(\pi n/4)$, for $0 \le n \le 31$.
(1) Plot this sequence. All the plots in this project should be performed with PLOT.
(2) Compute the 32-point DFT of $\{x(n)\}$ and plot the real and imaginary sequences $X_R(k)$ and $X_I(k)$, for $0 \le k \le 31$.
(3) Pad these sequences with 32 zeros to produce $Y_R(k)$ and $Y_I(k)$. To retain the symmetry properties of the real and imaginary sequences, the zero-padding is performed as follows:

$$Y_R(k) = \begin{cases} X_R(k) & \text{for } 0 \le k \le 15 \\ 0 & \text{for } 16 \le k \le 47 \\ X_R(k-32) & \text{for } 48 \le k \le 63 \end{cases}$$

and

$$Y_I(k) = \begin{cases} X_I(k) & \text{for } 0 \le k \le 15 \\ 0 & \text{for } 16 \le k \le 47 \\ X_I(k-32) & \text{for } 48 \le k \le 63 \end{cases}$$

Plot $\{Y_R(k)\}$ and $\{Y_I(k)\}$.
(4) Compute the 64-point IDFT of $\{Y(k)\}$. Plot the real and imaginary components and compare with $\{x(n)\}$.
(5) Determine the equivalent time-domain interpolation function by letting $\{x(n)\} = \{d(n)\}$ and repeat the steps above. Describe the characteristics of the corresponding interpolation filter.

4.7. Errors when employing the DFT for infinite-duration sequences
Object. Observe the errors that are produced when applying the DFT to infinite-duration sequences.
Let $h(n) = a^n$ for $n \ge 0$ with $a = 0.95$. Then

$$H(e^{j\omega}) = 1/(1 - a\,e^{-j\omega})$$

Computer results. Generate the 16-point DFT sequence $\{H(k)\}$ by evaluating $H(e^{j\omega})$ at the appropriate frequency points. Compute the 16-point IDFT of $\{H(k)\}$ and compare with $h(n)$. Repeat for $N = 32$. Compare these results with those of Example 4.3.

CHAPTER
5

THE
z-TRANSFORM

5.1 INTRODUCTION

The z-transform is the primary mathematical tool for the analysis and synthesis of digital filters. Procedures for determining the z-transform of a discrete-time sequence are described. The *system function* of a digital filter is defined as the z-transform of its unit-sample response. The complex z *plane* is employed to display the configuration of the poles and zeros of the system function. The circle centered at the origin of the z plane and having radius equal to unity is called the *unit circle*. The *region of convergence* (ROC) in the z plane determines the values of z for which the z-transform is defined. The form of the ROC is also related to the properties of the discrete-time sequence.

The location of the system function singularities in the z plane is exploited in several ways. The stability of the discrete-time system can be determined by inspection, since stable causal systems have all their poles inside the unit circle. Stable noncausal systems are allowed to have poles at infinity. The Fourier transform is shown to be the evaluation of the z-transform along the unit circle in the z plane. The discrete Fourier transform can be determined by evaluating the z-transform at a set of equally-spaced points on the unit circle. Finally, the digital filter coefficients can be determined from the locations of the poles and zeros. The chapter concludes by considering three methods for performing the inverse operation of determining the discrete-time sequence from its z-transform and the region of convergence.

5.2 THE DEFINITION OF THE z-TRANSFORM

The z-transform of a sequence $\{h(n)\}$, denoted by $H(z)$, is defined as

$$H(z) = \sum_{n=-\infty}^{\infty} h(n)z^{-n} \tag{5.1}$$

where z is a complex-valued variable. The z-transform of a sequence is said to exist for all values of z for which the sum is finite. When the sequence is the unit-sample response of a digital filter, as implied by the sequence $\{h(n)\}$ in Eq. (5.1), then its z-transform is called the *system function*. The z-transform of a signal sequence $\{x(n)\}$, denoted by $X(z)$, is similarly given by

$$X(z) = \sum_{n=-\infty}^{\infty} x(n)z^{-n} \tag{5.2}$$

Like the Fourier transform, the z-transform will be shown to convert a convolution in the time domain into a multiplication in the z domain. Its primary use, however, is for the analysis and synthesis of digital filters in the z domain. We illustrate the computation of the z-transform in the following examples.

Example 5.1.   The z-transform of the 3-sample averager. Let

$$h(n) = \begin{cases} 1/3 & \text{for } -1 \leq n \leq 1 \\ 0 & \text{otherwise} \end{cases}$$

Applying the definition of the z-transform, we have

$$H(z) = \frac{z}{3} + \frac{1}{3} + \frac{z^{-1}}{3}.$$

Example 5.2.   The z-transform of the first-order recursive filter. Let

$$h(n) = \begin{cases} a^n & \text{for } n \geq 0 \\ 0 & \text{otherwise} \end{cases}$$

The z-transform is given by

$$H(z) = \sum_{n=0}^{\infty} a^n z^{-n} = \sum_{n=0}^{\infty} (az^{-1})^n$$

Applying the infinite geometric sum formula, we have

$$H(z) = \frac{1}{1 - az^{-1}} \qquad \text{for } |az^{-1}| < 1$$

It is common to encounter sequences whose elements are expressed in terms of $\sin(\omega n)$ and $\cos(\omega n)$. To apply the geometric sum formulae, the sequences should be transformed by applying the Euler identities, as illustrated in the following examples.

Example 5.3.   The z-transform of a second-order recursive filter. Let

$$h(n) = \begin{cases} r^n \cos(\omega_0 n), & \text{for } n \geq 0 \\ 0 & \text{otherwise} \end{cases}$$

The z-transform is given by

$$H(z) = \sum_{n=0}^{\infty} r^n \cos(\omega_0 n) z^{-n}$$

Applying the Euler identity and the infinite geometric sum formula, we get

$$H(z) = \sum_{n=0}^{\infty} r^n \frac{(e^{j\omega_0 n} + e^{-j\omega_0 n})}{2} z^{-n}$$

$$= \frac{1}{2} \left[\sum_{n=0}^{\infty} (r\, e^{j\omega_0} z^{-1})^n + \sum_{n=0}^{\infty} (r\, e^{-j\omega_0} z^{-1})^n \right]$$

$$= \frac{1}{2} \left[\frac{1}{1 - r\, e^{j\omega_0} z^{-1}} + \frac{1}{1 - r\, e^{-j\omega_0} z^{-1}} \right] \qquad \text{for } |rz^{-1}| < 1$$

$$= \frac{1}{2} \left[\frac{2 - r(e^{j\omega_0} + e^{-j\omega_0}) z^{-1}}{(1 - r\, e^{j\omega_0} z^{-1})(1 - r\, e^{-j\omega_0} z^{-1})} \right]$$

$$= \frac{1 - r \cos(\omega_0) z^{-1}}{1 - 2r \cos(\omega_0) z^{-1} + r^2 z^{-2}}$$

Example 5.4.   The z-transform of another second-order system. Let

$$h(n) = \begin{cases} r^n \sin(\omega_0 n) & \text{for } n \geq 0 \\ 0 & \text{otherwise} \end{cases}$$

The z-transform is given by

$$H(z) = \sum_{n=0}^{\infty} r^n \sin(\omega_0 n) z^{-n}$$

Applying the Euler identity and breaking up the sum, we have

$$H(z) = \frac{1}{2j} \left[\sum_{n=0}^{\infty} (r\, e^{j\omega_0} z^{-1})^n - \sum_{n=0}^{\infty} (r\, e^{-j\omega_0} z^{-1})^n \right]$$

$$= \frac{1}{2j} \left[\frac{1}{1 - r\, e^{j\omega_0} z^{-1}} - \frac{1}{1 - r\, e^{-j\omega_0} z^{-1}} \right]$$

$$= \frac{r \sin(\omega_0) z^{-1}}{1 - 2r \cos(\omega_0) z^{-1} + r^2 z^{-2}}$$

5.3 PROPERTIES OF THE z-TRANSFORM

In this section, we describe the properties of the z-transform that are relevant to describing discrete-time signals and systems. These are listed in Table 5.1 and discussed below.

LINEARITY OF THE z-TRANSFORM. To demonstrate that the z-transform is a linear transformation, we apply the superposition principle to the definition of the z-transform. Consider the sequence $\{x(n)\} = \{ah_1(n) + bh_2(n)\}$, where a and b are constants. The z-transform of $\{x(n)\}$ is equal to

$$
\begin{aligned}
X(z) &= \sum_{n=-\infty}^{\infty} x(n)z^{-n} \\
&= \sum_{n=-\infty}^{\infty} [ah_1(n) + bh_2(n)]z^{-n} \\
&= a \sum_{n=-\infty}^{\infty} h_1(n)z^{-n} + b \sum_{n=-\infty}^{\infty} h_2(n)z^{-n} \\
&= aH_1(z) + bH_2(z)
\end{aligned}
\tag{5.3}
$$

The linearity of the z-transform allows us to implement the digital filter in the z domain using the same four step procedure as that presented in the time domain in Chapter 2.

THE z-TRANSFORM OF A DELAYED SEQUENCE. Since one of the main operations of a digital filter involves delaying the elements in a sequence, let us

TABLE 5.1
Properties of the z-transform

---

1. Definition:

$$H(z) = \sum_{n=-\infty}^{\infty} h(n)z^{-n}$$

2. Region of convergence (ROC): values of z for which

$$\sum_{n=-\infty}^{\infty} |h(n)z^{-n}| < \infty$$

3. Linearity:

$$\text{if} \quad \{x(n)\} = \{ah_1(n)\} + \{bh_2(n)\} \quad \text{then} \quad X(z) = aH_1(z) + bH_2(z)$$

4. z-Transform of delayed sequence:

$$\text{if} \quad \{y(n)\} = \{x(n - n_0)\} \quad \text{then} \quad Y(z) = z^{-n_0}X(z)$$

5. z-Transform of the convolution of two sequences:

$$\text{if} \quad \{y(n)\} = \{h(n)\} * \{x(n)\} \quad \text{then} \quad Y(z) = H(z)X(z)$$

---

consider the z-transform of a delayed sequence. Let $\{h_k(n)\} = \{h(n - k)\}$ represent a sequence that is delayed by k samples and its z-transform by $H_k(z)$. From the definition of the z-transform, we have

$$H_k(z) = \sum_{n=-\infty}^{\infty} h_k(n)z^{-n} = \sum_{n=-\infty}^{\infty} h(n-k)z^{-n} \qquad (5.4)$$

Changing the index to $m = n - k$, we have

$$H_k(z) = \sum_{m=-\infty}^{\infty} h(m)z^{-(m+k)} = z^{-k} \sum_{m=-\infty}^{\infty} h(m)z^{-m} = z^{-k}H(z) \qquad (5.5)$$

In words, delaying the sequence by k samples changes the z-transform of the sequence only by an additional multiplication by z^{-k}. It is for this reason that the unit-delay component in the digital filter is denoted by a box containing a z^{-1}.

THE z-TRANSFORM OF THE CONVOLUTION OF TWO SEQUENCES. As introduced in Chapter 2, the convolution operation that relates the input and output values of a linear time-invariant system is given by

$$y(n) = \sum_{k=-\infty}^{\infty} h(k) x(n - k) \qquad (5.6)$$

Applying the definition of the z-transform to $\{y(n)\}$, we have

$$Y(z) = \sum_{n=-\infty}^{\infty} y(n)z^{-n}$$

$$= \sum_{n=-\infty}^{\infty} \left[\sum_{k=-\infty}^{\infty} h(k) x(n - k) \right] z^{-n} \qquad (5.7)$$

From this last equation, we want to determine $Y(z)$ in terms of $H(z)$ and $X(z)$. The required infinite sums are present, but we need a z^{-k} for the $h(k)$ terms and an additional z^k for the $x(n - k)$ terms. Multiplying Eq. (5.7) by $z^k z^{-k}$ and changing the order of the summations, we get

$$Y(z) = \sum_{k=-\infty}^{\infty} \left[h(k)z^{-k} \sum_{n=-\infty}^{\infty} x(n - k)z^{-(n-k)} \right] \qquad (5.8)$$

In the sum over n, the variable k represents a delay of k samples in $\{x(n)\}$. Also, the same delay occurs in the exponent of z. Since the sum is taken over the entire range of $\{x(n)\}$, this delay of k samples is inconsequential in evaluating the sum. By substituting $m = n - k$, this sum is found to be equal to $X(z)$. Since $X(z)$ is not a function of k, it can be factored out of the sum over k, leaving the definition of $H(z)$. Hence, we have the desired result:

$$Y(z) = H(z) X(z) \qquad (5.9)$$

5.4 THE SYSTEM FUNCTION OF A DIGITAL FILTER

The previous result provides a simple description of the behavior of a linear system in the z domain. The system acts through the z-transform of its unit-sample response, or its *system function* $H(z)$. The z-transform of the output $Y(z)$ is commonly called the *response transform*, while that of the input $X(z)$ is the *excitation transform*. In words, the response transform is equal to the product of the excitation transform and the system function. We illustrate the application of this equation in the following example.

> **Example 5.5.** **An application of the system function.** Let us determine the filter that converts the input sequence
>
> $$x(n) = \begin{cases} r^n \cos(\omega_0 n) & \text{for } n \geq 0 \\ 0 & \text{otherwise} \end{cases}$$
>
> into
>
> $$y(n) = \begin{cases} r^n \sin(\omega_0 n) & \text{for } n \geq 0 \\ 0 & \text{otherwise} \end{cases}$$
>
> From Example 5.3, we have
>
> $$X(z) = \frac{1 - r\cos(\omega_0)z^{-1}}{1 - 2r\cos(\omega_0)z^{-1} + r^2 z^{-2}}$$
>
> From Example 5.4, we have
>
> $$Y(z) = \frac{r\sin(\omega_0)z^{-1}}{1 - 2r\cos(\omega_0)z^{-1} + r^2 z^{-2}}$$
>
> The required system function is then given by
>
> $$H(z) = Y(z)/X(z)$$
>
> $$= \frac{r\sin(\omega_0)z^{-1}}{1 - r\cos(\omega_0)z^{-1}}$$
>
> It is left as an exercise for the student to determine the unit-sample response of this filter.

The system function was defined above as the z-transform of the unit-sample response. We can also determine the system function in terms of the coefficients in the difference equation. In Chapter 2, the finite-order difference equation describing a linear time-invariant system was given by

$$y(n) = \sum_{k=1}^{M} a_k y(n-k) + \sum_{k=-N_F}^{N_P} b_k x(n-k) \qquad (5.10)$$

Taking the z-transform of the output sequence, we have

$$Y(z) = \sum_{n=-\infty}^{\infty} y(n)z^{-n}$$

$$= \sum_{n=-\infty}^{\infty} \left[\sum_{k=1}^{M} a_k y(n-k) + \sum_{k=-N_F}^{N_P} b_k x(n-k) \right] z^{-n}$$

Changing the order of summation, we get

$$Y(z) = \sum_{k=1}^{M} a_k \left[\sum_{n=-\infty}^{\infty} y(n-k)z^{-n} \right] + \sum_{k=-N_F}^{N_P} b_k \left[\sum_{n=-\infty}^{\infty} x(n-k)z^{-n} \right] \quad (5.12)$$

Applying the relation for the z-transform of a delayed sequence, we have

$$Y(z) = \sum_{k=1}^{M} a_k z^{-k} Y(z) + \sum_{k=-N_F}^{N_P} b_k z^{-k} X(z) \quad (5.13)$$

Since $Y(z)$ and $X(z)$ are not functions of k, they can be factored out to give

$$Y(z) \left[1 - \sum_{k=1}^{M} a_k z^{-k} \right] = X(z) \sum_{k=-N_F}^{N_P} b_k z^{-k} \quad (5.14)$$

Applying the definition for the system function $H(z)$, we have our desired result, or

$$H(z) = \frac{Y(z)}{X(z)} = \sum_{k=-N_F}^{N_P} b_k z^{-k} \Bigg/ 1 - \sum_{k=1}^{M} a_k z^{-k} \quad (5.15)$$

The numerator allows both positive and negative exponent values of z, corresponding to future and past values of the input. The denominator, however, does not contain positive exponent values, indicating that only current and past values of the output are accessible. To implement a digital filter from the system function, it must be expressed in an analytic form, such that the denominator contains terms in z with exponent values that are either zero or negative.

Example 5.6. System function of the 3-sample averager. For the 3-sample averager, the difference equation is

$$y(n) = \tfrac{1}{3}[x(n+1) + x(n) + x(n-1)]$$

Taking the z-transform, we have

$$Y(z) = \tfrac{1}{3}[zX(z) + X(z) + z^{-1}X(z)]$$

The system function is then

$$H(z) = Y(z)/X(z) = z/3 + \tfrac{1}{3} + z^{-1}/3$$

Example 5.7. System function of the first-order recursive filter. For the first-order recursive filter, the difference equation is

$$y(n) = ay(n-1) + x(n)$$

Taking the z-transform, we have

$$Y(z) = az^{-1} Y(z) + X(z) \quad \text{or} \quad (1 - az^{-1}) Y(z) = X(z)$$

The system function is then

$$H(z) = \frac{1}{1 - az^{-1}}$$

Example 5.8 System function of a second-order recursive filter. Consider the second-order recursive filter whose difference equation is given by

$$y(n) = a_1 y(n-1) + a_2 y(n-2) + b_0 x(n) + b_1 x(n-1)$$

Taking the z-transform, we have

$$Y(z) = a_1 z^{-1} Y(z) + a_2 z^{-2} Y(z) + b_0 X(z) + b_1 z^{-1} X(z)$$
$$(1 - a_1 z^{-1} - a_2 z^{-2}) Y(z) = (b_0 + b_1 z^{-1}) X(z)$$

The system function is then

$$H(z) = \frac{b_0 + b_1 z^{-1}}{1 - a_1 z^{-1} - a_2 z^{-2}}$$

5.5 COMBINING FILTER SECTIONS TO FORM MORE COMPLEX FILTERS

It is not uncommon for a digital filter to be implemented by combining simpler component digital filters to form a larger composite filter. There are two common ways of combining additional filters, either in *cascade* (also called *series*) or in *parallel*. In this section, we will consider the system function of the composite filter in terms of the system functions of its components.

CASCADE CONNECTION. Let us first consider the series combination of two filters shown in Fig. 5.1(*a*) Writing the system function explicitly for each

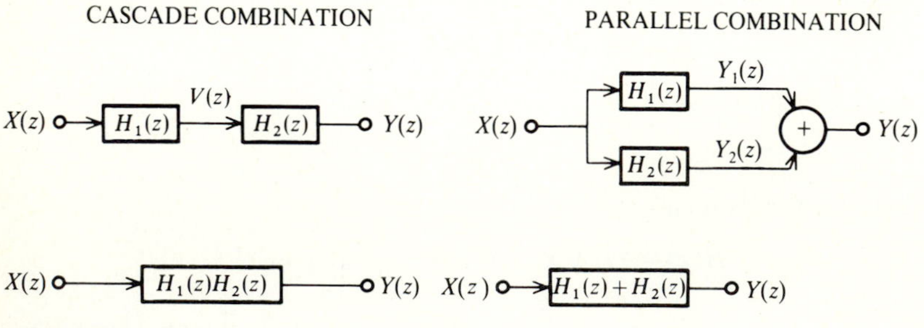

FIGURE 5-1
Cascade and parallel combinations of linear systems.

component filter, we have

$$V(z) = H_1(z) X(z) \qquad Y(z) = H_2(z) V(z) \tag{5.16}$$

Combining these two equations, the output $Y(z)$ can be expressed in the terms of $X(z)$ as

$$Y(z) = H_2(z) H_1(z) X(z) \tag{5.17}$$

Since the multiplication operation is commutative, we can also write

$$Y(z) = H_1(z) H_2(z) X(z) \tag{5.18}$$

This result can be generalized to the cascade connection of any finite number N of component filters. If $H_n(z)$ is the system function of the nth cascade section, then the response transform is given by

$$
\begin{aligned}
Y(z) &= H_1(z) H_2(z) \cdots H_N(z) X(z) \\
&= \prod_{n=1}^{N} H_n(z) X(z)
\end{aligned}
\tag{5.19}
$$

Thus the system function of the total filter composed of cascaded components, denoted by $H_C(z)$, is simply the *product* of the system functions of each of the individual series components, or

$$H_C(z) = \prod_{n=1}^{N} H_n(z) \tag{5.20}$$

Example 5.9. Cascade combination of digital filters. Let us cascade two 3-sample averagers to obtain an enhanced lowpass filtering effect. The total system function is then given by

$$H(z) = (z/3 + \tfrac{1}{3} + z^{-1}/3)^2 = (z^2 + 2z + 3 + 2z^{-1} + z^{-2})/9$$

Notice that in combining the two second-order filters in series, a fourth-order filter is obtained. In general, the number of delays required in the total filter is equal to the sum of the delays in the individual components.

It is useful to apply the previous result to the general filter structure, which can be viewed as the cascade combination of two simpler filters. The response transform can be expressed as

$$Y(z) = H(z) X(z) = H_1(z) H_2(z) X(z) \tag{5.21}$$

where

$$H_1(z) = \sum_{k=-N_F}^{N_P} b_k z^{-k}$$

and

$$H_2(z) = 1 \Big/ 1 - \sum_{k=1}^{M} a_k z^{-k}$$

$H_1(z)$ is the nonrecursive part of the general filter, while $H_2(z)$ is the recursive part. This result will be useful later for deriving an alternate filter structure that reduces the number of delays required for implementing a digital filter.

Example 5.10.   Two ways of implementing a second-order recursive filter. From Example 5.3, let us consider the following system function:

$$H(z) = \frac{1 - r\cos(\omega_0)z^{-1}}{1 - 2r\cos(\omega_0)\,z^{-1} + r^2z^{-2}}$$

which we can write as

$$H(z) = H_1(z)H_2(z)$$

where

$$H_1(z) = 1 - r\cos(\omega_0)z^{-1}$$

and

$$H_2(z) = (1 - 2r\cos(\omega_0)z^{-1} + r^2z^{-2})^{-1}$$

The desired filter can be implemented by either $H(z) = H_1(z)\,H_2(z)$ or $H(z) = H_2(z)\,H_1(z)$, as shown in Fig. 5.2.

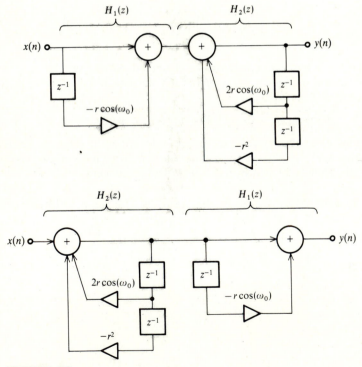

FIGURE 5-2
Two digital filter structures that implement the same system function.

PARALLEL CONNECTION. When filters are connected in parallel, each component has the same input, and the output is the sum of the component filter outputs. Let us consider N component filters that are connected in parallel. $N = 2$ is shown in Fig. 5.1(b). The kth filter system function is denoted by $H_k(z)$, for $1 \le k \le N$, and $Y_k(z)$ is the output of the kth filter, equal to

$$Y_k(z) = H_k(z) X(z) \tag{5.22}$$

The output response for each of the parallel sections is added to produce the composite filter output response, given by

$$Y(z) = \sum_{k=1}^{N} Y_k(z) = \sum_{k=1}^{N} H_k(z) X(z) = X(z) \sum_{k=1}^{N} H_k(z) \tag{5.23}$$

From this last equation, it is apparent that the system function of the total filter composed of parallel components, denoted by $H_P(z)$, is given by

$$H_P(z) = \sum_{k=1}^{N} H_k(z) \tag{5.24}$$

Example 5.11. Parallel connection of digital filters. Let us combine two first-order recursive filters, having coefficients a and b, respectively, in parallel. The overall system function is equal to

$$H(z) = \frac{1}{1 - az^{-1}} + \frac{1}{1 - bz^{-1}} = \frac{1 - bz^{-1} + 1 - az^{-1}}{(1 - az^{-1})(1 - bz^{-1})} = \frac{2 - (a + b)z^{-1}}{1 - (a + b)z^{-1} + abz^{-2}}$$

When combining the two first-order recursive filters, the overall system is second-order and has a nonrecursive component, if $a \ne -b$.

5.6 DIGITAL FILTER IMPLEMENTATION FROM THE SYSTEM FUNCTION

Since the z-transform is a linear transformation, the filter implementation procedure is similar to that in the time domain. The most convenient form for filter synthesis is the z-transform of the general difference equation given by

$$Y(z) = \sum_{k=1}^{M} a_k z^{-k} Y(z) + \sum_{k=-N_F}^{N_P} b_k z^{-k} X(z) \tag{5.25}$$

This form is convenient because the response transform is expressed in terms of the past response transforms as well as the future, current and past excitation transforms. The delay operator acts as a multiplier with coefficient z^{-1}. As in the time domain, a digital filter structure can be determined in four steps, as shown in Fig. 5.3.

Step 1. Two points are drawn, one corresponding to the excitation transform $X(z)$ and the other to the response transform $Y(z)$.

STEP 1

$X(z)\,o$———   ———$o\,Y(z)$

STEP 2

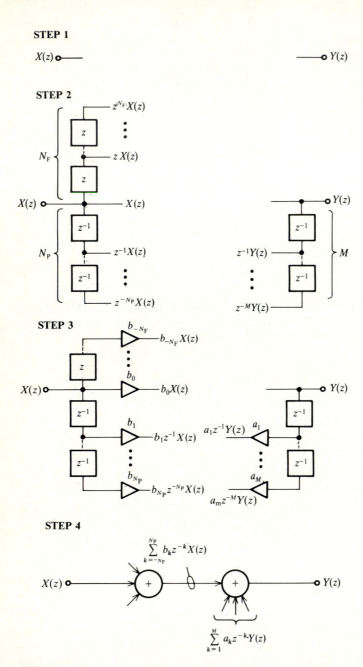

STEP 3

STEP 4

FIGURE 5-3
Steps in implementing a digital filter from the system function $H(z)$.

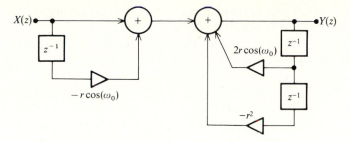

FIGURE 5-4
Implementation of a second-order filter from the system function.

Step 2. Delayed versions of the excitation and response are obtained by connecting delays, denoted by z^{-1}, to the respective points. Advances, denoted by z, are connected only to the excitation transform.

Step 3. Multipliers are connected to the outputs of the delays and advances to obtain the desired products.

Step 4. The products are combined through adders to obtain the desired sums.

Example 5.12. **Digital filter structure of second-order recursive filter.** Let

$$y(n) = 2r \cos(\omega_0) y(n-1) - r^2 y(n-2) + x(n) - r \cos(\omega_0) x(n-1)$$

Taking z-transforms we get

$$Y(z) = 2r \cos(\omega_0)z^{-1} Y(z) - r^2 z^{-2} Y(z) + X(z) - r \cos(\omega_0)z^{-1} X(z)$$

Applying the four steps above, we arrive at the filter structure shown in Fig. 5.4.

5.7 THE COMPLEX z PLANE

The z-transform $H(z)$ is a function of the complex-valued variable z. The complex z plane is shown in Fig. 5.5 with the horizontal axis defining the real values and the vertical axis the imaginary values. In this section, we consider $H(z)$ in terms of its singularities in the z plane. The configuration of the

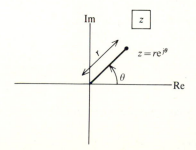

FIGURE 5-5
The complex z plane.

singularities determines the type of the digital filter, either recursive or nonrecursive, and can be used to interpret its frequency behavior.

To define the singularities of a z-transform, let us consider the system function of the general digital filter given by

$$H(z) = \sum_{k=-N_F}^{N_P} b_k z^{-k} \bigg/ 1 - \sum_{k=1}^{M} a_k z^{-k} \qquad (5.26)$$

The roots of the numerator, or the values of z for which $H(z) = 0$, define the locations of the *zeros* in the z plane. The roots of the denominator, or the values of z for which $H(z)$ becomes infinite, define the locations of the *poles*. The locations of the poles and zeros can be shown explicitly by factoring the above sum-of-products form into the following product-of-terms form:

$$H(z) = A z^{N_F} \prod_{k=1}^{N_P + N_F} (1 - c_k z^{-1}) \bigg/ \prod_{k=1}^{M} (1 - d_k z^{-1}) \qquad (5.27)$$

where A is a real-valued gain constant; c_k, for $1 \le k \le N_P + N_F$, is the location (possibly complex) of the kth zero and d_k, for $1 \le k \le M$, is the location of the kth pole.

Note the following important points about Eq. (5.27).

1. Taken individually, each numerator term $(1 - c_k z^{-1})$ generates a zero at $z = c_k$ and also a pole at $z = 0$.
2. Each denominator term generates a pole at $z = d_k$ and a zero at $z = 0$.
3. If $N_F > 0$, the filter is noncausal, having N_F elements in the unit-sample response that have negative index values. The term z^{N_F} generates N_F poles at $z = \infty$ and N_F zeros at $z = 0$.
4. There is a pole/zero cancellation that occurs at $z = 0$, in which a numerator pole will typically cancel a denominator zero. When the singularities at $z = \infty$ are also counted, the net result is that **the number of poles will equal the number of zeros.**
5. For the digital filter to have real-valued coefficients, the complex-valued singularities must occur in complex-conjugate pairs.
6. The filter type can easily be recognized from the pole/zero configuration. The system function of a *nonrecursive* filter can be expressed as a numerator polynomial. Hence, all the singularities in the z plane, except at $z = 0$ and $z = \infty$, will be *zeros*, whereas *recursive* filters will have at least one pole in this region.

These points are summarized in Table 5.2 and illustrated in the following examples.

Example 5.13. Pole/zero pattern of the 3-sample averager. From Example 5.1, the system function is given by

$$H(z) = z/3 + \tfrac{1}{3} + z^{-1}/3$$

TABLE 5.2
Location of singularities in the z plane

Consider the z-transform given in the following form:

$$H(z) = Az^{N_F} \prod_{k=1}^{N_P+N_F} (1 - c_k z^{-1}) \Big/ \prod_{k=1}^{M} (1 - d_k z^{-1})$$

The locations of the singularities are given by
  zeros at $z = c_k$, for $1 \le k \le N_P + N_F$, and $N_P + N_F$ poles at $z = 0$
  poles at $z = d_k$, for $1 \le k \le M$, and M zeros at $z = 0$
  and
  N_F poles at $z = \infty$ and N_F zeros at $z = 0$, due to the non-causal part of $\{h(n)\}$.
(Pole/zero cancellation will usually occur at $z = 0$.)

Converted into the product-of-terms form, we get

$$H(z) = z(1 + z^{-1} + z^{-2})/3 = z(1 - c_1 z^{-1})(1 - c_2 z^{-1})/3$$

where

$$A = 1/3, \qquad c_1 = \frac{-1 + j\sqrt{3}}{2} = e^{j2\pi/3} \qquad \text{and} \qquad c_2 = \frac{-1 - j\sqrt{3}}{2} = e^{-j2\pi/3}$$

The first term, z, generates a pole at $z = \infty$ and a zero at $z = 0$. The two terms containing z^{-1} generate zeros at $z = c_1$ and $z = c_2$ and two poles at $z = 0$. One of these poles cancels the zero at $z = 0$. The pole/zero pattern is shown in Fig. 5.6. The number of poles is equal to the number of zeros. Notice that the gain term A does not affect the locations of the singularities.

Example 5.14.   Pole/zero pattern of the causal 3-sample averager. Let

$$H(z) = \tfrac{1}{3} + z^{-1}/3 + z^{-2}/3 = (1 - c_1 z^{-1})(1 - c_2 z^{-1})/3$$

where c_1 and c_2 are the same as in Example 5.13. There are two poles at $z = 0$ and none at $z = \infty$. The pole/zero pattern is shown in Fig. 5.7. For a causal filter there are no poles at $z = \infty$.

Example 5.15.   Pole/zero pattern of the first-order recursive filter. From

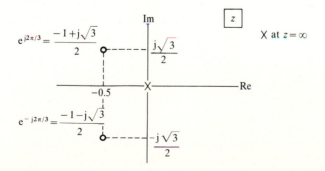

$e^{j2\pi/3} = \dfrac{-1 + j\sqrt{3}}{2}$

$e^{-j2\pi/3} = \dfrac{-1 - j\sqrt{3}}{2}$

X at $z = \infty$

FIGURE 5-6
Pole/zero pattern for
$H(z) = \tfrac{1}{3}(z + 1 + z^{-1})$.

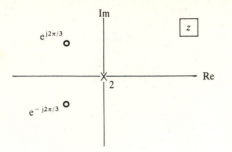

FIGURE 5-7
Pole/zero pattern for
$H(z) = \frac{1}{3}(1 + z^{-1} + z^{-2})$.

Example 5.2, the system function is given by

$$H(z) = [1 - az^{-1}]^{-1} = z/(z - a)$$

In this case, there is a pole at $z = a$ and a zero at $z = 0$, as shown in Fig. 5.8.

The system function in the previous example is expressed with positive powers of z in the denominator. This latter form tends to be less confusing for determining the locations of the poles and zeros. However, it should not be used for implementing a filter structure, since only zero and negative exponents of z are allowed in the denominator of the system function.

Example 5.16. Pole/zero pattern of a second-order recursive filter. From Example 5.3, the system function is given by

$$H(z) = \frac{1 - r\cos(\omega_0)z^{-1}}{1 - 2r\cos(\omega_0)\,z^{-1} + r^2 z^{-2}}$$

$$= \frac{1 - r\cos(\omega_0)\,z^{-1}}{1 - r(e^{j\omega_0} + e^{-j\omega_0})z^{-1} + r^2 z^{-2}}$$

$$= \frac{1 - r\cos(\omega_0)\,z^{-1}}{(1 - r\,e^{j\omega_0}z^{-1})(1 - r\,e^{-j\omega_0}z^{-1})}$$

The numerator generates a zero at $z = r\cos(\omega_0)$ and a pole at $z = 0$, while the denominator generates two zeros at $z = 0$ and poles at $z = r\,e^{\pm j\omega_0}$, as shown in Fig. 5.9.

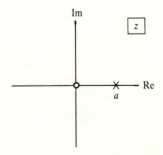

FIGURE 5-8
Pole/zero pattern for $H(z) = 1/(1 - az^{-1})$.

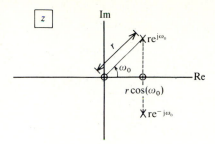

FIGURE 5-9
Pole/zero pattern for $H(z) = (1 - r\cos(\omega_0)z^{-1})/(1 - 2r\cos(\omega_0)z^{-1} + r^2 z^{-2})$.

Example 5.17. Multiple-order singularities in the z plane. Let

$$H(z) = z^2 + 2z + 3 + 2z^{-1} + z^{-2}$$

This can be factored to give

$$H(z) = (z + 1 + z^{-1})^2 = z^2(1 - c_1 z^{-1})^2(1 - c_2 z^{-1})^2$$

where

$$c_1 = \frac{-1 + j\sqrt{3}}{2} = e^{-j2\pi/3} \quad \text{and} \quad c_2 = \frac{-1 - j\sqrt{3}}{2} = e^{-j2\pi/3}$$

The first term, z^2, generates two poles at $z = \infty$ and two zeros at $z = 0$. The two terms containing z^{-1} each generate two zeros at $z = c_1$ and $z = c_2$ and two poles at $z = 0$. The net pole/zero pattern is shown in Fig. 5.10.

Example 5.18. Zeros lying on a circle in the z plane. In designing filters, we commonly encounter the system function having the form

$$H(z) = 1 - z^{-N}$$

The zeros of $H(z)$ occur at values of z for which $z^{-N} = 1$. Since $H(z)$ is an Nth order polynomial, there are N roots. To find these roots, let

$$z^N = r\, e^{j(\theta + 2m\pi)} \quad \text{where } m \text{ is an integer}$$

Then the Nth root of z^N is equal to

$$z = r^{1/N}\, e^{j(\theta + 2m\pi)/N}$$

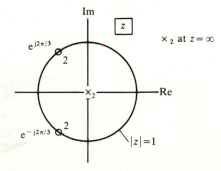

FIGURE 5-10
Pole/zero pattern for $H(z) = z^2 + 2z + 3 + 2z^{-1} + z^{-2}$.

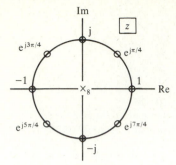

FIGURE 5-11
Pole/zero pattern for $H(z) = 1 - z^{-N}$. Case is shown for $N = 8$.

The roots have distinct values for $m = 0, 1, 2, \ldots, N-1$. In this example,

$$z^{-N} = 1 = e^{j2m\pi}$$

Hence, the zeros occur at $z = e^{j2m\pi/N}$, for $m = 0, 1, 2, \ldots, N-1$. These zeros lie on a circle having unit radius. They start at $z = 1$ and are separated by angles equal to $2\pi/N$. The case for $N = 8$ is shown in Fig. 5.11.

5.8 THE REGION OF CONVERGENCE IN THE z PLANE

In the definition of the z-transform, Eq. (5.1), the limits of the summation index can take on both positive and negative infinite values. The concept of a z-transform is only useful for values of z for which the sum is finite. The values of z for which $H(z)$ is finite lie within a region, called the *region of convergence* (*ROC*), in the z plane. Given a sequence $\{h(n)\}$, the ROC is defined as the set of z values for which

$$\sum_{n=-\infty}^{\infty} |h(n)z^{-n}| < \infty \qquad (5.28)$$

Since $H(z)$ becomes infinite at the locations of the poles, the poles cannot lie in the ROC. In fact, it is the location of the poles that determine the boundaries of the ROC, as illustrated in the following examples. The ROC is important for determining whether the filter is stable, and is related to the type of the time sequence, causal or noncausal, finite-duration or infinite-duration. These relationships are listed in Table 5.3 and discussed below.

Example 5.19. Region of convergence for 3-sample averager. Let

$$h(n) = \begin{cases} \frac{1}{3} & \text{for } -1 \leq n \leq 1 \\ 0 & \text{otherwise} \end{cases}$$

Then

$$\sum_{n=-\infty}^{\infty} |h(n)z^{-n}| = \frac{|z|}{3} + \frac{1}{3} + \frac{|z^{-1}|}{3}$$

TABLE 5.3
Relationship of the ROC to the discrete-time sequence type

Let us consider $h(n)$, for $-N_F \leq n \leq N_P$, where N_P may be infinite. Let ρ be the largest finite pole radius of the system function $H(z)$.

For finite-duration sequences ($N_P < \infty$), $\rho = 0$.
For infinite-duration sequences (N_P is infinite), $\rho > 0$.

	Region of convergence	
For causal, stable systems ($N_F \leq 0$) (all poles inside the unit circle)	$\lvert z \rvert > \rho$	$(\rho < 1)$
For causal, unstable systems ($N_F \leq 0$) (at least one pole outside the unit circle)	$\lvert z \rvert > \rho$	$(\rho > 1)$
For noncausal, stable systems ($N_F > 0$) (N_F poles at $z = \infty$)	$\rho < \lvert z \rvert < \infty$	$(\rho < 1)$

This sum is finite for $\lvert z \rvert < \infty$ and $\lvert z \rvert > 0$. Hence, the ROC contains all values of z for which $0 < \lvert z \rvert < \infty$, or the entire z plane except for $z = 0$ and $z = \infty$.

Example 5.20.   Region of convergence for the first-order recursive filter. Let

$$h(n) = \begin{cases} a^n & \text{for } n \geq 0 \\ 0 & \text{otherwise} \end{cases}$$

Then

$$\sum_{n=-\infty}^{\infty} \lvert h(n)z^{-n} \rvert = \sum_{n=0}^{\infty} \lvert az^{-1} \rvert^n = [1 - \lvert az^{-1} \rvert]^{-1} \quad \text{for } \lvert az^{-1} \rvert < 1$$

The sum converges for $\lvert az^{-1} \rvert < 1$ or $\lvert z \rvert > \lvert a \rvert$. Hence, the ROC is the outside of a circle of radius $\lvert a \rvert$ in the z plane as shown in Fig. 5.12.

DETERMINING THE STABILITY OF A DISCRETE-TIME SYSTEM IN THE z-PLANE. The definition of a stable system was given in Chapter 2 in terms of the unit-sample response sequence as

$$\sum_{n=-\infty}^{\infty} \lvert h(n) \rvert < \infty \tag{5.29}$$

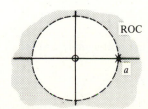

FIGURE 5-12
Region of convergence (ROC) for first-order recursive filter.

Systems having a finite-duration unit-sample response are always stable because the sum of a finite number of finite-valued coefficients cannot become infinite. Hence, the stability question is addressed to the infinite-duration component, if any, of the unit-sample response. This component is generated by the *poles* of the system function. The question of stability can then be addressed in terms of the pole locations in the z plane.

Let us consider the simplest form of the general digital filter that contains only finite-valued poles and all the zeros at $z = 0$, given by

$$H(z) = 1 \bigg/ 1 - \sum_{k=1}^{M} a_k z^{-k} \tag{5.30}$$

To show explicitly the locations of the poles, we factor the denominator into the product-of-terms form and perform a partial fraction expansion to obtain

$$H(z) = \sum_{k=1}^{M} \frac{\alpha_k}{1 - d_k z^{-1}} \tag{5.31}$$

where both α_k and d_k may be complex-valued. Since the z-transform is linear, the unit-sample response is equal to the sum of that produced by each term in the sum, or

$$h(n) = \sum_{k=1}^{M} h_k(n) \qquad \text{for } n \geq 0 \tag{5.32}$$

The unit-sample response of the kth first-order recursive filter is given by

$$h_k(n) = \alpha_k d_k^n u(n). \tag{5.33}$$

For the system to be stable, each component sequence $\{h_k(n)\}$ must satisfy the infinite sum condition, or

$$\sum_{n=0}^{\infty} |h_k(n)| < \infty \tag{5.34}$$

Applying this condition, we have

$$\sum_{n=0}^{\infty} |h_k(n)| = \sum_{n=0}^{\infty} |\alpha_k d_k^n| = |\alpha_k| \sum_{n=0}^{\infty} |d_k|^n \tag{5.35}$$

For the last sum, composed of a geometric series, to be finite, the magnitude of each term must be less than unity, or $|d_k| < 1$. Hence, we have shown that each pole of a stable system must lie *within the unit circle* in the z plane. This is illustrated in the following examples of two unstable systems.

Example 5.21. First-order unstable system. Consider the unit-sample response given by

$$h(n) = \begin{cases} 1 & \text{for } n \geq 0 \\ 0 & \text{otherwise} \end{cases}$$

Testing for stability, we compute the sum given by

$$\sum_{n=0}^{\infty} |h(n)| = \sum_{n=0}^{\infty} 1$$

which is not finite by inspection. Hence, the system is not stable. The system function is given by

$$H(z) = \sum_{n=0}^{\infty} h(n)z^{-n} = \sum_{n=0}^{\infty} z^{-n}$$

Applying the infinite geometric sum formula, we get

$$H(z) = \frac{1}{1 - z^{-1}} \qquad \text{for } |z^{-1}| < 1 \quad \text{or} \quad |z| > 1$$

The pole of $H(z)$ lies *on* the unit circle, at $z = 1$.

Example 5.22. Unstable second-order system. Consider the unit-sample response given by

$$h(n) = \begin{cases} \cos(\beta n) & \text{for } n \geq 0 \\ 0 & \text{otherwise} \end{cases}$$

Testing for stability, we compute the sum given by

$$\sum_{n=0}^{\infty} |h(n)| = \sum_{n=0}^{\infty} |\cos(\beta n)|$$

which is not finite by inspection. Hence, the system is not stable. The system function is given by

$$H(z) = \sum_{n=0}^{\infty} h(n)z^{-n} = \sum_{n=0}^{\infty} \cos(\beta n)z^{-n}$$

Applying the Euler identities, we have

$$H(z) = \frac{1}{2} \sum_{n=0}^{\infty} e^{j\beta n} z^{-n} + \frac{1}{2} \sum_{n=0}^{\infty} e^{-j\beta n} z^{-n}$$

Applying the infinite geometric sum formula to each term, we get

$$H(z) = \frac{1/2}{1 - e^{j\beta} z^{-1}} + \frac{1/2}{1 - e^{-j\beta} z^{-1}} \qquad \text{for } |z^{-1}| < |e^{j\beta}| = 1, \text{ and } |z^{-1}| < |e^{-j\beta}| = 1$$

The poles of $H(z)$ lie *on* the unit circle, at $z = e^{\pm j\beta}$.

DETERMINING THE ROC FROM THE UNIT-SAMPLE RESPONSE. The ROC can also be determined directly from the properties of the unit-sample response. The filter having a finite-duration unit-sample response has a system function that can be expressed as a polynomial given by

$$H(z) = \sum_{n=-N_F}^{N_P} h(n) z^{-n} \tag{5.36}$$

For such a polynomial system function, poles can exist only at $z = 0$ and $z = \infty$, with those at $z = \infty$ being present only for sequences that are noncausal, i.e. for $N_F > 0$. For infinite-duration sequences extending for $n \geq 0$, typified by the unit-sample response of a recursive filter, the system function will contain at least one denominator term producing a finite pole in the z plane in the region $0 < |z| < \infty$. Then the ROC is outside the pole having the largest radius. Infinite-duration sequences extending into negative time result when the ROC is inside the smallest pole in the z plane. Since these have little practical utility, they will not be considered in this book.

5.9 DETERMINING THE FILTER COEFFICIENTS FROM THE SINGULARITY LOCATIONS

For the filter design procedures described in the following chapters, it will be useful to determine the filter coefficients from the singularities in the z plane. As indicated by Eq. (5.19), any filter can be implemented as the cascade combination of first- and second-order sections. The types of strictly recursive and nonrecursive elementary filters that are needed for this application are shown in Fig. 5.13, along with their respective z plane patterns. Let us define r_0 to be a real number and ω_0 a frequency value, where $0 \leq \omega_0 \leq \pi$. Then the elementary filters generate singularities of the following four types.

1. A single zero on the real axis at $z = r_0$, where $0 < |r_0| < \infty$, with the corresponding pole at $z = 0$. The system function of the nonrecursive filter that generates this pattern is given by

$$H_1(z) = 1 - r_0 z^{-1} \tag{5.37}$$

2. A single pole on the real axis at $z = r_0$, where $0 < |r_0| < 1$, with the corresponding zero at $z = 0$. The system function of the recursive filter that generates this pattern is given by

$$H_2(z) = [1 - r_0 z^{-1}]^{-1} \tag{5.38}$$

3. A pair of complex-conjugate zeros at $z = r_0 e^{\pm j\omega_0}$, where $0 < r_0 < \infty$, with the two corresponding poles at $z = 0$. The system function of the filter that generates this pattern is given by

$$H_3(z) = (1 - r_0 e^{j\omega_0} z^{-1})(1 - r_0 e^{-j\omega_0} z^{-1})$$
$$= 1 - 2r_0 \cos(\omega_0) z^{-1} + r_0^2 z^{-2} \tag{5.39}$$

4. A pair of complex-conjugate poles at $z = r_0 e^{\pm j\omega_0}$, where $0 < r_0 < 1$, with the two corresponding zeros at $z = 0$. The system function of the filter that generates this pattern is given by

$$H_4(z) = [(1 - r_0 e^{j\omega_0} z^{-1})(1 - r_0 e^{-j\omega_0} z^{-1})]^{-1}$$
$$= [1 - 2r_0 \cos(\omega_0) z^{-1} + r_0^2 z^{-2}]^{-1} \tag{5.40}$$

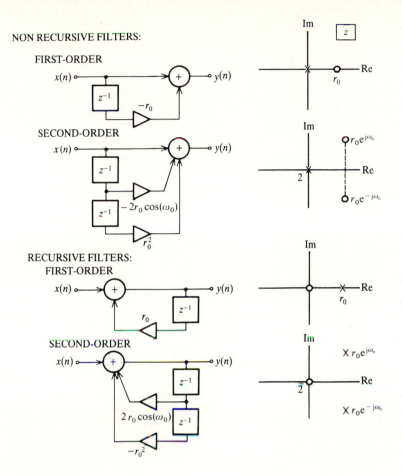

FIGURE 5-13
Relationships of simple filters to their pole/zero patterns.

Having implemented the singularities lying in $0 < |z| < \infty$, those remaining at $z = 0$ and $z = \infty$ are then considered. There will *always* remain an equal number of opposite singularities at $z = 0$ and $z = \infty$. If the system is noncausal, recognized by the unit-sample response starting at $n = -N_F$, then there will be N_F poles at $z = \infty$. The elementary system function component is then

$$H_5(z) = z^{N_F} \qquad (5.41)$$

If the system is causal, but has a delay in the unit-sample response, starting at $n = M > 0$, then there will be M zeros at $z = \infty$. The elementary system function component is then

$$H_5(z) = z^{-M} \qquad (5.42)$$

The system function of the total filter is then proportional to the product of these elementary system functions, or

$$H(z) = A \prod_{k=1}^{K} H_k(z) \tag{5.43}$$

where A is the filter gain constant, which *cannot* be determined from the pole/zero pattern, and K is the number of elementary systems. By breaking up a given z plane pole/zero pattern into these elementary types, the coefficients of the filters generating these singularities can be determined by inspection.

If a pole or zero is needed at $z = 0$, but not available, a pole/zero pair can be created there. Then the singularity that is needed is taken, leaving the opposite singularity at $z = 0$. The singularities remaining at $z = 0$, after all those located in $0 < |z| < \infty$ have been taken, have mates at $z = \infty$. If there are N excess poles at $z = 0$, then these cause a net delay of N samples in the system function, or z^{-N}. If there are N excess zeros at $z = 0$, then these cause a net advance, or z^N.

Example 5.23. Filter coefficients from the z plane singularities. Consider the pole/zero pattern shown in Fig. 5.14. The system function will be represented by

$$H(z) = A \prod_{k=1}^{K} H_k(z)$$

We first collect the finite, real-valued singularities.

(1) The zero at $z = d$ is paired with a pole at $z = 0$ to produce

$$H_1(z) = 1 - dz^{-1}$$

Since there was no existing pole at $z = 0$, a pole/zero pair was created at $z = 0$, the pole was taken and the zero remained, leaving two zeros at $z = 0$.

(2) The pole at $z = a$ is paired with a zero at $z = 0$ to produce

$$H_2(z) = [1 - az^{-1}]^{-1}$$

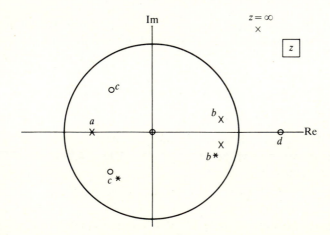

FIGURE 5-14
Pole/zero pattern to be implemented by a digital filter.

The finite, complex-conjugate singularities are then collected.

(3) The zero at $z = c$ and $z = c^*$ and two poles at $z = 0$ produce

$$H_3(z) = (1 - cz^{-1})(1 - c^*z^{-1}) = 1 - (c + c^*)z^{-1} + |c|^2 z^{-2}$$

Since there were no existing poles at $z = 0$, two pole/zero pairs were created at $z = 0$, the poles were taken and the zeros remained, leaving three zeros at $z = 0$.

(4) The poles at $z = b$ and $z = b^*$ and two zeros at $z = 0$ produce

$$H_4(z) = [(1 - bz^{-1})(1 - b^*z^{-1})]^{-1} = [1 - (b + b^*)z^{-1} + |b|^2 z^{-2}]^{-1}$$

(5) The remaining pole at $z = \infty$ indicates a noncausal system. The pole at $z = \infty$ and the zero at $z = 0$ produces

$$H_5(z) = z$$

The system function gain factor A cannot be determined from the pole/zero pattern, but must be given separately.

PROGRAMS TO DETERMINE THE FILTER COEFFICIENTS FROM THE z-PLANE SINGULARITIES. To determine the filter coefficients from the locations of the singularities, the following two subroutines are helpful. The first determines the nonrecursive filter coefficients, and is given by

```
ZERO(R, DEG, BCOEF, NB)
```

where

R	(real input) is the radius of the zero ($R > 0$);
DEG	(real input) is the angle in degrees of the zero with respect to the real axis ($0 \leq DEG \leq 180$);
BCOEF	(real output array) is the coefficient array for the nonrecursive filter that implements the zero or zeros; and
NB	(integer output) is the size of the array BCOEF.

The angle is specified in degrees, rather than radians, since it is easier to specify the singularity location in degrees.

The second program determines the recursive filter coefficients, and is given by

```
POLE(R, DEG, ACOEF, NA)
```

where

R	(real input) is the radius of the pole ($0 < R < 1$);
DEG;	(real input) is the angle in degrees of the pole with respect to the real axis ($0 \leq DEG \leq 180$);
ACOEF	(real output array) is the coefficient array for the recursive filter that implements the pole or poles; and
NA	(integer output) is the size of the array ACOEF.

If $DEG = 0$ or 180, then the singularity lies on the real axis, at $z = R$ or $z = -R$, respectively. If $DEG \neq 0$ or 180, then the singularity is complex-valued. In this case, the subroutines should then also include the corresponding complex-conjugate singularity to produce real-valued coefficients. The singularities at both $z = R\,e^{\pm j\pi(DEG)/180}$ are then implemented, as demonstrated in the following examples.

The coefficients produced by ZERO can be used directly by the NONREC filter subroutine, while those produced by POLE are used by REC. The complete pole/zero pattern can be generated by cascading these first- and second-order filter sections. In this case, the output of the previous section becomes the input to the next.

Example 5.24. Coefficients for a real-valued zero. We want the filter coefficients that implement a zero at $z = r_0$. Then $R = |r_0|$. If $r_0 > 0$, then DEG $= 0$, or if $r_0 < 0$, then DEG $= 180$. The subroutine should determine that the zero is produced by the filter having the system function

$$H(z) = 1 - r_0 z^{-1}$$

The call to ZERO(R, DEG, BCOEF, NB) should generate BCOEF(1) $= 1$, BCOEF(2) $= -r_0$ and NB $= 2$. The corresponding pole is generated at $z = 0$.

Example 5.25. Coefficient for a real-valued pole. We want the filter coefficients that implement a pole at $z = r_0$. Then $R = |r_0|$. If $r_0 > 0$, then DEG $= 0$, or if $r_0 < 0$, then DEG $= 180$. The subroutine should determine that the pole is produced by the filter having the system function

$$H(z) = [1 - r_0 z^{-1}]^{-1}$$

The call to POLE(R, DEG, ACOEF, NA) should generate ACOEF(1) $= r_0$ and NA $= 1$. The corresponding zero is generated at $z = 0$.

Example 5.26. Coefficients for a complex-valued zero. We want the filter coefficients that implement a zero at $z = r_0\,e^{j\omega_0}$, where $r_0 > 0$. Since we want a filter having real-valued coefficients, the filter should also implement the complex-conjugate zero at $z = r_0\,e^{-j\omega_0}$. Then $R = r_0$ and DEG $= 180\omega_0/\pi$. The desired system function is then given by

$$H(z) = (1 - r_0\,e^{j\omega_0}z - 1)(1 - r_0\,e^{-j\omega_0}z^{-1})$$
$$= 1 - 2r_0\cos(\omega_0)z^{-1} + r_0^2 z^{-2}$$

The call to ZERO(R, DEG, BCOEF, NB) should generate BCOEF(1) $= 1$, BCOEF(2) $= -2r_0\cos(\omega_0)$, BCOEF(3) $= r_0^2$ and NB $= 3$. The two corresponding poles are generated at $z = 0$.

Example 5.27. Coefficients for a complex-valued pole. We want the filter coefficients that implement a pole at $z = r_0\,e^{j\omega_0}$, where $0 < r_0 < 1$. Since we want a filter having real-valued coefficients, the filter should also implement the complex-conjugate pole at $z = r_0\,e^{-j\omega_0}$. Then $R = r_0$ and DEG $= 180\omega_0/\pi$. The desired system function is then given by

$$H(z) = [(1 - r_0\,e^{j\omega_0}z^{-1})(1 - r_0\,e^{-j\omega_0}z^{-1})]^{-1}$$
$$= [1 - 2r_0\cos(\omega_0)z^{-1} + r_0^2 z^{-2}]^{-1}$$

The call to POLE(R, DEG, ACOEF, NA) should generate ACOEF(1) $= 2r_0\cos(\omega_0)$, ACOEF(2) $= -r_0^2$ and NA $= 2$. The two corresponding zeros are generated at $z = 0$.

5.10 GEOMETRIC EVALUATION OF THE *z*-TRANSFORM IN THE *z* PLANE

In this section, we indicate how to evaluate $H(z)$ at any point in the z plane using geometric principles. This procedure is then extended to the evaluation of the Fourier transform in the next section.

First, let us define a *finite singularity* as one that is located at a finite value of z, i.e. not at $z = \infty$. The general system function, given by

$$H(z) = Az^{N_F} \prod_{k=1}^{N_P+N_F} (1 - c_k z^{-1}) \bigg/ \prod_{k=1}^{M} (1 - d_k z^{-1}) \tag{5.44}$$

has $N_F + N_P$ finite zeros located at c_k, $1 \le k \le N_P + N_F$, M finite poles located at d_k, $1 \le k \le M$, and $M - N_P$ zeros at $z = 0$. The value of $H(z)$ at some point $z = z_0$ in the z plane is a complex number that can be expressed as

$$H(z_0) = |H(z_0)| \, e^{j\mathrm{Arg}[H(z_0)]} \tag{5.45}$$

As shown in Fig. 5.15, the magnitude can be expressed in terms of the distances from the finite singularities to the point $z = z_0$ and the phase in terms of the relative angles.

To provide a formula for the evaluation of the magnitude and phase, consider the system function having K_p finite poles located at $z = p_k$ for $1 \le k \le K_p$, and K_z finite zeros located at $z = q_k$ for $1 \le k \le K_z$. Then the magnitude is given by

$$|H(z_0)| = |A| \prod_{k=1}^{K_z} D[q_k, z_0] \bigg/ \prod_{k=1}^{K_p} D[p_k, z_0] \tag{5.46}$$

where $D[x, y]$ is the distance between the points x and y in the complex z plane. The square of this distance is computed by

$$D^2[x, y] = (x - y)(x - y)^*$$
$$= [\mathrm{Re}(x) - \mathrm{Re}(y)]^2 + [\mathrm{Im}(x) - \mathrm{Im}(y)]^2 \tag{5.47}$$

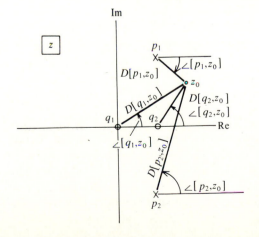

FIGURE 5-15
Geometric evaluation of the system function $H(z)$ at point $z = z_0$.

The phase of the complex number is equal to

$$\text{Arg}[H(z_0)] = \sum_{k=1}^{K_z} \angle[q_k, z_0] - \sum_{k=1}^{K_p} \angle[p_k, z_0] \qquad (5.48)$$

where $\angle[x, y]$ is the angle between the positive real axis and the line connecting points x and y. This angle can be computed by

$$\angle[x, y] = \arctan\left(\frac{\text{Im}(y) - \text{Im}(x)}{\text{Re}(y) - \text{Re}(x)}\right) \qquad (5.49)$$

The value of A cannot be determined from the pole/zero pattern, since the pattern is not affected by a constant multiplier of the system function. We will apply this evaluation procedure to determine the Fourier transform in the z plane in the next section.

5.11 RELATIONSHIP BETWEEN THE FOURIER TRANSFORM AND THE z-TRANSFORM

In this section we show the relationship between the transfer function $H(e^{j\omega})$ and the system function $H(z)$ of a filter. This relationship indicates the procedure to evaluate the Fourier transform in the z plane.

The complex variable z can be expressed in the complex-vector form given by $z = r\,e^{j\omega}$. Including this form in the definition of the z-transform, we have

$$H(r\,e^{j\omega}) = \sum_{n=-\infty}^{\infty} h(n)r^{-n}\,e^{-j\omega n} \qquad (5.50)$$

Note the similarity of this analytic form and the definition of the Fourier transform of $\{h(n)\}$ given by

$$H(e^{j\omega}) = \sum_{n=-\infty}^{\infty} h(n)\,e^{-j\omega n} \qquad (5.51)$$

In fact, the two definitions are identical when $r = 1$. In the z plane, this corresponds to the locus of points for which $|z| = 1$. This locus defines a circle about the origin with radius equal to 1, called the *unit circle*. Hence, $H(e^{j\omega})$ is equal to $H(z)$ evaluated along the unit circle, or

$$H(e^{j\omega}) = H(z)\big|_{z=e^{j\omega}} \qquad (5.52)$$

For $H(e^{j\omega})$ to exist, the ROC of $H(z)$ must include the unit circle. As shown previously, the ROC for a causal filter system function is the region outside the pole having the largest radius. Since the Fourier transform is evaluated on the unit circle, this implies that all the poles of a causal, stable filter must lie inside the unit circle.

To determine the Fourier transform at a particular frequency $\omega = \omega_0$, we must evaluate the system function at the point $z = e^{j\omega_0}$. In the z plane, this point is located at the intersection of the unit circle and the line passing through the origin and making an angle ω_0 with respect to the positive real axis. The evaluation is illustrated in the following examples.

Example 5.28. **Evaluation of the Fourier transform in the z plane for the first-order recursive filter.** Let

$$H(z) = [1 - az^{-1}]^{-1} \qquad \text{ROC: } |z| > |a|$$

There is a zero at $z = 0$ and a pole at $z = a$. For $H(e^{j\omega})$ to exist, the ROC must include the unit circle, which implies that $|a| < 1$. For any frequency ω, we evaluate the z-transform at the point on the unit circle $z = e^{j\omega}$.

Let us consider the particular frequency ω_0, as shown in Fig. 5.16. The distance from the zero is

$$D^2[0, e^{j\omega_0}] = 1$$

and that from the pole is

$$D^2[a, e^{j\omega_0}] = (a - \cos(\omega_0))^2 + \sin^2(\omega_0)$$

Hence,

$$|H(e^{j\omega_0})| = [(a - \cos(\omega_0))^2 + \sin^2(\omega_0)]^{-1/2}$$
$$= [1 - 2a \cos(\omega_0) + a^2]^{-1/2}$$

This is the same result obtained in Example 3.7.

The angle from $z = 0$ to $e^{j\omega_0}$ is equal to ω_0, while the angle from the pole is

$$\angle[a, e^{j\omega_0}] = \arctan[\sin(\omega_0)/(\cos(\omega_0) - a)]$$

Hence,

$$\text{Arg}[H(e^{j\omega_0})] = \omega_0 - \arctan[\sin(\omega_0)/(\cos(\omega_0) - a)].$$

Example 5.29. **Evaluation of the Fourier transform in the z-plane for the 3-sample averager.** Let

$$H(z) = \tfrac{1}{3}(z + 1 + z^{-1})$$

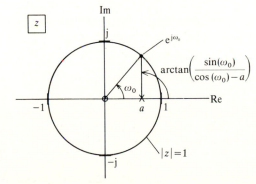

FIGURE 5-16
Geometric evaluation of the Fourier transform along the unit circle in the z plane for the first-order recursive filter.

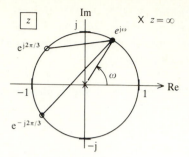

FIGURE 5-17
Geometric evaluation of the Fourier transform along the unit circle in the z plane for the 3-sample averager.

From Example 5.13, there are zeros at $z = e^{\pm j2\pi/3}$ and a finite pole at $z = 0$, as shown in Fig. 5.17. The distance from the pole to $e^{j\omega}$ is

$$D^2[0, e^{j\omega}] = \cos^2(\omega) + \sin^2(\omega) = 1$$

The product of the distances from each of the zeros to $e^{j\omega}$ is

$$D^2[e^{j2\pi/3}, e^{j\omega}]\, D^2[e^{-j2\pi/3}, e^{j\omega}]$$
$$= [2 - e^{j(\omega - 2\pi/3)} - e^{-j(\omega - 2\pi/3)}][2 - e^{j(\omega + 2\pi/3)} - e^{-j(\omega + 2\pi/3)}]$$
$$= 3 + 2\cos(2\omega) + 4\cos(\omega) = 1 + 4([1 + \cos(2\omega)]/2) + 4\cos(\omega)$$
$$= 1 + 4\cos^2(\omega) + 4\cos(\omega) = (1 + 2\cos(\omega))^2.$$

This result is 9 times the value for $|H(e^{j\omega})|^2$ in Example 5.13. Hence, $A = \frac{1}{3}$.

The angle from the pole at $z = 0$ to $e^{j\omega}$ is equal to ω. The angle from each of the zeros is equal to

$$\angle[e^{j2\pi/3}, e^{j\omega}] = \arctan\left(\frac{\sin(2\pi/3) - \sin(\omega)}{\cos(2\pi/3) - \cos(\omega)}\right)$$

$$\angle[e^{-j2\pi/3}, e^{j\omega}] = \arctan\left(\frac{-\sin(2\pi/3) - \sin(\omega)}{\cos(2\pi/3) - \cos(\omega)}\right)$$

Then

$$\text{Arg}[H(e^{j\omega})] = \angle[e^{j2\pi/3}, e^{j\omega}] + \angle[e^{-j2\pi/3}, e^{j\omega}] - \omega$$

Evaluating the phase, we find

$$\text{Arg}[H(e^{j\omega})] = \begin{cases} 0 & \text{for } -2\pi/3 < \omega < 2\pi/3 \\ -\pi & \text{for } 2\pi/3 < \omega \le \pi \\ \pi & \text{for } -\pi < \omega < -2\pi/3 \end{cases}$$

From the previous examples, note the following general points.

Point 1. Since the distance from the origin to the unit circle is unity, the singularities at the origin do not contribute to the magnitude response. They do, however, contribute to the phase response.

Point 2. The magnitude response is zero only when there is at least one zero of the system function lying on the unit circle at that frequency location.

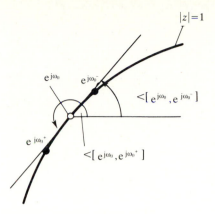

FIGURE 5-18
Evaluation of the phase response around a zero located on the unit circle.

Point 3. The phase jump of π radians results as the frequency passes through each zero lying on the unit circle. Let the zero exist at $z = e^{j\omega_0}$, and let $\omega_0^- = \omega_0 - \varepsilon$ and $\omega_0^+ = \omega_0 + \varepsilon$, for some small $\varepsilon > 0$. As shown in Fig. 5.18, if a single zero exists at $z = e^{j\omega_0}$, then the phase value at $\omega = \omega_0^-$ is π radians less than that at $\omega = \omega_0^+$. If there are multiple zeros, say M, at $z = e^{j\omega_0}$, then there is a jump of $M\pi$ in the phase as the ω passes from ω_0^- to ω_0^+. If M is an even-valued integer, then the jump is an integral number of 2π. In this case, when the phase is plotted within the principal value range of $-\pi$ to π, then no net change of the phase value is observed.

Point 4. As ω increases from 0 to π, the phase generated by each zero that lies strictly inside the unit circle increases by π radians. The phase generated by each pole inside the unit circle decreases by π. If the number of such zeros is N_z and N_p is the number of poles, the net increase in phase as ω goes from 0 to π is equal to $(N_z - N_p)\pi$.

The same analytic results could have been obtained with much less effort by merely substituting $e^{j\omega}$ for z in $H(z)$ to obtain the Fourier transform. The justification for this geometric evaluation is that the approximate shape of the magnitude response of a complicated Fourier transform can be readily inferred from the pole/zero pattern. For example, if a singularity is located close to the unit circle, it will dominate the frequency response in the frequency range adjacent to the singularity. This is true because the magnitude response is determined by the ratio of distances, given in Eq. (5.46). If the singularity is close to the unit circle, then the distances from it can change radically for small changes in frequency, which are increments along the unit circle, as shown in Fig. 5.19. This *dominant singularity* idea is helpful not only for determining the approximate shape of the frequency response, but also for the placement of poles and zeros to achieve a desired transfer function, as demonstrated in the following chapters.

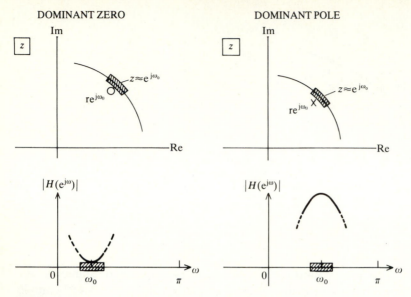

FIGURE 5-19
Magnitude response features for dominant zeros and poles. ($r = 0.99$).

5.12 THE z-TRANSFORM OF SYMMETRIC SEQUENCES

In this section, we consider the pole/zero pattern for linear, time-invariant systems having unit-sample response sequences that exhibit a point of symmetry. In Chapter 3, symmetric and antisymmetric sequences were discussed, with special cases being the even and odd sequences. To simplify the analysis, we first consider these special cases and then extend these results to the general symmetric and antisymmetric cases.

EVEN SEQUENCES. Let us consider an even sequence, denoted by $\{h_E(n)\}$, having duration $N = 2M + 1$, for some integer M. By the definition of an even sequence, $h(n) = h(-n)$. The z-transform of $\{h_E(n)\}$ is given by

$$H_E(z) = \sum_{n=-M}^{M} h(n)z^{-n}$$

$$= \sum_{n=-M}^{-1} h(n)z^{-n} + h(0) + \sum_{n=1}^{M} h(n)z^{-n} \qquad (5.53)$$

Substituting $k = -n$ in the first sum, we get

$$H_E(z) = \sum_{k=1}^{M} h(k)z^{k} + h(0) + \sum_{n=1}^{M} h(n)z^{-n}$$

$$= h(0) + \sum_{n=1}^{M} h(n)[z^{n} + z^{-n}] \qquad (5.54)$$

$H_E(z)$ is a function of $(z^n + z^{-n})$. Hence, if z were replaced by $1/z$, there is no change in the function. This condition can be expressed analytically as

$$H_E(z) = H_E(1/z)$$

In terms of the pole/zero pattern, if a zero is located at $z_0 = r_0 e^{j\omega_0}$ then another zero is also located at $1/z_0 = e^{-j\omega_0}/r_0$. Hence, real zeros occur in reciprocal pairs and the complex zeros occur in reciprocal, complex-conjugate quadruplets. When the zeros lie on the unit circle, they form their own reciprocals and occur only as complex conjugates. Since the number of poles is always equal to the number of zeros, half of the poles of $H_E(z)$ are located at $z = 0$ and the other half lie at $z = \infty$. These latter poles correspond to the noncausal elements in $\{h_E(n)\}$.

Example 5.30.   Pole/zero pattern for an even sequence. Let

$$h(-2) = 1 \qquad h(-1) = -2.5 \qquad h(0) = 5.25 \quad h(1) = -2.5 \qquad h(2) = 1$$
$$\text{and } h(n) = 0, \text{ otherwise}$$

The system function is equal to

$$H(z) = z^2 - 2.5z + 5.25 - 2.5z^{-1} + z^{-2}$$
$$= z^2(1 - 0.5\,e^{j2\pi/3}z^{-1})(1 - 0.5\,e^{-j2\pi/3}z^{-1})(1 - 2\,e^{j2\pi/3}z^{-1})(1 - 2\,e^{-j2\pi/3}z^{-1})$$

The pole/zero pattern in shown in Fig. 5.20. Note the complex-conjugate reciprocal pattern of the zeros and the two poles at $z = \infty$.

Example 5.31.   Pole/zero pattern for an even sequence with zeros on the unit circle. Let
$$h(-1) = 1 \qquad h(0) = -\tfrac{1}{2} \qquad h(1) = 1 \qquad \text{and } h(n) = 0, \text{ otherwise}$$

The system function is given by

$$H(z) = z - \tfrac{1}{2} + z^{-1}$$
$$= z(1 - e^{j\pi/4}z^{-1})(1 - e^{-j\pi/4}z^{-1})$$

The pole/zero pattern is shown in Fig. 5.21. Note the zeros on the unit circle form their own reciprocals.

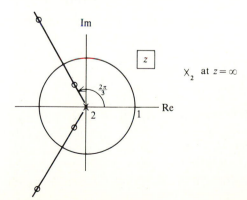

FIGURE 5-20
Pole/zero pattern for an even sequence.

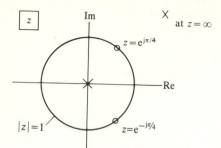

FIGURE 5-21
Pole/zero pattern for an even sequence having zeros on the unit circle.

CAUSAL SYMMETRIC SEQUENCES. For these sequences we must consider two cases, depending on whether the point of symmetry falls on a sample point or between sample points. If the point of symmetry falls on a sample point, this causal sequence, denoted by $\{h_S(n)\}$, can be obtained from an even sequence by applying sufficient delay that the first nonzero element starts at $n = 0$. For the even sequence above, a delay of M samples is needed, or

$$h_S(n) = h_E(n - M) \qquad \text{for } 0 \le n \le 2M$$

Since delaying a sequence by k samples is equivalent to multiplying the z-transform by z^{-k}, we get

$$H_S(z) = z^{-M}H_E(z) \tag{5.55}$$

This last equation indicates that delaying a sequence by M time units generates M poles at $z = 0$ and M zeros at $z = \infty$. These M zeros at $z = \infty$ cancel the M poles of the even sequence that were there. Hence, all the poles of the z-transform for a causal finite-duration sequence are located at $z = 0$. The ROC is then defined by the entire z-plane, except $z = 0$, or $|z| > 0$.

> **Example 5.32. Pole/zero configuration for a symmetric sequence having an odd number of elements.** Let
>
> $$h(0) = 1 \qquad h(1) = -2.5 \qquad h(2) = 5.25 \qquad h(3) = -2.5 \qquad h(4) = 1$$
>
> $$\text{and } h(n) = 0, \text{ otherwise}$$
>
> Then the system function is equal to
>
> $$H(z) = 1 - 2.5z^{-1} + 5.25z^{-2} - 2.5z^{-3} + z^{-4}$$
> $$= (1 - 0.5e^{j2\pi/3}z^{-1})(1 - 0.5e^{-j2\pi/3}z^{-1})(1 - 2e^{j2\pi/3}z^{-1})(1 - 2e^{-j2\pi/3}z^{-1})$$
>
> The pole/zero pattern is shown in Fig. 5.22.

If the point of symmetry falls between sample points, the analysis becomes more complicated. However, we can arrive at the desired result by the following argument. Let the symmetric sequence $h_S(n)$, for $0 \le n \le N - 1$, where N is an even-valued integer, have the z-transform $H_S(z)$. The point of symmetry is then $N_S = (N - 1)/2$, which does not fall on a sample point, as

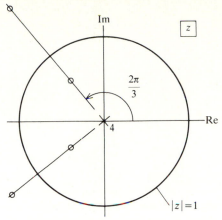

FIGURE 5.22
Pole/zero pattern for a causal symmetric sequence.'

shown in Fig. 5.23. Let us now consider the square of the system function, or

$$H'(z) = H_S(z) H_S(z) \qquad (5.56)$$

$H'(z)$ has the same singularity locations as $H_S(z)$, although there are now twice as many at each point. The equivalent time-domain operation is the linear convolution of $\{h(n)\}$ with itself, or

$$\{h'(n)\} = \{h_S(n)\} * \{h_S(n)\} \qquad (5.57)$$

By performing the convolution of an N-sample duration, symmetric function with itself, the resulting sequence $\{h'(n)\}$ has length $2N - 1$ and is symmetric about the sample point $n = N - 1$, as shown in Fig. 5.23. Hence, $H'(z)$ has zeros that occur in complex-conjugate reciprocal quadruplets. By the squaring operation given in Eq. (5.56), the singularities of $H'(z)$ occur an even number of times at each singularity location. Half of these singularities belong to $H_S(z)$ when we factor $H'(z)$. Hence, we have shown that $H_S(z)$ also has zeros that occur in complex-conjugate reciprocal quadruplets.

Example 5.33. Pole/zero pattern for sequences having symmetry between sample points. Let

$$h_S(n) = \begin{cases} 1 & \text{for } 0 \le n \le 3 \\ 0 & \text{otherwise} \end{cases}$$

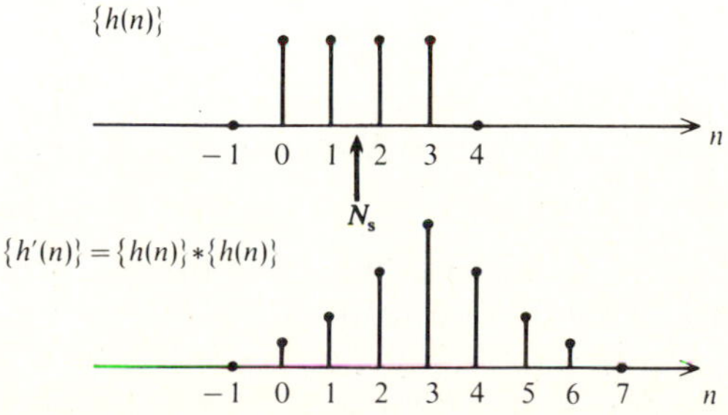

$\{h(n)\}$

N_s

$\{h'(n)\} = \{h(n)\} * \{h(n)\}$

FIGURE 5-23
The self-convolution of a symmetric sequence that has its point of symmetry between sampling points produces a sequence that has its point of symmetry on a sampling point.

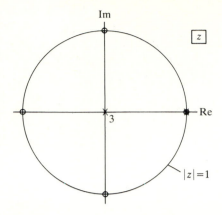

FIGURE 5-24
Pole/zero pattern for N-point symmetric sequence, where N is even integer.

shown in Fig. 5.23. The self-convolution sequence is then equal to

$$h'(n) = \begin{cases} n+1 & \text{for } 0 \le n \le 3 \\ 7-n & \text{for } 4 \le n \le 6 \\ 0 & \text{otherwise} \end{cases}$$

Note that $\{h'(n)\}$ is symmetric about $n = 3$. Hence, $H'(z)$ has zeros that are complex-conjugate quadruplets with an even number of zeros at each singularity location. The z-transform of $\{h'(n)\}$ is equal to

$$H'(z) = \frac{1 - 2z^{-4} + z^{-8}}{1 - 2z^{-1} + z^{-2}} = \left(\frac{1 - z^{-4}}{1 - z^{-1}}\right)^2$$

Hence,

$$H_S(z) = \frac{1 - z^{-4}}{1 - z^{-1}}$$

The numerator and denominator terms produce the pole/zero pattern shown in Fig. 5.24. Note that a pole/zero cancellation occurs at $z = 1$.

The previous example illustrates a general result when N is even. Let us consider a causal symmetric sequence $h_S(n)$, for $0 \le n \le N - 1$, when N is an even integer. In this case, the highest power of z^{-1} in $H_S(z)$ is $z^{-(N-1)}$. In the previous example $N = 4$. The terms in $H_S(z)$ having the same coefficient value can be grouped together as

$$H_S(z) = \sum_{n=0}^{(N/2)-1} h(n)[z^{-n} + z^{-(N-1-n)}] \tag{5.58}$$

Note that when n is even, $(N - 1 - n)$ is odd. Hence, when $z = -1$, the term in the brackets is zero. This demonstrates that when N is even, $H_S(z)$ will always have a zero at $z = -1$.

ODD SEQUENCES. Applying the same analysis as for the even sequence to an odd sequence, and recalling that $h_O(0) = 0$, we obtain

$$H_O(z) = \sum_{n=1}^{M} h_O(n)[z^{-n} - z^n] \tag{5.59}$$

If z is replaced by $1/z$, we get

$$H_O(1/z) = \sum_{n=1}^{M} h_O(n)[z^n - z^{-n}] = -H_O(z) \qquad (5.60)$$

Since multiplying the z-transform by the constant $A = -1$ does not change the locations of the singularities in the z plane, the result here is the same as the case for the even sequence: if there is a zero at $z = z_0$, then there is also one at $z = 1/z_0$. Hence, the zeros for an odd sequence also occur in complex-conjugate quadruplets.

Example 5.34. **Pole/zero pattern for an odd sequence.** Let

$$h(-3) = 1 \quad h(-2) = 2 \quad h(-1) = 3 \quad h(0) = 0 \quad h(1) = -3 \quad h(2) = -2$$
$$h(3) = -1 \quad \text{and } h(n) = 0, \text{ otherwise}$$

The z-transform is then given by

$$H(z) = z^3 + 2z^2 + 3z - 3z^{-1} - 2z^{-2} - z^{-3}$$

The locations of the zeros are at

$$z = 1, \; -1, \; 1.7e^{\pm j0.644\pi}, \; \text{and } (1/1.7)e^{\pm j0.644\pi}$$

and three poles are at $z = 0$ and three (due to the noncausal component) are at $z = \infty$, as shown Fig. 5.25.

CAUSAL ANTI-SYMMETRIC SEQUENCES. As above, an odd sequence can be converted into a causal antisymmetric sequence, denoted by $\{h_A(n)\}$, by

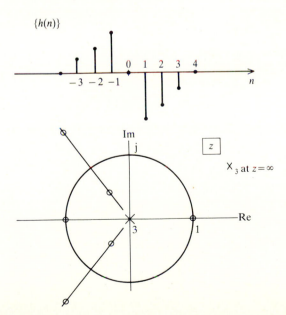

FIGURE 5-25
Pole/zero pattern for an odd-symmetric sequence.

applying sufficient delay:

$$h_A(n) = h_O(n - M) \qquad \text{for } 0 \leq n \leq 2M$$

Then

$$H_A(z) = z^{-M} H_O(z)$$

The M zeros at $z = \infty$ produced by z^{-M} cancel all the poles of the odd sequence that were there. As expected, all the poles of a causal finite-duration sequence are located at $z = 0$.

Again, we must consider the two cases when the point of antisymmetry falls on an index value and when it falls between index values. The results are exactly analogous to the causal symmetric results above. We present an example of the latter, more complicated, case.

Example 5.35. **Pole/zero pattern for an antisymmetric sequence.** Let

$$h(0) = 1 \qquad h(1) = 2 \qquad h(2) = -2 \qquad h(3) = -1 \qquad \text{and } h(n) = 0, \quad \text{otherwise}$$

The self-convolution sequence is then equal to

$$h'(0) = 1 \qquad h'(1) = 4 \qquad h'(2) = 0 \qquad h'(3) = -10 \qquad h'(4) = 0 \qquad h'(5) = 4$$
$$h'(6) = 1 \qquad \text{and } h'(n) = 0, \text{ otherwise}$$

and is shown in Fig. 5.26. The sequence $\{h'(n)\}$ is symmetric about $n = 3$. Hence, $H'(z)$ has zeros that are complex-conjugate quadruplets with an even number of zeros at each singularity location. The z-transform of $\{h'(n)\}$ is equal to

$$H'(z) = 1 + 4z^{-1} - 10z^{-3} + 4z^{-5} + z^{-6}$$

The locations of the zeros were found by finding the roots of the polynomial. The

{h(n)}

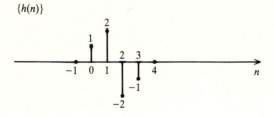

{h'(n)} = {h(n)} * {h(n)}

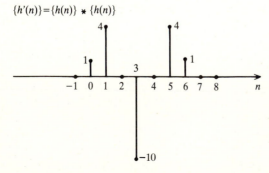

FIGURE 5-26
The self-convolution of an antisymmetrc sequence that has its point of symmetry between sampling points produces a sequence that has its point of symmetry on a sampling point.

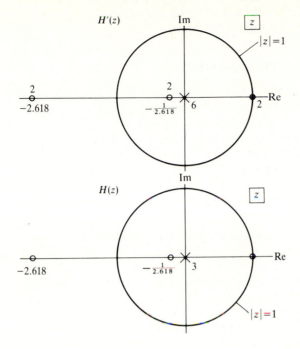

FIGURE 5-27
Pole/zero pattern for an antisymmetric sequence whose point of symmetry falls between sample points.

locations of the double zeros are at

$$z = 1 \qquad -2.618 \qquad \text{and} \qquad -1/2.168$$

and six poles are at $z = 0$, as shown in Fig. 5.27. Hence, the z-transform of the original sequence, given by

$$H_A(z) = 1 + 2z^{-1} - 2z^{-2} - z^{-3}$$

has zeros that occur in reciprocal pairs (zeros on the unit circle form their own reciprocals).

One important consequence of the above result is that we can determine by inspection of the pole/zero pattern whether a finite-duration sequence is symmetric. For the z-transform of a finite-duration symmetric sequence, the singularities in the region $0 < |z| < \infty$ are all zeros and occur in complex-conjugate reciprocal quadruplets, and all the poles lie at $z = 0$ or at $z = \infty$. Zeros at $z = 1$ and $z = -1$ form their own complex-conjugates and reciprocals.

APPLICATION TO LINEAR-PHASE FILTERS. In Chapter 3 it was shown that both symmetric and antisymmetric unit-sample responses produce linear-phase responses. The above results indicate the pole/zero patterns corresponding to symmetric or antisymmetric unit-sample responses can easily be recognized. Hence, the system function of a linear-phase filter has a pole/zero pattern that consists of zeros that occur in complex-conjugate quadruplets. The poles lie at

$z = 0$ and $z = \infty$, the number at $z = \infty$ being equal to N_F, the number of noncausal elements in the unit sample response.

5.13   THE INVERSE z-TRANSFORM

In this section, we consider the converse problem: Given $H(z)$ and the ROC, how do we determine the discrete-time sequence $\{h(n)\}$? This question is important for determining the unit-sample response of a system from its pole/zero pattern in the z plane. From the definition of the z-transform, we note that $h(n)$ is the coefficient of z^{-n}. Hence, if $H(z)$ is expressed as a polynomial in powers of z, determining the sequence is a trivial matter, since we need only to equate $h(n)$ with the coefficient of z^{-n}, as shown in the following example.

>**Example 5.36.   Inverse operation on a polynomial z-transform.** Let
>
>$$H(z) = az + b + cz^{-1}$$
>
>From the definition of the z-transform, we find by inspection that $h(-1) = a$, $h(0) = b$ and $h(1) = c$, and $h(n) = 0$ for $n \geq 2$ and $n \leq -2$.

For more complicated functions of the z-transform, three methods are usually employed to determine the values of the discrete-time sequence: long division, Taylor expansion and application of the residue theorem from complex-variable theory. The first method is the simplest for determining the first few terms of the sequence. The second and third methods are more complicated, but provide the functional form of the sequence. This functional form then allows the value of any term of the sequence to be computed and also indicates the trends that are present in the sequence. Each method will be described separately below and applied to the same z-transforms for comparison.

LONG-DIVISION METHOD. Since $h(n)$ is defined as the coefficient of z^{-n}, we need only to expand $H(z)$ into a polynomial in powers of z^{-1}. The coefficient of z^{-n} is then equal to $h(n)$. We illustrate the application of this method to various types of sequences in the following examples.

>**Example 5.37.   Long-division method for causal sequences.** Let
>
>$$H(z) = \frac{1}{1 - az^{-1}} = \frac{z}{z - a} \qquad \text{ROC: } |z| > a$$
>
>Since the ROC is outside the pole, we should obtain a right-sided sequence, i.e. a polynomial in z^{-n} for $n \geq 0$. To accomplish this, we perform the following division:
>
>$$\frac{z}{z - a} = 1 + az^{-1} + a^2z^{-2} + \cdots$$

Since $h(n)$ is the coefficient of z^{-n}, we find

$$h(n) = \begin{cases} a^n & \text{for } n \geq 0 \\ 0 & \text{otherwise} \end{cases}$$

Example 5.38. **Long-division method for noncausal sequence.** Let

$$H(z) = \frac{z^K}{1 - bz^{-1}} = \frac{z^{K+1}}{z - b} \qquad \text{ROC: } b < |z| < \infty$$

Since the ROC is outside the pole, we should obtain a right-sided sequence, but because there are K poles at $z = \infty$, the sequence starts at $n = -K$. To accomplish this, we perform the following division:

$$\frac{z^{K+1}}{z - b} = z^K + bz^{K-1} + b^2 z^{K-2} + \cdots + b^K + b^{K+1} z^{-1} + b^{K+2} z^{-2} + \cdots$$

Since $h(n)$ is the coefficient of z^{-n}, we find

$$h(n) = \begin{cases} b^{n+K} & \text{for } n \geq -K \\ 0 & \text{otherwise} \end{cases}$$

Example 5.39. **Long-division method for a second-order system.** Let

$$H(z) = \frac{1 - b^2 z^{-2}}{1 + a^2 z^{-2}} \qquad \text{ROC: } |z| > a$$

Since the ROC defines the region outside the pole, we should expect an infinite duration sequence extending for positive values of n. To obtain such a right-sided sequence, we perform the division to obtain

$$H(z) = 1 - (a^2 + b^2)z^{-2} + a^2(a^2 + b^2)z^{-4} - \cdots$$

Equating the coefficient of z^{-n} with $h(n)$, we have

$$h(0) = 1$$
$$h(1) = 0$$
$$h(2) = -(a^2 + b^2)$$
$$h(3) = 0$$
$$h(4) = a^2(a^2 + b^2)$$
$$\cdots$$

This result can be expressed analytically as

$$h(n) = \begin{cases} 1 & \text{for } n = 0 \\ (-1)^{n/2} a^{n-2}(a^2 + b^2) & \text{for } n > 0 \text{ and even} \\ 0 & \text{otherwise} \end{cases}$$

The long-division method is robust in that it can always be applied to any rational function of z. By observing several terms of the sequence the analytic form of the series can often be guessed, as in the examples above, or approximated by some simple functional form. The two methods described below provide analytic expressions for $h(n)$, from which any element of the sequence can then be determined.

TAYLOR SERIES EXPANSION METHOD. Recall that for a complex number c, where $|c| < 1$, then the Taylor series expansion is given by

$$\frac{1}{1-c} = \sum_{n=0}^{\infty} c^n \tag{5.61}$$

The trick in successfully applying this method is to express the z-transform in the appropriate form. To apply the Taylor series method, the denominator of the z-transform must be factored to find the locations of the poles. Then a partial fraction expansion can be performed to obtain the sum of first-order terms with possibly complex coefficients. The series expansion can then be calculated for each term. Since the z-transform is a linear transformation, the coefficients of the z^{-n} term from each expansion can be added to find $h(n)$. This is illustrated in the following examples.

Example 5.40. Taylor series method for determining the inverse z-transform. Let

$$H(z) = \frac{1}{1 - az^{-2}} \qquad \text{ROC: } |z| > a^{1/2}$$

In the ROC, $|az^{-2}| < 1$. Hence, the formula can be applied directly to express $H(z)$ in the following form:

$$H(z) = \sum_{n=0}^{\infty} (az^{-2})^n = \sum_{n=0}^{\infty} a^n z^{-2n}$$

Since $h(n)$ is the coefficient of z^{-n}, we can write

$$h(n) = \begin{cases} a^{n/2} & \text{for } n \geq 0 \text{ and } n \text{ even} \\ 0 & \text{otherwise} \end{cases}$$

Example 5.41. Taylor series method applied to a second-order system. Let

$$H(z) = \frac{1 - b^2 z^{-2}}{1 + a^2 z^{-2}} \qquad \text{ROC: } |z| > a$$

The z-transform above can be factored into

$$H(z) = \frac{1}{1 + a^2 z^{-2}} - \frac{b^2 z^{-2}}{1 + a^2 z^{-2}}$$

$$= \sum_{n=0}^{\infty} (-a^2 z^{-2})^n - b^2 z^{-2} \sum_{n=0}^{\infty} (-a^2 z^{-2})^n$$

Combining terms having the same exponent value, we get

$$H(z) = 1 - (a^2 + b^2)z^{-2} + a^2(a^2 + b^2)z^{-4} - a^4(a^2 + b^2)z^{-6} + \cdots$$

This is the same result produced by the long-division method.

The following example shows that a partial fraction expansion can simplify the application of the Taylor series method.

Example 5.42.   The inverse *z*-transform of a second-order system. Let

$$H(z) = \frac{r \sin(\omega_0) z^{-1}}{1 - 2r \cos(\omega_0) z^{-1} + r^2 z^{-2}}$$

Factoring the denominator into the product of two terms and then performing the partial fraction expansion, we get

$$H(z) = \frac{1}{2j}[(1 - r e^{j\omega_0}z^{-1})^{-1} - (1 - r e^{-j\omega_0}z^{-1})^{-1}]$$

Applying the Taylor series expansion to each term, we have

$$H(z) = \frac{1}{2j}\left[\sum_{n=0}^{\infty}(r e^{j\omega_0}z^{-1})^n - \sum_{n=0}^{\infty}(r e^{-j\omega_0}z^{-1})^n\right]$$

Combining the coefficients of z^{-n}, we get

$$H(z) = \frac{1}{2j}\sum_{n=0}^{\infty} r^n (e^{j\omega_0 n} - e^{-j\omega_0 n})z^{-n}$$

$$= \sum_{n=0}^{\infty} r^n \sin(\omega_0 n)z^{-n}$$

Equating the coefficient of z^{-n} with $h(n)$, we finally get

$$h(n) = \begin{cases} r^n \sin(\omega_0 n) & \text{for } n \geq 0 \\ 0 & \text{otherwise} \end{cases}$$

RESIDUE THEOREM METHOD. This method is analytically the most complicated of the three, using results from complex-variable theory. In a first-level course, this section can be skipped without loss in continuity. From complex-variable theory, it can be shown that a direct relationship exists between the *n*th element of a discrete-time sequence $\{h(n)\}$ and its *z*-transform $H(z)$. This relationship is given by

$$h(n) = \frac{1}{2\pi j}\oint_C H(z)z^{n-1}\,dz \tag{5.62}$$

where the integration is taken in a counterclockwise direction along the contour C that lies in the region of convergence of $H(z)$ and completely encloses the origin, as shown in Fig. 5.28. The direct evaluation of the above integral is usually difficult, and so a second result, called the *Residue Theorem*, is usually employed. This theorem is stated without proof, which can be found in the references.

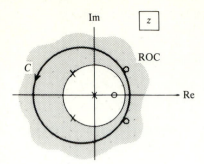

FIGURE 5-28
Contour of integration in the region of convergence
for determining the inverse z-transform.

Residue theorem. If $H(z)z^{n-1}$ has N poles, denoted by $p_1, p_2, \ldots p_N$, then

$$h(n) = \sum_{i=1}^{N} \text{Res}[H(z)z^{n-1} \quad \text{at pole } p_i] \tag{5.63}$$

where $\text{Res}[\cdot]$ is the *residue*.

To define the residue, let $H(z)z^{n-1}$ be a rational function (the ratio of two polynomials) of z, having an mth-order pole at $z = z_0$. We can then express this mth-order pole explicitly as

$$H(z)z^{n-1} = \frac{P(z)}{(z - z_0)^m} \tag{5.64}$$

where $P(z)$ does not have any poles at $z = z_0$. The residue of $H(z)z^{n-1}$ at $z = z_0$ is then defined to be

$$\text{Res}[H(z)z^{n-1} \quad \text{at } z = z_0] = \frac{1}{(m-1)!} \frac{d^{m-1}P(z)}{dz^{m-1}}\bigg|_{z=z_0} \tag{5.65}$$

If $m = 1$, there is only one pole at $z = z_0$. The evaluation then simplifies to

$$\text{Res}[H(z)z^{n-1} \quad \text{at } z = z_0] = P(z_0) \tag{5.66}$$

The following examples illustrate the application of the Residue Theorem.

Example 5.43.   Residue theorem method for a noncausal sequence. Let

$$H(z) = z + a + z^{-1}$$

The contour C is located in the ROC, which is the entire z plane except for $z = 0$ and $z = \infty$. Then

$$H(z) z^{n-1} = z^n + az^{n-1} + z^{n-2}$$

For $n = -2$

$$H(z) z^{-3} = z^{-2} + az^{-3} + z^{-4} = \frac{z^2 + az + 1}{z^4} = \frac{P(z)}{z^4}$$

There is only a fourth-order pole at $z = 0$. Evaluating the residue, we get

$$h(-2) = \frac{1}{3!} \frac{d^3 P(z)}{dz^3}\bigg|_{z=0} = 0$$

The same results would be obtained for $n < -2$.

For $n = -1$

$$H(z) z^{-2} = z^{-1} + az^{-2} + z^{-3} = \frac{z^2 + az + 1}{z^3} = \frac{P(z)}{z^3}$$

There is only a third-order pole at $z = 0$. Hence

$$h(-1) = \frac{1}{2} \frac{d^2 P(z)}{dz^2}\bigg|_{z=0} = 1$$

For $n = 0$

$$H(z) z^{-1} = 1 + az^{-1} + z^{-2} = \frac{z^2 + az + 1}{z^2} = \frac{P(z)}{z^2}$$

There is only a second-order pole at $z = 0$. Hence

$$h(0) = \frac{dP(z)}{dz}\bigg|_{z=0} = a$$

For $n = 1$

$$H(z) = z + a + z^{-1} = \frac{z^2 + az + 1}{z} = \frac{P(z)}{z}$$

There is only a first-order pole at $z = 0$ that is inside C. Hence $h(1) = P(z)|_{z=0} = 1$

For $n = 2$

$$H(z) z = z^2 + az + 1$$

There are no poles inside C. Hence $h(2) = 0$. This is also true for $n \geq 2$.

Example 5.44.   Residue theorem method for a causal sequence. Let

$$H(z) = \frac{1}{1 - az^{-1}} \qquad \text{ROC: } |z| > a$$

To enclose all the poles, the contour C is situated in the region of convergence, as shown in Fig. 5.29. Then

$$H(z) z^{n-1} = \frac{z^n}{z - a}$$

For $n > 0$, C encloses only one pole, at $z = a$. Then

$$P(z) = H(z) z^{n-1}(z - a) = z^n$$

Evaluating the residue, we get

$$h(n) = P(a) = a^n \qquad \text{for } n > 0$$

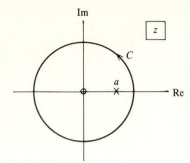

FIGURE 5-29
Contour of integration for the first-order recursive filter.

For $n < 0$, C also encloses additional poles at $z = 0$. For $n = -1$:

$$H(z)z^{n-1} = \frac{1}{z(z-a)}$$

Then,

$$h(-1) = \text{Res}[H(z)z^{n-1} \quad \text{at } z = 0] + \text{Res}[H(z)z^{n-1} \quad \text{at } z = a]$$

The residue at $z = 0$ is equal to

$$P(z) = \frac{1}{z-a} \qquad P(0) = -1/a$$

The residue at $z = a$ is equal to

$$P(z) = 1/z \qquad P(a) = 1/a$$

Adding the results of the two residues, we get

$$h(-1) = 0$$

In a similar manner, it can be shown that $h(n) = 0$ for $n < 0$.

Example 5.45.   Residue theorem method for a second-order system. Let

$$H(z) = \frac{1 - b^2 z^{-2}}{1 + a^2 z^{-2}} \qquad \text{ROC: } |z| > a$$

If $H(z)$ contains N_P poles, then

$$h(n) = \sum_{k=1}^{N_P} \text{Res}[H(z)\, z^{n-1} \quad \text{at pole } p_k]$$

We first start with

$$H(z)z^{n-1} = \frac{(z^2 - b^2)z^{n-1}}{(z+ja)(z-ja)}$$

For $n < 0$, $h(n) = 0$, because the sequence is right-sided.

For $n = 0$, $H(z)\, z^{n-1}$ has poles at $z = 0$, ja and $-ja$. Hence, the sum of the residues is

$$h(0) = \frac{z^2 - b^2}{z^2 + a^2}\bigg|_{z=0} + \frac{(z^2 - b^2)}{z(z + ja)}\bigg|_{z=ja} + \frac{(z^2 - b^2)}{z(z - ja)}\bigg|_{z=-ja}$$

$$= -\frac{b^2}{a^2} + \frac{a^2 + b^2}{2a^2} + \frac{a^2 + b^2}{2a^2} = 1$$

For $n > 1$, $H(z)\,z^{n-1}$ has poles only at $z = ja$ and $-ja$. Hence, for $n > 1$

$$h(n) = \frac{(z^2 - b^2)z^{n-1}}{(z + ja)}\bigg|_{z=ja} + \frac{(z^2 - b^2)z^{n-1}}{(z - ja)}\bigg|_{z=-ja}$$

$$= \frac{(a^2 + b^2)}{2ja}[(-ja)^{n-1} - (ja)^{n-1}] \qquad \text{for } n > 1$$

Evaluating this equation produces the same values for $\{h(n)\}$ as given in Example 5.39.

5.14 SUMMARY

In this chapter, procedures for determining the z-transform of a discrete-time sequence and for performing the inverse operation have been described. The system function of a digital filter has been defined as the z-transform of its unit-sample response. Cascade and parallel combinations of elemental digital filters can be formed by factoring the system function.

The complex z plane was defined. The configuration of the poles and zeros of the system function in the complex z plane were used to determine whether the filter is recursive or nonrecursive. Stable causal filters were found to have all their poles inside the unit circle, while stable noncausal filters also had poles located at $z = \infty$. The region of convergence (ROC) was employed to determine the type of the discrete-time sequence, either causal or noncausal, finite-duration or infinite-duration. The Fourier transform was shown to be the evaluation of the z-transform along the unit circle in the z plane. Three methods to determine the discrete-time sequence from the z-transform were also presented.

FURTHER READING

Antoniou, Andreas: *Digital Filters: Analysis and Design*, McGraw-Hill, New York, 1979.

Hamming, R. W.: *Digital Filters*, 2d ed., Prentice-Hall, Englewood Cliffs, NJ, 1983.

Jong, M. T.: *Methods of Discrete Signal and System Analysis*, McGraw-Hill, New York, 1982.

Jury, E. I.: *Theory and Application of the z-Transform Method*, Wiley, New York, 1964.

Oppenheim, Alan V., and Ronald W. Schafer: *Digital Signal Processing*, Prentice-Hall, Englewood Cliffs, NJ, 1975.

Peled, A., and B. Liu: *Digital Signal Processing: Theory, Design and Implementation*, Wiley, New York, 1976.

Rabiner, Lawrence R., and Bernard Gold: *Theory and Application of Digital Signal Processing*, Prentice-Hall, Englewood Cliffs, NJ, 1975.

REFERENCE TO TOPIC

Complex variables

Churchill, R. V.: *Complex Variables and Applications*, McGraw-Hill, New York, 1960.

PROBLEMS

5.1. Given the following unit-sample response sequences, compute the z-transform, sketch the pole/zero pattern in the z plane and the filter block diagram.
(a) $h(-1) = 1$, $h(1) = 1$, $h(n) = 0$, otherwise
(b) $h(0) = 1$, $h(1) = -1$, $h(n) = 0$, otherwise
(c) $h(-1) = 1$, $h(0) = -2$, $h(1) = 1$, $h(n) = 0$, otherwise
(d) $h(n) = \begin{cases} r^n \sin(\omega_0 n), & \text{for } n \geq 0, \\ 0 & \text{otherwise.} \end{cases}$ Let $\omega_0 = \pi/8$ and $r = 0.9$

5.2. For the following difference equations, compute the z-transform of $\{h(n)\}$, sketch the pole/zero pattern in the z plane and the filter block diagram.
(a) $y(n) = x(n) - x(n - N)$. Let $N = 5$
(b) $y(n) = x(n) + x(n - N)$. Let $N = 5$
(c) $y(n) = 2r \cos(\theta) y(n - 1) + r^2 y(n - 2) + x(n)$. Let $\theta = \pi/4$ and $r = 0.9$

5.3. For the digital filter structures shown in Fig. P2.11, compute the system function and sketch the pole/zero pattern in the z plane.

5.4. For the following magnitude and phase functions, compute the system function and sketch the pole/zero pattern in the z plane and the filter block diagram.
(a) $|H(e^{j\omega})| = 2 \sin(\omega/2)$ for $0 \leq \omega \leq 2\pi$

$\quad \text{Arg}[H(e^{j\omega})] = \pi/2 - \omega/2$ for $0 < \omega < 2\pi$

(b) $|H(e^{j\omega})| = [5 + 4 \cos(2\omega)]^{1/2}$,

$\quad \text{Arg}[H(e^{j\omega})] = \arctan\left[\dfrac{-\sin(2\omega)}{2 + \cos(2\omega)}\right]$, $\text{Arg}[H(e^{j0})] = 0$

(c) $|H(e^{j\omega})| = [17 - 8 \cos(2\omega)]^{1/2}$,

$\quad \text{Arg}[H(e^{j\omega})] = \arctan\left[\dfrac{4 \sin(2\omega)}{1 - 4 \cos(2\omega)}\right]$, $\text{Arg}[H(e^{j0})] = 0$

(d) $|H(e^{j\omega})| = [1.81 - 1.8 \cos(\omega)]^{1/2}$,

$\quad \text{Arg}[H(e^{j\omega})] = \omega + \arctan\left[\dfrac{0.9 \sin(\omega)}{1 - 0.9 \cos(\omega)}\right]$, $\text{Arg}[H(e^{j0})] = 0$

5.5. For the z plane pole/zero patterns shown in Fig. P5.5, compute the z-transform, sketch the filter block diagram, determine the transfer function and sketch the unit-sample response sequence. Let the filter gain $A = 1$.

5.6. Determine the pole locations of the systems having the following unit-sample response and determine whether they are stable.

(a) $h(n) = \begin{cases} (-1)^n & \text{for } 0 \leq n \leq N - 1, \text{ for any finite value of } N \\ 0 & \text{otherwise} \end{cases}$

(b) $h(n) = \begin{cases} (-1)^n & \text{for } n \geq 0 \\ 0 & \text{otherwise} \end{cases}$

(a)

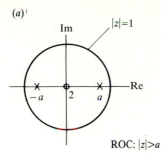

ROC: $|z|>a$

(b)

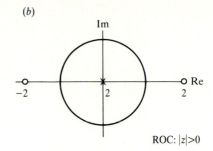

ROC: $|z|>0$

(c)

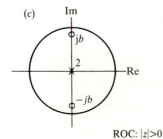

ROC: $|z|>0$

(d)

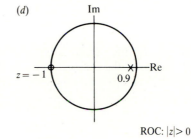

ROC: $|z|> 0$

(e)

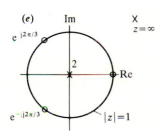

ROC: $0<|z|<\infty$

(f)

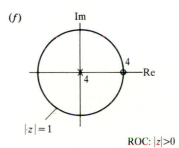

ROC: $|z|>0$

FIGURE P5.5
Pole/zero patterns.

5.7. Determine the possible values of the phase response at $\omega = 0$. Indicate the pole/zero pattern that can cause it. Repeat for $\omega = \pi$.

5.8. Demonstrate that the z-transform of any antisymmetric sequence has a zero at $z = 1$.

5.9. Consider the z-transform given by

$$X(z) = \sin(1/z)$$

Find a sequence $\{x(n)\}$ that has this z-transform by performing a Taylor series expansion.

5.10. Consider the sequence $\{x(n)\}$ and its z-transform $X(z)$.

(a) Find the z-transform of the sequence

$$\{y(n)\} = \{\alpha^n x(n)\} \qquad \text{where } \alpha \text{ is a positive real number}$$

in terms of $X(z)$.

(b) If $X(z)$ has a pole inside the unit circle at radius r, what is the value of α, such that the pole of $Y(z)$ falls on the unit circle?

5.11. Consider the filter with unit-sample response $\{h(n)\}$ and input $\{x(n)\}$ given by

$$h(n) = \begin{cases} 1 & \text{for } n = 0 \\ -1 & \text{for } n = 8 \\ 0 & \text{otherwise} \end{cases}$$

and

$$x(n) = 1 \qquad \text{for } n \geq 0$$
$$= 0 \qquad \text{otherwise}$$

(a) Determine the output $\{y(n)\}$ by performing the discrete-time convolution.
(b) Determine the output $\{y(n)\}$ from the inverse z-transform $Y(z) = H(z) X(z)$. Note that the Fourier transform of $\{x(n)\}$ does not exist.

5.12. Consider the system function given by

$$H(z) = \frac{1 - \rho^{-1} z^{-1}}{1 - \rho z^{-1}} \qquad \text{for } 0 < \rho < 1$$

(a) Draw the pole/zero pattern of this system function.
(b) Show that $|H(e^{j\omega})|$ has a constant value for all ω.

5.13. Consider a second-order filter having a finite-duration unit-sample response. Let its magnitude response $|H(e^{j\omega})|$, over $-\pi \leq \omega < \pi$, equal zero at two frequency points $\omega = \omega_0$ and $\omega = -\omega_0$.

(a) Show that the magnitude response for all frequencies can be determined to within a gain constant from this information.
(b) Draw the pole/zero pattern for the filter that has both poles at $z = 0$.

COMPUTER PROJECTS

5.1. Determining the filter coefficients from the z plane singularity location
Object. To determine the filter coefficients and the time-domain sequence from the singularities in the z plane.

Write the two subroutines described in this chapter given by

        ZERO(R,DEG,BCOEF,NB)

and

        POLE(R,DEG,ACOEF,NA)

Verify the operation of your subroutines by generating the coefficients for the following causal filters. For each filter provide the following:
(i) the sketch of the block diagram of the digital filter
(ii) the sketch of the z plane pole/zero pattern

(iii) the computer graph of the unit-sample response
(iv) the computer graphs of the first 65 points of the magnitude and phase response computed from the 128-point DFT.

The filters are:

(a) the causal 3-sample averager
(b) the first-order recursive filter, with $a = -0.9$,
(c) the second-order filter described by

$$y(n) = 2r \cos(\omega_0) y(n-1) - r^2 y(n-2) + x(n) - r \cos(\omega_0) x(n-1)$$

with $\omega_0 = \pi/10$ and $r = 0.95$.

Generate 128 points of $\{h(n)\}$, but plot only the first 55.

5.2. Magnitude and phase response curves for simple poles and zeros

Object. To determine the magnitude and phase response curves for simple poles and zeros in the z plane.

Analytic results

(a) Consider a filter having a real zero located at $z = r_0$ and a pole at $z = 0$. Show that the magnitude response of this filter is identical to within a gain constant to that of the filter having a real zero at $z = 1/r_0$ and pole at $z = 0$. Repeat the analysis for a pair of complex-conjugate zeros located at $z = r_0 e^{\pm j\omega_0}$ and two poles at $z = 0$.

(b) Determine the magnitude response of a filter having a zero at $z = r_0$ and a pole at $z = (1/r_0)$, where $|r_0| > 1$.

Computer results

Use ZERO and POLE to determine the filter coefficients for the following singularity locations. Use NONREC or REC to determine the unit-sample response (since all the filters are causal, use the unit-sample sequence without any delay). Use FFT to find the 128-sample Fourier transform, followed by MAGPHS to find the magnitude and phase response over the 65 points covering $0 \le \omega \le \pi$. Use PLOTMR and PLOTPR to plot the curves.

For each part below include the following:
sketch of the filter block diagram
plot of the unit-sample response
analytic form of the transfer function and system function
sketch of the pole/zero pattern in the z plane
the magnitude and phase response plots

(1) Real zero at $z = \frac{3}{4}$ and pole at $z = 0$. Repeat for zero at $z = \frac{4}{3}$. Explain the difference in the two phase responses.

(2) Complex-conjugate zeros at $z = 0.9 e^{\pm j\pi/8}$ and two poles at $z = 0$. Repeat for zeros at $z = 1/(0.9 e^{\pm j\pi/8})$ and two poles at $z = 0$. Explain the difference in the two phase responses.

(3) Real pole at $z = -0.9$ and zero at $z = 0$. Repeat by moving the zero from $z = 0$ to $z = -1/0.9$.

(4) Complex-conjugate poles at $z = 0.95 e^{\pm j\pi/8}$ and two zeros at $z = 0$. Repeat by moving one of the zeros from $z = 0$ to $z = 0.95 \cos(\pi/8)$.

CHAPTER
6

DIGITAL
FILTER
STRUCTURES

6.1 INTRODUCTION

In this chapter, we describe the most commonly employed methods to implement a digital filter structure from either its difference equation, its unit-sample response, or its z-transform. We start by defining the terms conventionally used for describing digital filters and relating them to the system function coefficients. The digital filter structure determined directly from either the difference equation or the system function is called the *Direct Form I*. An alternate view of the same equations results in the memory-efficient structure, called the *Direct Form II*. Digital filters structured as cascade and parallel combinations of second-order components are shown to have advantages in terms of hardware implementation and numerical sensitivity. The above methods are employed to implement the general filter structure, one that contains both poles and zeros. An *all-zero filter* having a linear-phase response is shown to have a special nonrecursive structure that reduces the number of multiplications by approximately one-half. The *frequency-sampling structure* is presented as an alternative to the tapped-delay line structure for all-zero filters.

6.2 SYSTEM-DESCRIBING EQUATIONS

The equations that describe the input/output relations in the time and z-transform domains have been defined in the previous chapters. They are repeated here for reference.

In the time domain, the convolution equation is given by

$$y(n) = \sum_{k=-\infty}^{\infty} h(k)\, x(n-k) \tag{6.1}$$

and the difference equation is given by

$$y(n) = \sum_{k=1}^{M} a_k\, y(n-k) + \sum_{k=-N_F}^{N_P} b_k\, x(n-k) \tag{6.2}$$

In the z-transform domain, the system function can be expressed in two useful forms. The first is the sum-of-products form given by

$$H(z) = \sum_{k=-N_F}^{N_P} b_k\, z^{-k} \bigg/ 1 - \sum_{k=1}^{M} a_k\, z^{-k} \tag{6.3}$$

The second is the product-of-terms form given by

$$H(z) = A z^{N_F} \prod_{k=1}^{N_P+N_F} (1 - c_k z^{-1}) \bigg/ \prod_{k=1}^{M} (1 - d_k z^{-1}) \tag{6.4}$$

We describe several different structures that can be used to implement the digital filter from any one of these defining equations. First, we discuss the terms commonly used for classifying the filter types.

6.3 FILTER CATEGORIES

Digital filters are often categorized either by the duration of their unit-sample response or by their structure. When a filter produces a unit-sample response that has an infinite duration, it is called an *infinite-impulse response (IIR) filter*. This term originated when it was popular to refer to the unit-sample response as the impulse response, the conventional term for analog filters. If the digital filter has a unit-sample response having a finite duration, then it is called a *finite-impulse response (FIR) filter*.

Digital filter classification can also be based on the filter structure. In general, the output of a filter can be a function of future, current and past input values, as well as past output values. If the output is a function of the past outputs, then a feedback, or *recursive*, mechanism from the output must exist. Hence, such a filter is referred to as a *recursive filter*. A recursive filter can be recognized from the defining equations, since at least one a_k coefficient, for $1 \le k \le M$, in Eqs. (6.2) and (6.3) is nonzero, and at least one d_k coefficient, for $1 \le k \le M$, in Eq. (6.4) is nonzero.

If the filter output value is a function of only the input sequence values, it is called a *nonrecursive filter*. A nonrecursive filter can be easily recognized from the defining equations, for which $a_k = 0$ in Eqs. (6.2) and (6.3), or $d_k = 0$ in Eq. (6.4), for all k. As shown in the previous chapter, all the poles of a nonrecursive filter are at either $z = 0$ or $z = \infty$. In the geometric evaluation of the transfer function, they did not contribute to the magnitude response. Hence, they are said not to be *relevant*, and the nonrecursive filter is also called an *all-zero* filter. The poles at $z = 0$, however, do contribute to the phase response, by introducing linear phase.

To achieve an infinite-duration unit-sample response, some form of recursive structure is necessary. Hence, the terms IIR and recursive are commonly accepted as being interchangeable. Similarly, finite-duration unit-sample responses are typically implemented with nonrecursive filters. Hence, FIR and nonrecursive are also usually interchangeable. An exception to this latter case is the frequency-sampling structure for FIR filters, described later in this chapter. In this filter, a recursive structure is employed to generate a finite-duration unit-sample response.

Several additional terms, used mostly in signal modeling, statistics and system identification, are *autoregressive (AR)*, *moving-average (MA)* and *autoregressive/moving-average (ARMA)*. The FIR filter can be viewed as an MA system, while the general IIR filter, having both poles and zeros, is an ARMA system. An AR system is an IIR filter whose system function has all its zeros at $z = 0$, that is $b_k = 0$ for $k \neq 0$ and $c_k = 0$ for all k. An AR filter is also commonly called an *all-pole filter*, since all the zeros are located at $z = 0$, and hence are not relevant, in that they do not contribute to the filter magnitude response.

6.4 THE DIRECT FORM I AND II STRUCTURES

The digital filter structures implemented with the procedures described in the previous chapters provide separate delay buffers for both the input and output sequences, as shown in Fig. 6.1. This implementation is called the *Direct Form*

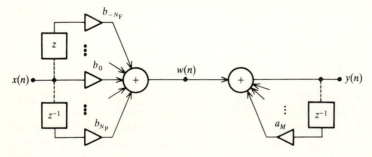

FIGURE 6-1
Direct Form I implementation of a digital filter.

I. Since it shows explicitly the delayed values of the input and output, the Direct Form I structure is a convenient way to implement a digital filter with a computer program. One subroutine (NONREC) is used to implement the zeros of the system function, followed by a second (REC) to implement the poles. However, this form does not make the most efficient use of memory. A second structure that performs the same transformation but uses fewer memory elements is considered next.

   This alternate structure is most easily achieved by considering the system function expression given in Eq. (6.3). The system function is then divided into two parts connected in cascade, the first part containing only the zeros, followed by the part containing only the poles. As shown in Fig. 6.1, we can introduce an intermediate sequence $\{w(n)\}$ representing the output of the first part and the input to the second. Hence, we write

$$w(n) = \sum_{k=-N_F}^{N_P} b_k x(n - k) \tag{6.5}$$

and

$$y(n) = \sum_{k=1}^{M} a_k y(n - k) + w(n) \tag{6.6}$$

In the z-transform domain, the equivalent relationships are given by

$$W(z) = X(z) \sum_{k=-N_F}^{N_P} b_k z^{-k} \tag{6.7}$$

and

$$Y(z) = W(z) \bigg/ 1 - \sum_{k=1}^{M} a_k z^{-k} \tag{6.8}$$

Since we are dealing with linear systems, the order of these two parts can be interchanged, as shown in Fig. 6.2(*a*), to generate the same overall system function, but through a different intermediate sequence $\{u(n)\}$. The equations for this second system are then given in the time-domain by

$$u(n) = \sum_{k=1}^{M} a_k u(n - k) + x(n) \tag{6.9}$$

and

$$y(n) = \sum_{k=-N_F}^{N_P} b_k u(n - k) \tag{6.10}$$

In the z-transform domain, we have

$$U(z) = X(z) \bigg/ 1 - \sum_{k=1}^{M} a_k z^{-k} \tag{6.11}$$

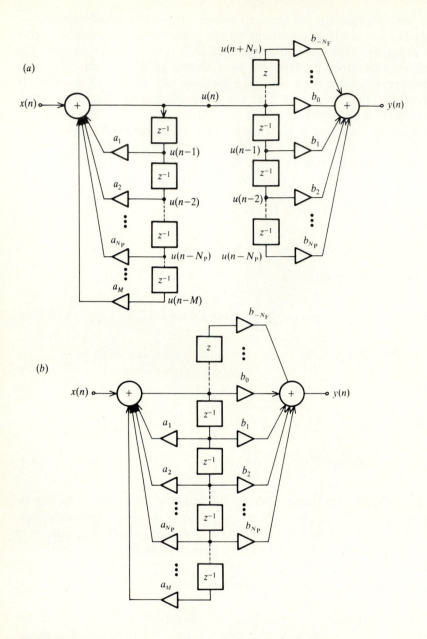

FIGURE 6-2
Two step procedure to obtain the Direct Form II digital filter structure. a) Interchange recursive and nonrecursive parts. b) Combine common delays.

and

$$Y(z) = U(z) \sum_{k=-N_F}^{N_P} b_k z^{-k} \qquad (6.12)$$

By examining the outputs of the delays in Fig. 6.2(a), we note that their contents are identical. Hence, we can eliminate one set of delays, as indicated in Fig. 6.2(b), to obtain a filter structure that uses fewer delay elements. This filter structure, called the *Direct Form II*, uses a smaller number of delay elements, equal to the maximum of M or $N_F + N_P$, than the Direct Form I, which uses $M + N_P + N_F$ elements. Since we need K delays to implement a system function term containing z^{-K}, *the Direct Form II uses the minimum number of delays required to implement a given system function*. It is therefore said to be a *canonical structure*.

Example 6.1. Direct Form II second-order filter implementation. Let

$$y(n) = 2r \cos(\omega_0) y(n-1) - r^2 y(n-2) + x(n) - r \cos(\omega_0) x(n-1)$$

The corresponding system function is given by

$$H(z) = \frac{1 - r \cos(\omega_0) z^{-1}}{1 - 2r \cos(\omega_0) z^{-1} + r^2 z^{-2}} = \frac{Y(z)}{X(z)}$$

The Direct Form I structure is obtained by factoring the system function to give

$$Y(z) = H_1(z) H_2(z) X(z)$$

where $H_1(z)$ is the numerator of $H(z)$ and $H_2(z)$ is the remainder. The intermediate sequence $\{w(n)\}$, having z-transform $W(z)$, is introduced between these two sections as

$$H_1(z) = 1 - r \cos(\omega_0) z^{-1} = \frac{W(z)}{X(z)}$$

and

$$H_2(z) = (1 - 2r \cos(\omega_0) z^{-1} + r^2 z^{-2})^{-1} = \frac{Y(z)}{W(z)}$$

The order of the component filters is reversed to give

$$Y(z) = H_2(z) H_1(z) X(z)$$

A second intermediate sequence $\{u(n)\}$, with z-transform $U(z)$, is introduced between these two sections to give

$$H_2(z) = (1 - 2r \cos(\omega_0) z^{-1} + r^2 z^{-2})^{-1} = \frac{U(z)}{X(z)}$$

and

$$H_1(z) = 1 - r \cos(\omega_0) z^{-1} = \frac{Y(z)}{U(z)}$$

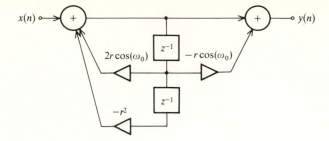

FIGURE 6-3
Implementation of second-order filter as a Direct Form II structure.

These latter two systems are implemented as two Direct Form I filters and the common delayed values of $\{u(n)\}$ are combined to eliminate one set of delays, as shown in Fig. 6.3. The result is the Direct Form II structure implementation of the system function $H(z)$.

The Direct Forms I and II are commonly employed to implement digital filters, both in hardware and in programs. However, they have two shortcomings. First, they lack hardware flexibility, in that filters of different orders, having a different number of multipliers and delay elements, require different hardware structures. Second, the sensitivity of the coefficients to quantization effects that occur when using finite-precision arithmetic, described in Chapter 10, increases with the order of the filter. This sensitivity is especially important when generating poles, since small deviations in the coefficient values may significantly change the form of the frequency response, or even cause the filter to become unstable, as shown in Chapter 10. To overcome these effects, we will next consider implementing a digital filter by combining filter building blocks consisting of second-order sections, either in cascade or in parallel. The cascade form has the advantage that the overall filter system function is easier to determine. Being the product of the components, the overall system function has the same zeros and poles as the individual components. The parallel implementation does not have this property with its zeros, but does provide an easier control of the dynamic range of the signal values. In the cascade form, intermediate results can exceed the range of the number system in some cases. The parallel implementation also allows the filter operation to be performed in parallel. This feature is important for high-speed filtering applications, in which a separate computer can be used to implement each parallel structure and the processing can be performed simultaneously.

6.5 CASCADE COMBINATION OF SECOND-ORDER SECTIONS

To obtain a cascade implementation, the system function $H(z)$ is factored into a product of second-order terms in both the numerator and denominator, after the noncausal part is factored out. If M, the number of delays in the denominator, is greater than $N_P + N_F$, the number in the numerator, we can write

$$H(z) = Az^{N_F} \prod_{k=1}^{[M/2]} H_k(z) \qquad (6.13)$$

where

$$H_k(z) = \frac{1 + b_{1k}z^{-1} + b_{2k}z^{-2}}{1 - a_{1k}z^{-1} - a_{2k}z^{-2}} \qquad (6.14)$$

The constant A is a gain term, z^{N_F} is the noncausal advance component and $[M/2]$ is the integer obtained by rounding up $M/2$ to the next integer value. For example, if $M = 5$, then $[M/2] = 3$. If $N_P + N_F > M$, then the upper limit would be $[(N_P + N_F)/2]$.

For the filter coefficients to be real, two complex-conjugate zeros of the system function $H(z)$ are combined to form each numerator term and two complex-conjugate poles are combined for each denominator term. Two real-valued singularities can also be combined to form a second-order term as well. If the number of poles or zeros is odd, the remaining single singularity is implemented by making $b_{2k} = 0$ or $a_{2k} = 0$ for one of the sections. Each $H_k(z)$ term describes a *cascade second-order section (CSOS)*, that can be efficiently implemented using the Direct Form II, as shown in Fig. 6.4. The defining equations for the kth CSOS are given by

$$y_k(n) = a_{1k}\,y_k(n-1) + a_{2k}\,y_k(n-2) + x_k(n) + b_{1k}\,x_k(n-1) + b_{2k}\,x_k(n-2) \qquad (6.15)$$

in the time domain, and in the z domain by

$$Y_k(z) = a_{1k}z^{-1}\,Y_k(z) + a_{2k}z^{-2}\,Y_k(z) + X_k(z) + b_{1k}z^{-1}\,X_k(z) + b_{2k}z^{-2}\,X_k(z) \qquad (6.16)$$

Configuring a digital filter in this cascade form allows us to put all the

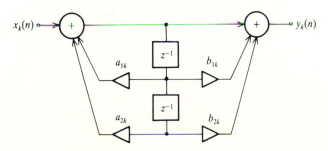

FIGURE 6-4
Cascade second-order section (CSOS) structure.

design effort into an efficient implementation of a single second-order section in hardware. Then, to implement any desired filter, the required number of these identical sections are simply cascaded. The multiplier coefficient values for a particular filter design can be stored and accessed from a read-only memory (ROM). Modifications in the filter design can then easily be accomplished by changing the contents of this ROM.

Example 6.2. Implementing a filter with CSOS components. Let

$$H(z) = \frac{z/3 + 5/12 + 5z^{-1}/12 + z^{-2}/12}{1 - z^{-1}/2 + z^{-2}/4}$$

This system function can be factored into

$$H(z) = \frac{z/3(1 + z^{-1}/4)(1 + z^{-1} + z^{-2})}{1 - z^{-1}/2 + z^{-2}/4}$$

The filter can be implemented in two ways, depending on which numerator term is combined with the denominator to form a CSOS. The first is

$$H(z) = \frac{1}{3} z(1 + z^{-1}/4) \frac{1 + z^{-1} + z^{-2}}{1 - z^{-1}/2 + z^{-2}/4}$$

which is shown in Fig. 6.5(a). The second is

$$H(z) = \tfrac{1}{3} z(1 + z^{-1} + z^{-2}) \frac{1 + z^{-1}/4}{1 - z^{-1}/2 + z^{-2}/4}$$

which is shown in Fig. 6.5(b).

In theory, there is no particular reason to prefer one structure over the other. In practice, however, the ordering is important when the limited dynamic range or the limited precision of the number system of the hardware is considered. These topics are discussed in Chapter 10.

HARDWARE SAVINGS WITH HIGH-SPEED CIRCUITS. Implementing a filter with CSOS components can also lead to savings in the amount of hardware required. Four multiplications and additions are performed to execute the general CSOS operation. Let the time for this procedure take T_P seconds. With current special-purpose signal processing chips, in which multiplication times are 200 nanoseconds, typical values for T_P are approximately 1 microsecond. If a new sample arrives every T_S seconds, *real-time operation,* in which the required processing is completed before the arrival of the next sample, occurs when $T_P < T_S$. To implement a high-order filter, we could use a large number of identical CSOS hardware sections.

If T_P is much smaller than T_S, the hardware performing the processing is very fast and is sitting idle most of the time. In this case, an alternate

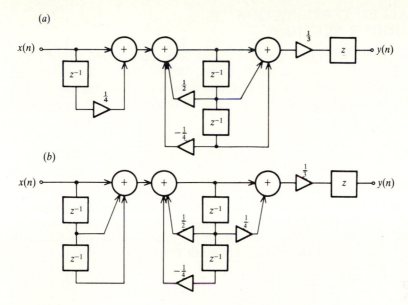

FIGURE 6-5
Two implementations of a digital filter using CSOS structures.

implementation is to have only one CSOS hardware structure that is used repeatedly. If $T_P \ll T_S$, the input data can be processed by the structure that has coefficients (obtained from a ROM) corresponding to the first section of the system function. The output of this section and the contents of the delays are then stored in a read/write memory (RAM). The output of this section then becomes the input for the next section of the system function. This second section is implemented by the *same* hardware CSOS with the appropriate coefficients read from ROM and whose delay contents are read from RAM. This procedure is repeated until the required number of sections have been employed to implement the filter. The process is again initiated upon the arrival of the next sample.

> **Example 6.3. Processing telephone-quality speech in real time.** Telephone-quality speech is defined to have frequency components between 300 and 3500 Hz. The sampling rate commonly employed is approximately 8 kHz, or $T_S = 125\ \mu s$. If the filter is implemented as a cascade of $[M/2]$ CSOS-sections, then for real-time operation $[M/2]$ must be less than T_S/T_P. When $T_P < 1\ \mu s$, T_S/T_P is greater than 100. Typical values for the filter order M rarely exceed 12. Hence, real-time operation with a single CSOS structure is readily achieved.
>
> When $T_S/T_P \gg [M/2]$, as above, then not only can the same CSOS hardware structure be used to implement the required filter, but it is fast enough to process the data on additional voice channels. In the case above, as many as 16 voice channels can be accommodated with a single CSOS structure.

6.6 PARALLEL COMBINATION OF SECOND-ORDER SECTIONS

An alternate configuration that is useful for implementing digital filters is the combination of *parallel second-order sections* (PSOS). This structure will be useful for the implementation of the frequency-sampling FIR filter structure, described later in this chapter. If $M \geqslant N_P + N_F$, then $H(z)$ can be factored into

$$H(z) = z^{N_F} \sum_{k=1}^{[M/2]} G_k(z) \tag{6.17}$$

where

$$G_k(z) = \frac{g_{0k} + g_{1k}z^{-1}}{1 - a_{1k}z^{-1} - a_{2k}z^{-2}}$$

and $[\cdot]$ indicates the next largest integer value. This filter configuration is shown in Fig. 6.6. For $N_P + N_F > M$, the upper limit of the sum becomes $[(N_P + N_F)/2]$. The defining equations for the kth PSOS filter are given by

$$y_k(n) = a_{1k} y_k(n-1) + a_{2k} y_k(n-2) + g_{0k} x_k(n) + g_{1k} x_k(n-1) \tag{6.18}$$

in the time domain, and in the z domain by

$$Y_k(z) = a_{1k}z^{-1} Y_k(z) + a_{2k}z^{-2} Y_k(z) + g_{0k} X_k(z) + g_{1k}z^{-1} X_k(z) \tag{6.19}$$

Since the input sequence $\{x(n)\}$ is applied simultaneously to all the PSOS sections, a scale factor is necessary for each section. This scale factor is included in the numerator coefficients g_{0k} and g_{1k}. The poles in the PSOS system function Eq. (6.17) are the same as those of the CSOS given in Eq. (6.14). However, there is only one zero in each PSOS, as opposed to a pair of complex-conjugate zeros in the CSOS. This occurs because additional zeros are generated in the total system function in making the parallel connections, as shown in the following example. In general, determining the locations of the zeros of a parallel combination of filters is not a simple problem.

Example 6.4. Parallel combination of two first-order IIR filters. Let

$$H_1(z) = \frac{1}{1 - az^{-1}} + \frac{1}{1 - bz^{-1}}$$

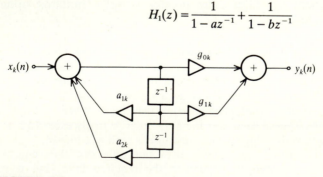

FIGURE 6-6
Parallel second-order section (PSOS) structure.

Combining over a common denominator, we have

$$H_2(z) = \frac{2 - (a + b)z^{-1}}{(1 - az^{-1})(1 - bz^{-1})}$$

$H_2(z)$ contains the same poles as the individual terms of $H_1(z)$, but there is a zero at $z = (a + b)/2$ that is explicitly expressed in $H_2(z)$. Of course, this zero is also present in $H_1(z)$, but is not obvious by inspection.

Example 6.5. Parallel structure from system function. Let us consider the implementation of the system function given in Example 6.2 as a combination of PSOS structures. Let

$$H(z) = \frac{z/3 + 5/12 + 5z^{-1}/12 + z^{-2}/12}{1 - z^{-1}/2 + z^{-2}/4}$$

This can be factored into the following

$$H(z) = \frac{z(1/3 + 5z^{-1}/12 + 5z^{-2}/12 + z^{-3}/12)}{1 - z^{-1}/2 + z^{-2}/4}$$

To obtain the form of $H(z)$ given in Eq. (6.17), we divide the denominator into the numerator, to get

$$H(z) = z\left[\frac{z^{-1}}{3} + \frac{7}{3} + \frac{5z^{-1}/4 - 2}{1 - z^{-1}/2 + z^{-2}/4}\right].$$

The PSOS implementation is shown in Fig. 6.7.

The above cascade and parallel forms can be employed to implement both nonrecursive and recursive filters. For filters having finite-duration unit-sample responses, alternate structures, described in the next section, also have some advantages.

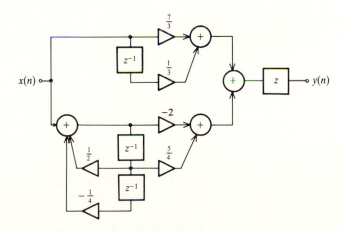

FIGURE 6-7
Digital filter implementation using PSOS structures.

6.7 LINEAR-PHASE FIR FILTER STRUCTURES

As shown in Chapter 3, for a digital filter to have linear phase, its unit-sample response sequence must be either symmetric or antisymmetric about some point in time. Since the FIR filter coefficients are identical to the elements of the unit-sample response, pairs of multipliers have coefficients with identical magnitudes. To save a multiplication operation, the inputs to these multipliers can be combined and then the product computed. Doing this reduces the number of multiplications by approximately one-half. Since performing a multiplication is the most time-consuming operation in a digital filter, this reduction in computation time is significant, possibly permitting real-time operation to occur.

To illustrate this procedure, let us consider a noncausal FIR filter having an even unit-sample response extending over $-M \le n \le M$. Then its system function is given by

$$H(z) = \sum_{n=-M}^{M} h(n) z^{-n} \tag{6.20}$$

The Direct Form I implementation of this filter shown in Fig. 6.8(a) requires $2M + 1$ multiplications. For an even unit-sample response, the equal-valued coefficients can be combined to give

$$H(z) = h(0) + \sum_{n=1}^{M} h(n) [z^n + z^{-n}] \tag{6.21}$$

This last equation illustrates that the coefficient $h(n)$ can be applied to the sum of the outputs of two delay elements, as shown in Fig. 6.8(b). This requires only $M + 1$ multiplications to accomplish the same result as before.

Example 6.6.   Linear-phase FIR structure. Let

$$H(z) = z/2 + 1 + z^{-1}/2$$

This linear interpolator is implemented in the direct form structure in Fig. 6.9. To exploit the even symmetry of the unit-sample response, we can write

$$H(z) = 1 + [z + z^{-1}]/2$$

and implement this filter with only one multiplication, as shown.

The same procedure can be applied to a causal FIR filter. The causal filter is first implemented as a Direct Form I structure. The contents of the delays leading to multipliers having the same coefficient magnitudes are then routed to summing junctions before the multiplication, as in the next example.

Example 6.7.   Causal linear-phase FIR filter structure. Let

$$H(z) = 1/2 + z^{-1} + z^{-2}/2$$

Combining the outputs of the delays that lead to multipliers having the same

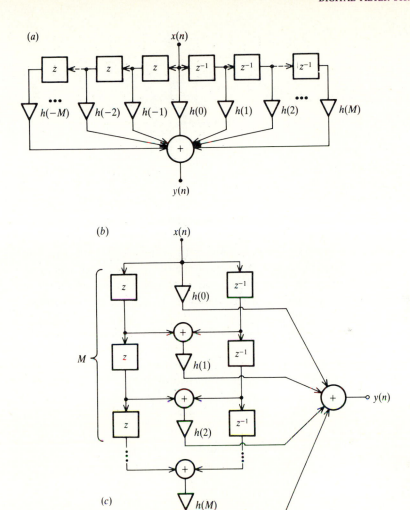

FIGURE 6-8
(*a*) Nonrecursive Direct Form I structure of a noncausal FIR filter; (*b*) efficient noncausal linear-phase FIR structure that reduces the multiplications by approximately one-half.

coefficient magnitudes, we can express the system function as

$$H(z) = z^{-1} + [1 + z^{-2}]/2$$

These equivalent structures are shown in Fig. 6.10.

6.8 FREQUENCY-SAMPLING STRUCTURE FOR THE FIR FILTER

In this section, we consider an alternate method of implementing an FIR filter. This structure is interesting in that it employs the cascade combination of a

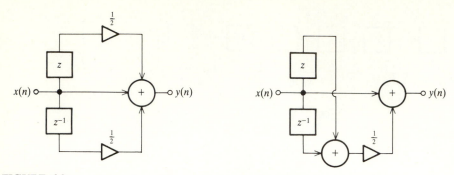

FIGURE 6-9
Two equivalent structures for a noncausal linear-phase FIR filter.

nonrecursive filter and a bank of recursive filters to generate a finite-duration unit-sample response. Since the unit-sample response of the final filter has finite duration, it is by definition an FIR filter, even though it uses a recursive structure. To illustrate the approach, we consider first the implementation of simple first- and second-order filter sections and then combine these to form more complicated filters. This approach is then generalized to be applicable to any FIR filter.

Let us consider the following FIR filter having a causal finite-duration unit-sample response containing N elements of constant value, or

$$h_0(n) = \begin{cases} G_0/N & \text{for } 0 \le n \le N-1 \\ 0 & \text{otherwise} \end{cases} \tag{6.22}$$

The corresponding filter system function is equal to

$$H_0(z) = \frac{G_0}{N} \sum_{n=0}^{N-1} z^{-n} \tag{6.23}$$

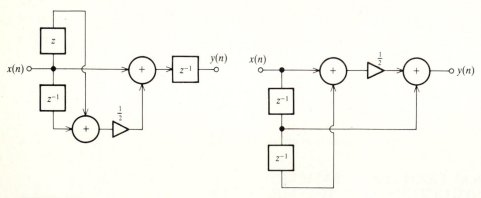

FIGURE 6-10
Two equivalent structures for a causal linear-phase FIR filter.

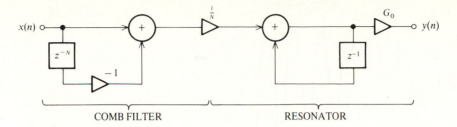

FIGURE 6-11
Comb filter followed by a resonator to implement an FIR filter.

Using the finite geometric sum formula, this sum can be evaluated to give

$$H_0(z) = \frac{G_0}{N} \frac{1 - z^{-N}}{1 - z^{-1}} \tag{6.24}$$

This analytic form of the system function suggests a novel way to implement the above filter: as the two-stage cascade structure shown in Fig. 6.11. The system function of this cascade combination can be written as

$$H_0(z) = H_C(z) H_{R,0}(z) \tag{6.25}$$

where

$$H_C(z) = \frac{1 - z^{-N}}{N}$$

and

$$H_{R,0}(z) = \frac{G_0}{1 - z^{-1}}$$

$H_C(z)$ is the system function of the *comb filter*; the reason for this name will become apparent shortly. The comb filter has the $(N + 1)$-point unit-sample response, only two elements of which are nonzero, $h(0) = 1/N$ and $h(N) = -1/N$, as shown in Fig. 6.12 for $N = 8$. These nonzero elements of

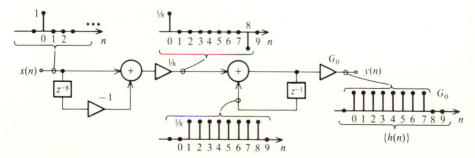

FIGURE 6-12
FIR filter implementation using the comb filter with resonator structure.

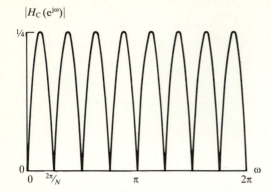

FIGURE 6-13
Magnitude response of comb filter ($N = 8$).

alternate signs produce the sinusoidal transfer function given by

$$H_C(e^{j\omega}) = \frac{1 - e^{-jN\omega}}{N}$$

$$= \frac{2j}{N} e^{-jN\omega/2} \sin(N\omega/2) \tag{6.26}$$

which is shown in Fig. 6.13 for $N = 8$. $H_C(e^{j\omega})$ has N zeros equally spaced over the frequency range $0 \le \omega < 2\pi$. Because the magnitude response resembles a comb, this filter has become known as a *comb filter*. This form of the transfer function can also be explained by considering the pole/zero pattern of the system function $H_C(z)$. This is most easily seen by expressing $H_C(z)$ as

$$H_C(z) = \frac{1}{N} \prod_{k=0}^{N-1} (1 - e^{j2\pi k/N} z^{-1}) \tag{6.27}$$

This last form explicitly indicates that there are N equally spaced zeros over the unit circle with the first zero at $z = 1$ and all the poles are at $z = 0$. This is shown in Fig. 6.14 for $N = 8$.

$H_{R,k}(z)$ is the system function of a *resonator*, or a filter that has poles on the unit circle at frequency $\omega_k = 2\pi k/N$. Here $H_{R,0}$ is the resonator at $\omega = 0$.

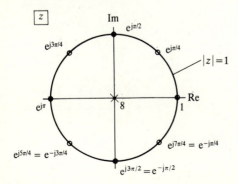

FIGURE 6-14
Pole/zero pattern for comb filter for $N = 8$.

The resonator $H_{R,0}(z)$ has a pole on the unit circle, at $z = 1$. Since this pole is not strictly within the unit circle, this filter is not stable, in that its unit-sample response does not decay to zero with time. One common application of such a filter is to implement an oscillator.

When this resonator is cascaded with the comb filter, however, the zero of the comb filter located at $z = 1$ cancels the resonator pole, making the combination stable, since the unit-sample response of the total filter has a finite duration.

The transfer function of this two-stage filter is equal to

$$H_0(e^{j\omega}) = \frac{G_0}{N} \frac{1 - e^{-j\omega N}}{1 - e^{-j\omega}}$$

$$= \frac{G_0}{N} e^{-j(N-1)\omega/2} \frac{\sin(N\omega/2)}{\sin(\omega/2)} \tag{6.28}$$

As shown in Fig. 6.15 for $N = 8$, $|H_0(e^{j\omega})|$ has a maximum at $\omega = 0$, equal to G_0, and is equal to zero at $\omega_k = 2\pi k/N$, for $1 \le k \le N - 1$. As a consequence, we can set the gain of the filter at $\omega = 0$ to any desired value G_0 without effecting the gains at the other ω_k points, since they are zero. We will exploit this result later by connecting additional resonators in parallel with $H_{R,0}(z)$, to set the gains at these other values of ω_k to arbitrary values.

The behavior of this filter combination can also be explained in the time domain. The comb filter energizes the resonator with the nonzero sample at time $n = 0$. The resonator is lossless, because the feedback coefficient has a magnitude equal to unity. Hence, it maintains the output value until the comb filter discharges the resonator with a nonzero sample of opposite sign at time $n = N$. This two-stage filter combination produces a rectangular pulse having N samples.

Let us extend the above two-stage filter by considering a complex-valued resonator whose pole is situated at the next zero location of the comb filter, located at frequency $\omega_1 = 2\pi/N$. The system function of this resonator is given

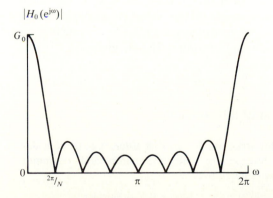

FIGURE 6-15
Magnitude response of comb filter plus resonator ($N = 8$).

by

$$H_{R,1}(z) = \frac{G_1}{1 - e^{j2\pi/N}z^{-1}} \tag{6.29}$$

where G_1 is now a *complex-valued gain*. This constant is set to the complex value of the desired transfer function $H(e^{j\omega})$ at $\omega = 2\pi/N$, or

$$G_1 = H(e^{j2\pi/N}) \tag{6.30}$$

To obtain real-valued coefficients in the filter, we need a second complex-valued resonator, whose pole is the complex conjugate of that of $H_{R,1}(z)$. This pole also cancels the $(N-1)$th zero of the comb filter. The system function of this second resonator is given by

$$H_{R,N-1}(z) = \frac{G_{N-1}}{1 - e^{-j2\pi/N}z^{-1}} \tag{6.31}$$

where G_{N-1} is complex-valued gain equal to the desired $H(e^{j\omega})$ at $\omega = 2\pi(N-1)/N = -2\pi/N$. In Chapter 3, the transfer function of a filter having a real-valued unit-sample response was shown to have the property that

$$H(e^{j\omega}) = H^*(e^{-j\omega}) \tag{6.32}$$

Hence, $G_{N-1} = G_1^*$. Connecting these two resonators in parallel, the resulting system function is given by

$$\begin{aligned}
H_{R,1}(z) + H_{R,N-1}(z) &= \frac{G_1}{1 - e^{j2\pi/N}z^{-1}} + \frac{G_{N-1}}{1 - e^{-j2\pi/N}z^{-1}} \\
&= \frac{G_1 + G_1^* - [G_1 e^{-j2\pi/N} + G_1^* e^{j2\pi/N}]z^{-1}}{1 - 2\cos(2\pi/N)z^{-1} + z^{-2}} \\
&= \frac{2\,\mathrm{Re}[G_1] - 2\,\mathrm{Re}[G_1 e^{-j2\pi/N}]z^{-1}}{1 - 2\cos(2\pi/N)z^{-1} + z^{-2}}
\end{aligned}$$

This combination forms a *second-order resonator,* whose coefficients are all real-valued.

This procedure can be generalized to include a complex-valued resonator at each of the N zero locations of the comb filter. These resonators are all connected in parallel and this parallel combination is then connected in cascade with the comb filter to produce the total filter $H_T(z)$, given by

$$H_T(z) = \frac{1 - z^{-N}}{N} \sum_{k=0}^{N-1} \frac{G_k}{1 - e^{j2\pi k/N}z^{-1}} \tag{6.34}$$

Example 6.8 illustrates the application of this technique.

Example 6.8. Comb and resonator structure for an FIR filter. Let us consider implementing the causal linear interpolator with the comb and resonator structure. The unit-sample response is given by

$$h(0) = \tfrac{1}{2} \qquad h(1) = 1 \qquad h(2) = \tfrac{1}{2} \qquad h(n) = 0, \quad \text{otherwise}$$

The transfer function is given by

$$H(e^{j\omega}) = (1 + \cos(\omega)) \, e^{-j\omega}$$

Since the number of elements in the unit-sample response is equal to 3, we choose $N = 3$. The comb filter system function is then

$$H_C(z) = (1 - z^{-3})/3$$

$H_C(z)$ has three zeros, located at $z = e^{j2\pi k/3}$, for $k = 0$, 1 and 2. The zero at $z = 1$ will be canceled by a real resonator, and the zeros at $z = e^{\pm j2\pi/3}$ will be canceled by a pair of complex resonators. The gains of the resonators are equal to

$$G_0 = H(e^{j0}) = 2$$
$$G_1 = H(e^{j2\pi/3}) = \tfrac{1}{2} \, e^{-j2\pi/3}$$
$$G_2 = H(e^{j4\pi/3}) = \tfrac{1}{2} \, e^{j2\pi/3}$$

Note that G_0 is real and $G_2 = G_1^*$. The real-valued first-order resonator system function is given by

$$H_{R,0}(z) = \frac{2}{1 - z^{-1}}$$

The system function of the parallel combination of the complex-valued first-order resonators is

$$
\begin{aligned}
H_{R,1}(z) + H_{R,2}(z) &= \frac{\tfrac{1}{2} \, e^{-j2\pi/3}}{1 - e^{j2\pi/3} z^{-1}} + \frac{\tfrac{1}{2} \, e^{j2\pi/3}}{1 - e^{-j2\pi/3} z^{-1}} \\
&= \frac{2 \operatorname{Re}[\tfrac{1}{2} \, e^{-j2\pi/3}] - 2 \operatorname{Re}[\tfrac{1}{2} \, e^{-j4\pi/3}] z^{-1}}{1 - 2 \cos(2\pi/3) z^{-1} + z^{-2}} \\
&= \frac{\cos(2\pi/3) - \cos(4\pi/3) z^{-1}}{1 - 2 \cos(2\pi/3) z^{-1} + z^{-2}} \\
&= \frac{-\tfrac{1}{2}(1 - z^{-1})}{1 + z^{-1} + z^{-2}}
\end{aligned}
$$

The coefficients of the resulting second-order resonator are real. The filter structure is shown in Fig. 6.16. The unit-sample response of the two resonators are shown in the figure. Their sum is equal to the desired unit-sample response.

This filter structure will be useful for the frequency-sampling procedure for FIR synthesis described in Chapter 9.

6.9 SUMMARY

In this chapter, we have described several common methods to implement a digital filter structure from either its difference equation, its unit-sample response or its z-transform. We started by defining the terms commonly used for describing digital filters and then considering the various structures. The structure determined directly from either the difference equation or the system function is called the Direct Form I. An alternate view of the same equations results in the more memory-efficient Direct Form II structure. Digital filters

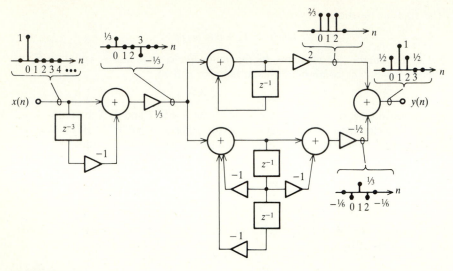

FIGURE 6-16
Implementation of FIR digital filter using the comb filter with a bank of resonators.

structured as cascade and parallel combinations of second-order components have advantages in terms of hardware implementation and numerical sensitivity. The above methods can be employed to implement the general filter structure, which contains both poles and zeros. FIR filters having a linear-phase response were shown to have a special structure that reduces the number of multiplications by approximately one-half. The frequency-sampling structure is shown to be an alternative to the tapped-delay line structure for FIR filters.

PROBLEMS

6.1. Implement the following system functions in the Direct Form I, Direct Form II, cascade and parallel structures. All coefficients are real.

(a) $H(z) = \dfrac{1}{(1 + az^{-1})(1 - bz^{-1})}$ (use first-order sections)

(b) $H(z) = \dfrac{1}{(1 - az^{-1})^3}$

(c) $H(z) = \dfrac{1 - b\cos(\theta) z^{-1}}{(1 - az^{-1})(1 - 2b\cos(\theta) z^{-1} + b^2 z^{-2})}$

(d) $H(z) = \dfrac{1}{(1 - az^{-1})^2} + \dfrac{1}{(1 - bz^{-1})^2}$

6.2. Implement the following system functions as Direct Form I, Direct Form II, cascade and parallel structures. All the coefficients are real.

(a) $H(z) = \dfrac{1 - 2r_1 \cos(\theta_1) \, z^{-1} + r_1^2 z^{-2}}{1 - 2r_2 \cos(\theta_2) \, z^{-1} + r_2^2 z^{-2}}$

(b) $H(z) = \dfrac{1 - z^{-4}}{1 - 2r \cos(\theta) \, z^{-2} + r^2 z^{-4}}$

(c) $H(z) = \dfrac{1 - z^{-2}}{(1 - 2r \cos(\theta) \, z^{-1} + r^2 z^{-2})^2}$

6.3. Implement the following difference equations in structures indicated by Direct Form I, Direct Form II, cascade of second-order sections and parallel of second-order sections.

(a) (See Example 7.5)

$$H(z) = \frac{A(1 + z^{-1})(1 + z^{-2})(1 + z^{-1} + z^{-2})}{(1 - 0.7z^{-1})(1 - 1.2z^{-1} + 0.8z^{-2})(1 - 1.2z^{-1} + 0.6z^{-2})(1 - 0.9z^{-1} + 0.4z^{-2})}$$

(b) (See Example 7.6)

$$H(z) = \frac{A(1 - z^{-2})(1 + 1.8z^{-1} + z^{-2})(1 - 1.8z^{-1} + z^{-2})}{(1 + 0.1z^{-1} + 0.5z^{-2})(1 - 0.6z^{-1} + 0.8z^{-2})(1 + 0.7z^{-1} + 0.8z^{-2})}$$

(c) (See Example 8.7)

$$H(z) = \frac{A}{(1 - 1.7z^{-1} + 0.7z^{-2})(1 - 1.7z^{-1} + 0.9z^{-2})}$$

(d) (See Example 8.8)

$$H(z) = A\frac{1 + z^{-1}}{1 - 0.8z^{-1}}\frac{1 + 2z^{-1} + z^{-2}}{1 - 1.6z^{-1} + 0.8z^{-2}}$$

(e) (See Example 8.11)

$$H(z) = A\frac{1 - z^{-2}}{1 - 0.4z^{-1} + 0.9z^{-2}}\frac{1 - z^{-2}}{1 + 0.1z^{-1} + 0.7z^{-2}}\frac{1 - z^{-2}}{1 + 0.6z^{-1} + 0.9z^{-2}}$$

6.4. Find the unit-sample response at the junction of the comb/first-order resonator (pole at $z = 1$) combination, when they are connected in the reverse order.

CHAPTER
7

FROM
ANALYSIS
TO SYNTHESIS

7.1 INTRODUCTION

In this chapter we extend the analytic techniques discussed in the previous chapters to be useful for the synthesis of digital filters. We then begin the filter design section of the book by considering some general issues that should be addressed at the initiation of a filter design. An alternative to designing the filter analytically is the intuitive approach of pole/zero placement considered in this chapter. Intuition dictates that the dips and peaks in the desired magnitude response can be accomplished by suitable placement of zeros and poles in the z plane. However, even though a filter can be designed in this manner, there is no guarantee that another filter, having fewer delays or operations, may not also satisfy the same requirement. This intuitive approach will form the starting point for the optimal filter synthesis procedures described in the following chapters. This intuitive and interactive procedure provides a fundamental understanding of the effects produced by poles and zeros on the frequency transfer function.

An additional benefit of this interactive procedure is that it solves two problems that exist with the conventional design procedures, described in the next chapter. The first is that the filter magnitude response resulting from conventional design approaches can often be improved by proper placement of additional poles and zeros. The intuitive procedure helps to place such additional singularities. The second problem occurs when the filter poles lie very close to the unit circle. Filter design computations require a high precision in the calculations. If the precision is not sufficient, the locations of the singularities deviate slightly from those desired and can significantly alter the magnitude response. The intuitive design procedure offers an alternative. It indicates how to shift the resulting singularities in the z plane to satisfy the specification without repeating the tedious computations.

7.2 INTER-RELATIONSHIPS OF ANALYTIC METHODS

As an overview, let us consider the flow of information given in Fig. 7.1, in which the five main categories of analytic techniques are shown interrelated by the solid arrows. In principle, the results presented in the previous chapters should allow us to start at any category, the starting point being the specification to be met by the digital filter, and to determine any other. The end point for synthesis is usually the digital filter structure and coefficient values. As the figure implies, the easiest path is usually through the z-transform. The numbers in the figure indicate examples in the book that illustrate the corresponding path.

 The important practical case in which only the desired magnitude response is specified is shown connected by a broken arrow to the z plane and to the complex frequency transfer function. In general, finding the z plane

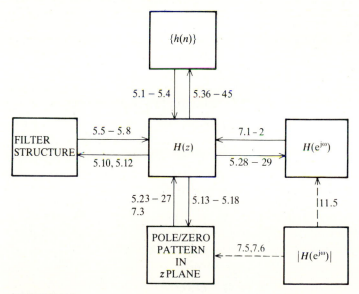

FIGURE 7-1
Relationship of analytic methods. Numbers refer to Examples in the book that illustrate each path.

pole/zero configuration from only the magnitude response, with no phase specification given, is not a well-defined analytical problem. The results of this chapter will demonstrate that it does not have a unique solution; that is, many filters can have the same magnitude response, but have different phase responses. In this chapter, poles and zeros will be placed in the z plane interactively to achieve a sufficiently accurate approximation to $|H(e^{j\omega})|$. If an appropriate phase response constraint, called the *minimum-phase* condition, is applied to the magnitude of the DFT sequence $\{|H(k)|\}$, then a unique phase response sequence $\{\text{Arg}[H(k)]\}$ can be determined. From these two, the complex DFT sequence $\{H(k)\}$ can be found. The explanation of this procedure is delayed until Chapter 11, since it involves the discrete Hilbert transform.

Let us now consider the paths indicated by the solid arrows in Fig. 7.1. The two paths from the Fourier transform to the z-transform and from the pole/zero pattern to the z-transform have not been explicitly described previously. Since these are important for filter synthesis, they are considered now.

FROM FOURIER TRANSFORM TO z-TRANSFORM. The Fourier transform is usually given in terms of the magnitude and phase response functions, which can be expressed as

$$|H(e^{j\omega})|^2 = H_R^2(e^{j\omega}) + H_I^2(e^{j\omega})$$
$$= H(e^{j\omega})H^*(e^{j\omega}) \tag{7.1}$$

and

$$\text{Arg}[H(e^{j\omega})] = \arctan[H_I(e^{j\omega})/H_R(e^{j\omega})] \tag{7.2}$$

Any common factor in $H_I(e^{j\omega})$ and $H_R(e^{j\omega})$, such as the filter gain, has usually been eliminated in the argument of the arctangent function through cancellation. The following procedure can then be employed to retrieve this common factor.

Let $N(e^{j\omega})$ and $D(e^{j\omega})$ represent the numerator and denominator polynomials in $\text{Arg}[H(e^{j\omega})]$ that have any common factors removed, or

$$\text{Arg}[H(e^{j\omega})] = \arctan[N(e^{j\omega})/D(e^{j\omega})] \tag{7.3}$$

Then the imaginary and real parts of $H(e^{j\omega})$ are given by

$$H_I(e^{j\omega}) = \alpha(\omega)\,N(e^{j\omega}),$$
$$H_R(e^{j\omega}) = \alpha(\omega)\,D(e^{j\omega})$$

where $\alpha(\omega)$ is the real-valued common factor, that may be a function of ω. We can find $\alpha(\omega)$ from the specified magnitude response by substituting these forms of the real and imaginary parts into Eq. (7.1), to find

$$\alpha^2(\omega) = \frac{|H(e^{j\omega})|^2}{N^2(e^{j\omega}) + D^2(e^{j\omega})} \tag{7.4}$$

To find the value of $\alpha(\omega)$, the square root must be taken, allowing both

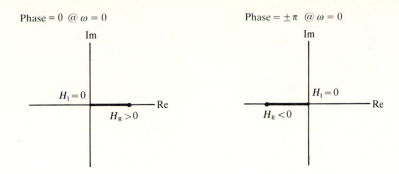

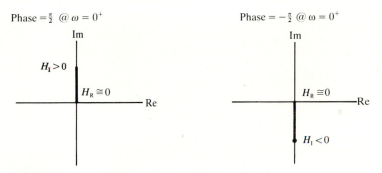

FIGURE 7-2
Possible values of phase at $\omega = 0$ or $\omega = 0^+$ and the corresponding relationship of the real and imaginary parts of the Fourier transform.

$\pm\alpha(\omega)$ as equally valid solutions. The correct sign must be determined from the value of the phase at $\omega = 0$ or $\omega = 0^+$, as shown in Fig. 7.2. If the phase is 0, then $H_I(e^{j0}) = 0$ and the sign of $\pm\alpha(\omega)$ should be taken to make $H_R(e^{j0}) > 0$. If the phase is $\pm\pi$, then $H_I(e^{j0}) = 0$ and the sign should be chosen to make $H_R(e^{j0}) < 0$. If the phase at $\omega = 0^+$ is $\pi/2$, $H_R(e^{j0}) \cong 0$ and the sign should be chosen to make $H_I(e^{j0})$ positive and if the phase is $-\pi/2$, $H_I(e^{j0})$ must be negative.

The complex-valued Fourier transform is then equal to

$$H(e^{j\omega}) = \alpha(\omega)D(e^{j\omega}) + j\alpha(\omega)N(e^{j\omega}) \tag{7.5}$$

where $\alpha(\omega)$ contains the correct sign. The Euler identities can then be applied to convert the sine and cosine functions in the analytic expression of $H(e^{j\omega})$ into complex exponentials. The z-transform is then obtained by replacing $a\,e^{-j\omega n}$ with az^{-n}.

Example 7.1.   From Fourier transform to z-transform. Let

$$|H(e^{j\omega})|^2 = \frac{4}{1 - 2a\cos(\omega) + a^2} \qquad \text{with } |a| < 1$$

and

$$\text{Arg}[H(e^{j\omega})] = \arctan\left[\frac{-a\,\sin(\omega)}{1 - a\,\cos(\omega)}\right] = \arctan[N(e^{j\omega})/D(e^{j\omega})]$$

with $\text{Arg}[H(e^{j0})] = \pi$.

Computing the squared magnitude with $N(e^{j\omega})$ and $D(e^{j\omega})$, we get

$$
\begin{aligned}
|H(e^{j\omega})|^2 &= \alpha^2(\omega)[N^2(e^{j\omega}) + D^2(e^{j\omega})] \\
&= \alpha^2(\omega)[(a\,\sin(\omega))^2 + (1 - a\,\cos(\omega))^2] \\
&= \alpha^2(\omega)[1 - 2a\,\cos(\omega) + a^2(\cos^2(\omega) + \sin^2(\omega))] \\
&= \alpha^2(\omega)[1 - 2a\,\cos(\omega) + a^2]
\end{aligned}
$$

This result must equal the specified magnitude response. Equating the two and taking the square root, we get

$$\alpha(\omega) = \pm\frac{2}{1 - 2a\,\cos(\omega) + a^2}$$

The real part is then equal to $\alpha(\omega)D(e^{j\omega})$, or

$$H_R(e^{j\omega}) = \pm\frac{2(1 - a\,\cos(\omega))}{1 - 2a\,\cos(\omega) + a^2}$$

Since the phase is equal to π at $\omega = 0$, $H_R(e^{j0})$ must be negative (note that $H_I(e^{j0}) = 0$). Hence, the negative square root is the correct choice, and

$$
\begin{aligned}
H(e^{j\omega}) &= \alpha(\omega)D(e^{j\omega}) + j\alpha(\omega)N(e^{j\omega}) \\
&= \frac{-2(1 - a\,\cos(\omega) - ja\,\sin(\omega))}{1 - 2a\,\cos(\omega) + a^2} = \frac{-2(1 - a\,e^{j\omega})}{(1 - a\,e^{j\omega})(1 - a\,e^{-j\omega})} \\
&= \frac{-2}{1 - a\,e^{-j\omega}}.
\end{aligned}
$$

Finally, the desired system function is given by

$$H(z) = \frac{-2}{1 - az^{-1}}$$

Occasionally, the evaluation of Eq. (7.4) results in a complicated trigonometric expression. An alternate procedure to determine $\alpha(\omega)$ follows from expressing the Fourier transform in the following form:

$$
\begin{aligned}
H(e^{j\omega}) &= \alpha(\omega)D(e^{j\omega}) + j\alpha(\omega)N(e^{j\omega}) \\
&= \alpha(\omega)[D(e^{j\omega}) + jN(e^{j\omega})]
\end{aligned} \tag{7.6}
$$

The term in the square brackets can then be simplified by converting the trigonometric terms into complex exponentials by applying the Euler identities. Then the squared magnitude of the right hand side of Eq. (7.6) can be computed and compared with the specified $|H(e^{j\omega})|^2$ to find $\alpha(\omega)$.

Example 7.2.   Alternate approach to solving Example 7.1. From the numerator and denominator terms of the phase response in Example 7.1, we have

$$H(e^{j\omega}) = \alpha(\omega)[1 - a\cos(\omega) - ja\sin(\omega)]$$
$$= \alpha(\omega)[1 - a\,e^{j\omega}]$$

Then the squared-magnitude response should have the following form:

$$|H(e^{j\omega})|^2 = \alpha^2(\omega)[1 - 2a\cos(\omega) + a^2]$$

This result must equal the specified function given by

$$|H(e^{j\omega})|^2 = \frac{4}{1 - 2a\cos(\omega) + a^2}$$

Equating the two and taking the square root, we find

$$\alpha(\omega) = \pm\frac{2}{1 - 2a\cos(\omega) + a^2}$$

which is equal to the result in the previous example.

FROM POLE/ZERO PATTERN IN THE z PLANE TO z-TRANSFORM.

In Chapter 5, the system function of a digital filter was given by

$$H(z) = Az^{N_F} \prod_{k=1}^{N_P+N_F} (1 - c_k z^{-1}) \Big/ \prod_{k=1}^{M} (1 - d_k z^{-1}) \tag{7.7}$$

where A is the gain constant. Since the pole/zero pattern is not effected by a constant gain factor in the system function, the value of A must be specified separately to determine a unique system from the pole/zero pattern.

To find the system function from the pole/zero pattern, the singularities in the z plane should be considered one by one. When the singularities at $z = 0$ and $z = \infty$ are included, the number of poles is equal to the number of zeros. Let us first consider the singularities in the z plane that are not located at $z = 0$ and $z = \infty$. A zero located at $z = c_k$, for $c_k \neq 0$ or ∞, produces a numerator term $(z - c_k)$ in the system function. Each such zero should be combined with a pole at $z = 0$, to produce the term $(z - c_k)/z$, or $(1 - c_k z^{-1})$, for one of the terms in the numerator of Eq. (7.7). If there is no pole at $z = 0$, then a pole/zero pair should be created there and the pole paired with the zero at $z = c_k$, leaving the zero at $z = 0$. A pole located at $z = d_k$, for $d_k \neq 0$ or ∞, produces a term $1/(z - d_k)$. Each such pole should be combined with a zero at $z = 0$, to produce the term $z/(z - d_k) = 1/(1 - d_k z^{-1})$ in Eq. (7.7). If there is no zero at $z = 0$, then a pole/zero pair should be created there and the zero paired with the pole at $z = d_k$, leaving the pole at $z = 0$.

After all the singularities lying in $0 < |z| < \infty$ have been eliminated, an equal number of singularities of opposite types will remain at $z = 0$ and $z = \infty$. If there are M_P poles at $z = 0$, each of them produces a z^{-1} term, or a z^{-M_P} term in the numerator of Eq. (7.7). Hence, $N_F = -M_P$, indicating that the

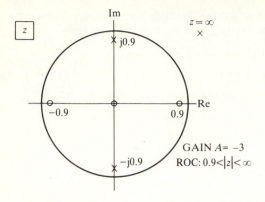

FIGURE 7-3
Pole/zero pattern to be implemented as a digital filter.

pole/zero pattern is that of a causal filter whose first nonzero term in the unit-sample response is $h(M_P)$. If there are M_Z zeros at $z = 0$, each of them produces a z term, or a z^{M_Z} term in the numerator of Eq. (7.7). Hence, $N_F = M_Z$, indicating a noncausal filter whose unit-sample response has its first nonzero term as $h(-M_Z)$.

> **Example 7.3. From pole/zero pattern to z-transform.** Let us consider the z plane pole/zero pattern shown in Fig. 7.3. Applying the procedure described above first to the zeros in $0 < |z| < \infty$, then to the poles in $0 < |z| < \infty$, and finally to the singularities at $z = 0$ and $z = \infty$, we have
>
> $$H(z) = \frac{Az(1 + 0.9z^{-1})(1 - 0.9z^{-1})}{(1 + j0.9z^{-1})(1 - j0.9z^{-1})}$$
>
> $$= -3\frac{z(1 - 0.81z^{-2})}{(1 + 0.81z^{-2})}$$
>
> Note that the gain $A = -3$ must be specified separately.

7.3 PRACTICAL MAGNITUDE RESPONSE SPECIFICATIONS

An important situation commonly arising in practice is one in which only the desired magnitude response $|H(e^{j\omega})|$ is specified and the phase response is left unspecified or arbitrary. Let us consider the ideal magnitude response specification of the lowpass filter characteristic having the *cutoff frequency* ω_C shown in Fig. 7.4(a). In the ideal filter, frequencies smaller than ω_C are transmitted with no attenuation and frequencies greater than ω_C are completely attenuated. We will see later that a characteristic having perfectly flat transmission and attenuation regions and a sharp discontinuity between the two cannot be realized in practice, but only approximated. This approximation procedure is the *art* of filter design, and the goal is to become a good artist.

In practice, the desired $|H(e^{j\omega})|$ is usually specified to lie within allowable regions. An example of an approximation to an ideal lowpass filter is shown in

(a)

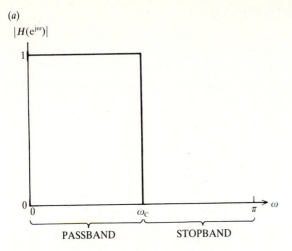

$$|H(e^{j\omega})|$$

PASSBAND STOPBAND

(b)

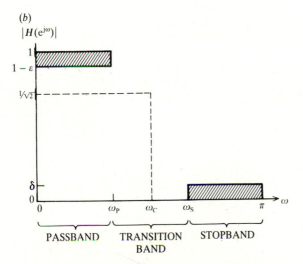

$$|H(e^{j\omega})|$$

PASSBAND TRANSITION BAND STOPBAND

FIGURE 7-4
(a) Ideal lowpass filter magnitude response; (b) practical specification to be satisfied by a digital filter.

Fig. 7.4(b). The cutoff frequency ω_C is replaced by a *passband frequency* ω_P and a *stopband frequency* ω_S. The passband is defined as the frequency range in which the transfer function magnitude is approximately equal to unity, or $1 - \varepsilon \leq |H(e^{j\omega})| \leq 1$, for some small $\varepsilon > 0$. The stopband is the range in which the transfer function magnitude is small, or $|H(e^{j\omega})| \leq \delta$, for some small $\delta > 0$. The location of ω_C is conventionally taken as the frequency at which the magnitude $|H(e^{j\omega})|$ is equal to $1/\sqrt{2}$. The power transfer response $|H(e^{j\omega})|^2$ is then equal to $\frac{1}{2}$ at ω_C. Hence, it is also referred to as the *half-power* frequency.

7.4 LOG-MAGNITUDE RESPONSE CURVES

When the magnitude response is plotted on a linear scale, the features in the stopband are difficult to observe when δ is small. This attenuation region can be expanded by plotting the magnitude response in logarithmic units, decibels (dB), defined by

$$|H(e^{j\omega})|_{dB} = 20 \log_{10} |H(e^{j\omega})| = 10 \log_{10} |H(e^{j\omega})|^2 \tag{7.8}$$

An example is shown in Fig. 7.5, in which the magnitude responses of two lowpass filters are compared. The first filter system function has a single pole at $z = 0.95$ and the second has a double pole. The difference is more apparent in the log-magnitude response curves.

It is conventional to normalize the log-magnitude response by vertically shifting it such that the maximum value of $20 \log_{10} |H(e^{j\omega})|$ over $0 \le \omega \le \pi$ is

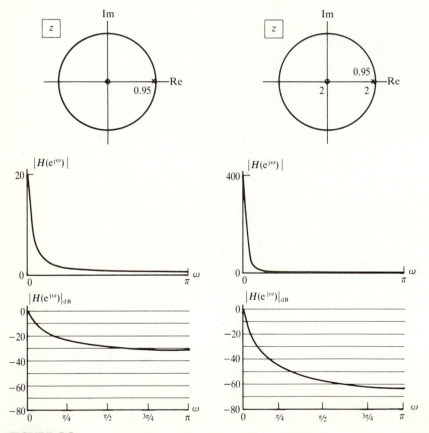

FIGURE 7-5
Comparison of magnitude and log-magnitude response curves for first- and second-order recursive filters.

equal to 0 dB. The curve then shows the degree of attenuation and is simple to interpret. When the gain A of the filter is included in the response, it produces only a vertical shift without otherwise changing the shape of the curve.

A problem in the evaluation of Eq. (7.8) arises when $|H(e^{j\omega})| = 0$, which results in an infinite logarithmic value and an execution error in the computer program. This can occur when a zero is located on the unit circle in the z plane. This problem can be avoided by checking for small values of $|H(e^{j\omega})|$ and limiting them to be no less than some suitably small positive number. A value equal to 10^{-6} times the maximum value of $|H(e^{j\omega})|$ over $0 \le \omega \le \pi$ is usually adequate. This procedure prevents the argument of the logarithm function from being zero without significantly affecting the response of most practical filters. This value of the constant allows the magnitude response to be plotted over an 80 dB range, from 0 dB to -80 dB, which is adequate for the filter designs in this book.

To show clearly both the passband and stopband characteristics, the log-magnitude response must usually be plotted on two different scales: an expanded dB scale to show the response in the passband, where the attenuation is small—typically between 0 and -4 dB—and a compressed dB scale to show the response in the stopband, where the attenuation is high—typically $|H(e^{j\omega})|_{dB} < -60$ dB.

The practical design utility of the log-magnitude response comes in cascading filter sections. The total filter log-magnitude response $|H_T(e^{j\omega})|_{dB}$ is then equal to the sum of the log-magnitudes of the individual sections $|H_i(e^{j\omega})|_{dB}$. If there are N sections, then the overall filter response is given by

$$|H_T(e^{j\omega})|_{dB} = \sum_{i=1}^{N} |H_i(e^{j\omega})|_{dB} \tag{7.9}$$

Transfer functions that have relatively simple forms, like those produced by single and complex-conjugate singularities, are more readily visualized when added than when they are multiplied. This observation will prove to be useful for the intuitive filter design procedure presented later in this chapter.

7.5 PROGRAMS TO COMPUTE THE LOG-MAGNITUDE RESPONSE OF POLES AND ZEROS

As shown in Chapter 5, the magnitude response of a first- or second-order filter section is a function of the locations of its singularities in the z plane. Viewing the magnitude response in this manner helps to perform efficiently the interactive procedure described below. In this section, we describe programs to compute and display the log-magnitude response. The first routine is LOGMAG that computes the log-magnitude response from the magnitude sequence generated by MAGPHS. To display the log-magnitude response, we describe the routine PLOTLM, which is given in Appendix A.

In Chapter 5, the coefficients of the filter section producing real and complex-conjugate singularities were determined with the subprograms ZERO and POLE. These coefficients are then used as inputs to NONREC, if ZERO is called, and to REC, if POLE is called. The unit-sample sequence is then used as the input to the filter to determine its unit-sample response. The FFT subprogram computes the discrete Fourier transform of the unit-sample response and MAGPHS calculates the magnitude and phase responses. What is still needed is a routine to compute the magnitude response in units of dB. This subprogram is called

> LOGMAG(HMAG,NH,HDB)

where
  HMAG   (input real array) is the magnitude response array
  NH   (integer input) is the number of spectral values over $0 \le \omega \le \pi$
           (typically, 65 when a 128-point FFT is computed)
  HDB   (output real array) is the log-magnitude array in units of dB.

This routine performs the transformation given by

$$HDB(N) = 20 \log_{10}[HMAG(N)] \qquad \text{for } 1 \le N \le NH \qquad (7.10)$$

This equation is invalid when $HMAG(N) = 0$ which can occur when a zero lies on the unit circle at $\omega_N = \pi(N-1)/(NH-1)$. Hence, values of $HMAG(N)$ less than some small number ε should be set to ε. The value $\varepsilon = 10^{-6}$ (equal to -120 dB) is adequately small.

This output of LOGMAG should be plotted twice, once in expanded vertical scale for the passband specification, and a second time in compressed scale for the stopband specification. This plot is performed with the subprogram given in Appendix A named

> PLOTLM(HTOT,DBVAL,SPEC,IF1,IF2,NPRNT)

where
  HTOT   (input real array) is the log-magnitude response sequence containing 65 elements to be plotted
  DBVAL   (input real) is the lower limit in dB of the plot (typical values are -4 for passband and -60 for stopband)
  SPEC   (input real) is the specification value in dB that is marked as a horizontal line on the plot (typical values are -1 for passband and -50 for stopband),
  IF1,IF2 (input integers) are the frequency points that are marked by vertical lines on the plot ($1 \le IF1,IF2 \le 65$, and is equal to frequency $\omega = 2\pi(k-1)/128$, where k is either IF1 or IF2. If IF1 = 0 or IF2 = 0, no mark is made), and
  NPRNT   (input integer) is set to 1 if a printout of the terminal display is desired.

The PLOTLM subroutine automatically shifts the log-magnitude re-

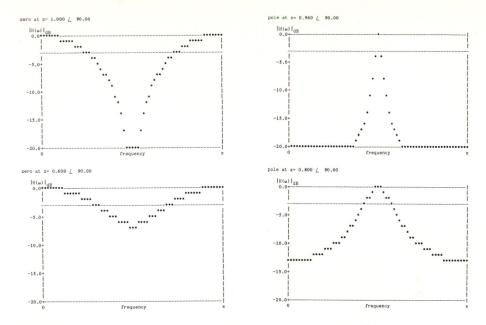

FIGURE 7-6
Log-magnitude response curves produced with PLOTLM subroutine for elemental sections. a) zeros. b) poles. DBVAL = −20, SPEC = −3, IF1 = IF2 = 0.

sponse curve, so that the maximum value of {HTOT(N)} is plotted to correspond to 0 dB.

To illustrate the effects of singularity position on the magnitude response, the log-magnitude response curves for complex-conjugate singularities have been computed with LOGMAG and plotted with PLOTLM. The results are shown in Fig. 7.6. The peaks produced by the poles and the troughs by the zeros have a sharpness that is related to the distance of the singularity from the unit circle. These response curves can be considered to be those of elemental filters.

With the above procedure, the log-magnitude response is computed for each first and second-order elemental filter section that corresponds to the real or complex-conjugate singularities, respectively. The total filter structure is then obtained by cascading these elemental sections. The log-magnitude response sequence of this total filter, denoted by {HTOT(N)} is then obtained by accumulating the log-magnitude responses of these elemental sections, as given in Eq. (7.9).

To complete the design procedure, the gain constant A of the filter must be determined. This procedure is described next.

Determining the value of the filter gain constant A. In Chapter 5, it was shown that the pole/zero pattern in the z plane is not affected by the gain constant of the system function. Conversely, when the digital filter is designed by placing poles and zeros in the z plane, the gain constant A of the filter must be computed to meet the specification. In this section, we describe how this gain constant is determined.

For illustration, let us consider how the gain constant of the digital filter is computed in order that the maximum value of its magnitude response is equal to one. We start with a digital filter pole/zero pattern whose magnitude response satisfies the specification to within a gain constant. For the pole/zero pattern having zeros at $z = c_k$, for $1 \le k \le N$, and poles at $z = d_k$, for $1 \le k \le M$, the transfer function is given by

$$H(e^{j\omega}) = A \frac{\prod\limits_{k=1}^{N} (1 - c_k e^{-j\omega})}{\prod\limits_{k=1}^{M} (1 - d_k e^{-j\omega})} \qquad (7.11)$$

$$= A H_{pz}(e^{j\omega})$$

where $H_{pz}(e^{j\omega})$ is the transfer function computed directly from the poles and zeros by assuming that the gain constant is equal to one. The gain constant A needs to be determined, such that the maximum value of $|H(e^{j\omega})|$ is equal to one. To find the exact value of A, we must find the frequency ω_m for which $|H_{pz}(e^{j\omega})|$ takes on its maximum value. Then the gain factor is equal to

$$A = 1/|H_{pz}(e^{j\omega_m})|. \qquad (7.12)$$

The standard way to find ω_m is to compute the derivative $d |H_{pz}(e^{j\omega})|/d\omega$, find the values of ω for which it is equal to zero, and evaluate $|H_{pz}(e^{j\omega})|$ at these values to find the global maximum over $0 \le \omega \le \pi$. In practice, however, the procedure is very tedious because of the complicated analytic form of the typical $|H_{pz}(e^{j\omega})|$ and of the presence of multiple maxima and minima.

Fortunately, there are two methods to find reasonably accurate approximations to the exact gain value that are usually satisfactory in practice. First, it may be sufficient to let the filter magnitude response be equal to one at some convenient frequency in the passband and disregard the maximum value, which may be slightly greater than one. For example, for lowpass filters, it is reasonable to specify that the magnitude response at $\omega = 0$ be equal to one. This can be easily accomplished by setting

$$A = 1/|H_{pz}(e^{j0})|. \qquad (7.13)$$

The gain is readily evaluated since

$$H_{pz}(e^{j0}) = \frac{\prod\limits_{k=1}^{N} (1 - c_k)}{\prod\limits_{k=1}^{M} (1 - d_k)} \qquad (7.14)$$

For highpass filters, the magnitude response at $\omega = \pi$ can be set to one in a similar manner. For bandpass filters the evaluation is slightly more complicated because the terms in Eq. (7.14) are complex-valued.

The second method of determining an approximate value for A involves computing the DFT sequence $\{H_{pz}(k)\}$, finding the maximum value of the magnitude response sequence $\{|H_{pz}(k)|\}$, denoted by $|H_{pz}|_{\max}$, and setting

$$A = 1/|H_{pz}|_{\max}. \qquad (7.15)$$

In the previous section, the log-magnitude response sequence of the filter $\{HTOT(N)\}$ was determined by accumulating the contributions of the elemental sections. If HMAX represents the maximum value of $\{HTOT(N)\}$ in dB, then the gain to make the maximum magnitude response unity is equal to

$$A = 10^{-\text{HMAX}/20} \qquad (7.16)$$

Determining the gain constant value from the DFT sequence is reasonably accurate if the DFT is computed at a sufficiently dense set of points, or such that the form of $|H_{pz}(e^{j\omega})|$ is readily apparent from a graph of the DFT sequence $\{|H_{pz}(k)|\}$. Then $|H_{pz}|_{\max}$ will be approximately equal to the maximum value of $|H_{pz}(e^{j\omega})|$. Since computing the DFT is readily accomplished on a computer, we will use this latter approximate method for determining the value of A in the filter designs in this book.

7.6 PHASE RESPONSE CONSIDERATIONS

The magnitude response of a filter is a description of the attenuation of the various frequency components in the signal that is produced by the filter. The phase response describes how the filter modifies the time relationship between the frequency components passing through the filter. When a filter design is performed to satisfy a magnitude response specification, the phase response is usually left arbitrary. In this section, we provide the motivation for specifying a linear phase response.

Let us start with the frequency domain description of the filtering operation given by

$$Y(e^{j\omega}) = H(e^{j\omega}) X(e^{j\omega}) \qquad (7.17)$$

Let us further consider the simple case of the bandpass magnitude response given by

$$|H(e^{j\omega})| = \begin{cases} 1, & \text{for } \omega_1 \le \omega \le \omega_2 \\ 0, & \text{for } 0 \le \omega < \omega_1 \text{ and } \omega_2 < \omega \le \pi \end{cases} \qquad (7.18)$$

Applying the superposition principle, we can separate the input signal into two parts

$$\{x(n)\} = \{x_1(n) + x_2(n)\} \qquad (7.19)$$

The corresponding spectra are then given by

$$X(e^{j\omega}) = X_1(e^{j\omega}) + X_2(e^{j\omega}) \qquad (7.20)$$

This separation is performed such that the desired part of $\{x(n)\}$, given by $\{x_1(n)\}$, lies in the passband of the filter, while the undesired part $\{x_2(n)\}$ lies outside the passband. That is, $X_1(e^{j\omega})$ is nonzero only for $\omega_1 \le \omega \le \omega_2$, and $X_2(e^{j\omega})$ is zero in this range. For example, $\{x_1(n)\}$ may be the signal on a desired radio channel and $\{x_2(n)\}$ represents the signals on the other channels.

Since only $X_1(e^{j\omega})$ passes unattenuated through the filter, the output spectrum is given by

$$Y(e^{j\omega}) = X_1(e^{j\omega})\, e^{j\,\mathrm{Arg}[H(e^{j\omega})]} \tag{7.21}$$

If the filter has a linear phase response in the passband, or

$$\mathrm{Arg}[H(e^{j\omega})] = -k\omega \qquad \text{for } \omega_1 \le \omega \le \omega_2 \tag{7.22}$$

then the output spectrum is given by

$$Y(e^{j\omega}) = X_1(e^{j\omega})\, e^{-jk\omega} \tag{7.23}$$

In Chapter 3, we showed that a linear phase response having a slope equal to $-k$ corresponds to a mere time delay of k samples. Hence, the output sequence is equal to

$$\{y(n)\} = \{x_1(n-k)\} \tag{7.24}$$

and is identical to the desired part of the input sequence, except for the time delay of k sample periods experienced by passing through the filter. That is, a linear phase response maintains the phase relationship of the frequency components of the input signal in the passband, as illustrated below in Example 7.4. In the stopband, the shape of the phase response is less important since these signal components are highly attenuated or even eliminated from the output. However, while a linear phase response is desirable, it is often difficult to achieve in practice when speed and cost considerations are applied. In these cases, a relatively smooth phase response that approximates a linear function over the passband is often satisfactory.

Example 7.4.   Effect of phase distortion. Let the desired input sequence be given by

$$\{x_1(n)\} = \{\cos(\omega_0 n) + \cos(2\omega_0 n)\}$$

If the filter has a transfer function with a linear phase response

$$H(e^{j\omega}) = e^{-jk\omega}$$

then

$$\{y(n)\} = \{\cos \omega_0(n-k)) + \cos(2\omega_0(n-k))\}$$

As shown in Fig. 7.7, the shape of the signal is maintained but only a delay of k samples is experienced.

Now consider a filter having the same magnitude response, equal to 1, but having the phase response given by

$$\mathrm{Arg}[H(e^{j\omega})] = \begin{cases} -\pi/4 & \text{for } 0 \le \omega \le 3\omega_0/2, \\ -\pi & \text{for } 3\omega_0/2 < \omega \le \pi \end{cases}$$

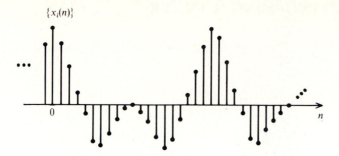

$\{x_i(n)\}$

LINEAR-PHASE OUTPUT

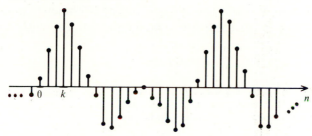

NON-LINEAR PHASE OUTPUT

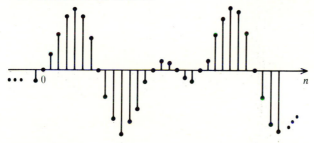

FIGURE 7-7
Illustration of phase distortion.

The output is then equal to

$$\{y(n)\} = \{\cos(\omega_0 n - \pi/4) + \cos(2\omega_0 n - \pi)\}$$

and is also shown in Fig. 7.7. The shape of the sequence has been altered appreciably, even though there was no amplitude distortion. This illustrates the effect of *phase distortion*.

7.7 STEPS IN PERFORMING A FILTER DESIGN

When designing a filter to meet a given specification, three steps must usually be taken:

1. The choice of the filter type (IIR or FIR).
2. The determination of the filter coefficients that meet the specification.
3. The verification that the resulting design indeed satisfies the specification.

We describe the implications of the first step in the next section. In this chapter, the second step is accomplished by an intuitive pole/zero placement. The conventional procedures for designing IIR filters are discussed in the next chapter, and for designing FIR filters in the following chapter. The problems encountered in the third step are due to finite accuracy arithmetic employed either in the calculations or in the hardware implementation. These effects are described in Chapter 10.

7.8 CHOICE OF FILTER TYPE

To illustrate the consequences of choosing between an IIR and FIR filter design, let us define a *relevant singularity* as one that contributes to the magnitude response of the filter. From the geometric evaluation of the magnitude response in the z plane, discussed in Chapter 5, the relevant singularities are those that are located in $0 < |z| < \infty$. The IIR filter system function can be expressed as a rational function in z^{-1}, or

$$H_{\text{IIR}}(z) = Az^{N_F} \prod_{k=1}^{N_P+N_F} (1 - c_k z^{-1}) \bigg/ \prod_{k=1}^{M} (1 - d_k z^{-1}) \tag{7.25}$$

This IIR system function has $N_P + N_F$ relevant zeros at $z = c_k$, for $1 \le k \le N_P + N_F$, that can lie anywhere in the z plane, and M relevant poles at $z = d_k$, for $1 \le k \le M$, that must be located within the unit circle for stability. The FIR filter system function is more restricted, in that it is expressed as only a polynominal in z^{-1}, or

$$H_{\text{FIR}}(z) = Az^{N_F} \prod_{k=1}^{N_P+N_F} (1 - c_k z^{-1}) \tag{7.26}$$

This FIR system function has $N_P + N_F$ relevant zeros that can lie anywhere in the finite z plane, and no relevant poles.

We now give some general observations regarding digital filter design. Let us consider a specification that can be satisfied either by an IIR filter having M poles and N zeros, where $N = N_P + N_F$ in (7.25), or by an FIR filter having K zeros, where $K = N_P + N_F$ in (7.26). Then the following general observations apply.

Observation 1. The IIR filter is usually the more efficient design in terms of computation time and memory requirements. The IIR filter is faster because $M + N$ will typically be less than K to satisfy a desired specification. The number of multiplications for an IIR filter is equal to $M + N$ compared with K for the general FIR filter. As shown in Chapter 6, for a linear-phase FIR filter, a special structure can be employed that requires only $(K/2) + 1$ multiplications. Since K delays are required to implement an FIR filter, while only the maximum of $N_F + M$ or $N_F + N_P$ in (7.25) for an IIR filter, the IIR filter also requires less memory.

Observation 2. The practical specification can approach that of the ideal filter by making ε, δ, and $(\omega_S - \omega_P)$ in Fig. 7.4 all approach zero. However, the values of both K and $M + N$ increase as the specification approaches the ideal. This, in turn, increases the computation time and memory requirements of the design. Hence, there is tradeoff between the accuracy of the approximation and the computation time.

Observation 3. FIR filters are always stable, since the poles of the FIR system function are located at $z = 0$ (and at $z = \infty$ for noncausal filters). The design of IIR filters with poles close to the unit circle must be performed with care to ensure that instabilities do not occur. It is not uncommon to have poles that were intended to be close to the unit circle actually fall outside the unit circle in the implementation, owing to round-off errors in the calculations. In this case, the implemented filter is unstable.

These general rules are illustrated in the design methods described in this and the following chapters. The first method is the interactive and intuitive procedure for designing a filter described in the next section. This procedure does not generally result in an optimal filter, or one that has the minimum number of delays or multiplications. However, it does lead to an intuitive understanding of the effects of poles and zeros, and provides a better appreciation of the conventional design procedures described in the following chapters. An additional benefit of this interactive procedure is that it solves a problem that exists with the conventional design procedures. The filter magnitude response resulting from conventional design approaches can be improved by proper placement of additional poles and zeros. The intuitive procedure helps in deciding where to place such additional singularities.

7.9 INTERACTIVE FILTER DESIGN BY INTUITIVE POLE/ZERO PLACEMENT

In this section, we determine the filter structure and its coefficients to meet a desired magnitude response by placing poles and zeros in the z plane. The filter design is accomplished interactively by successively adding real and complex-conjugate singularities in the z plane and observing whether the magnitude of the resulting filter is approaching the desired magnitude specification. The final filter structure is then the cascade connection of the elemental filter sections that generate these real and complex-conjugate singularities.

To assist in the design procedure, the following rules form the basis for pole/zero placement:

Rule 1. The magnitude response specifications will consist of passband and stopband regions. These bands are mapped onto the unit circle, as shown in Fig. 7.8, to assist in the placement of the singularities. If a complex singularity is specified, then it is understood that the complex-conjugate singularity is also required. The log-magnitude response of the singularities are computed and compared to the specification at 65 points equally spaced over $0 \le \omega \le \pi$. For easier interpretation, the angles of the singularities with respect to the positive real axis will be given in degrees, rather than radians.

Rule 2. All the poles in the design will have a radius r_P lying in the range $0.6 \le r_P < 0.96$. The upper limit guarantees that the unit-sample response of each elemental section will not have significant values beyond the 128th point. This prevents truncation errors from affecting the log-magnitude response of a particular elemental filter section. The lower limit is meant as a guide to avoid poles having anemic magnitude responses.

Rule 3. The filter design is performed interactively by examining the log-magnitude response after each additional singularity is added. In some cases, the singularity position must be modified or removed. The effect of an existing singularity can be canceled by specifying the opposite singularity at the same position. Of course, such pole/zero cancellation pairs are not included in the final filter design.

The design is initiated by placing a single pole in the middle of each passband at radius r_P, and a single zero in the middle of each stopband on the unit circle. The object of this design procedure is to satisfy the specification by pole/zero placement, not to find the filter having the minimum number of elements. The total number of singularities in the designs below will typically be less than 20, including the complex conjugates. It may be possible to find another filter, having a smaller number of elements, that also satisfies the specification. The procedure to find these simpler filters will be described in the next two chapters.

Based on the log-magnitude response of poles and zeros, the following hints may be useful:

Hint 1. The magnitude response over a narrow band of frequencies can be increased by placing a pole at the angle corresponding to the center of the band, but inside the unit circle. The radius of the pole determines both the size of the bandwidth and the amount of gain. These two effects, however, are inversely related, that is, for a single pole, high gains are usually achieved only over narrow bandwidths. For wider bandwidths, additional poles must be used.

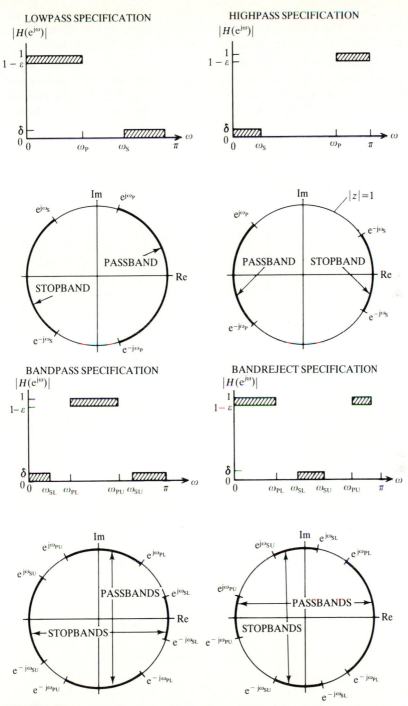

FIGURE 7-8
Mapping frequency regions in the specification onto the unit circle in the z plane.

Hint 2. The magnitude response over a narrow band of frequencies can be decreased by placing a zero at the angle corresponding to the center of the band. For maximum attenuation, the zero is placed on the unit circle.

Hint 3. When the log-magnitude responses of two singularities are added together, the location of the peak (for poles) or of the valley (for zeros) may shift owing to their interaction. This effect is shown in Fig. 7.9. In general, when two poles are close to the unit circle or distant from each other, there is

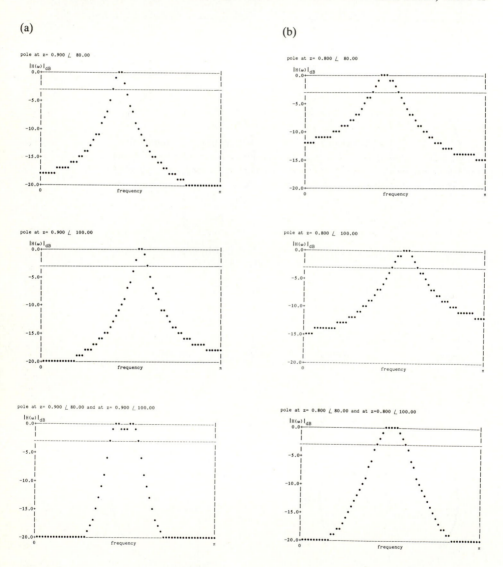

FIGURE 7-9
Interactions of two poles. a) Interaction produces two peaks when poles are close to the unit circle. b) Interaction produces one peak when poles are farther from the unit circle.

little interaction. When the poles are far from the unit circle or close together, their interaction is greater.

Two examples of performing the filter designs are given next. An algorithm for programming this procedure is suggested in the next section.

Example 7.5. Interactive design of a lowpass filter. Let us consider the following lowpass specification:

passband:

$$-1 < |H(e^{j\omega})|_{dB} \leq 0 \qquad \text{for } 0 \leq \omega \leq \pi/4$$

stopband:

$$|H(e^{j\omega})|_{dB} < -50 \qquad \text{for } \pi/2 \leq \omega \leq \pi$$

The pole/zero pattern indicating the placement of singularities is shown in Fig. 7.10a. The sequence of magnitude response curves are shown in Fig. 7.10b. These curves were produced by the log-magnitude plotting routine PLOTLM with DBVAL = −4, SPEC = −1, IF1 = 17 and IF2 = 0 for the passband plots, and for the stopband plots DBVAL = −60, SPEC = −50, IF1 = 33 and IF2 = 0.

The following sequence of poles and zeros resulted in a filter design that met the specification:

Singularities 1 and 2: A pole is placed in the middle of the passband at radius = 0.7 and a zero is placed in the middle of the stopband on the unit circle.

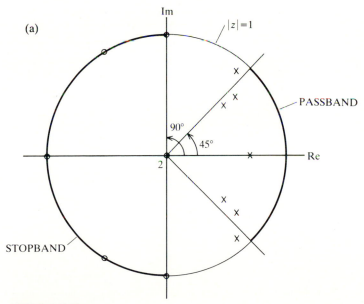

(a)

FIGURE 7-10
Interactive design of lowpass filter. a) Pole/zero pattern in z plane. b) Sequence of log-magnitude response curves. c) Digital filter structure.

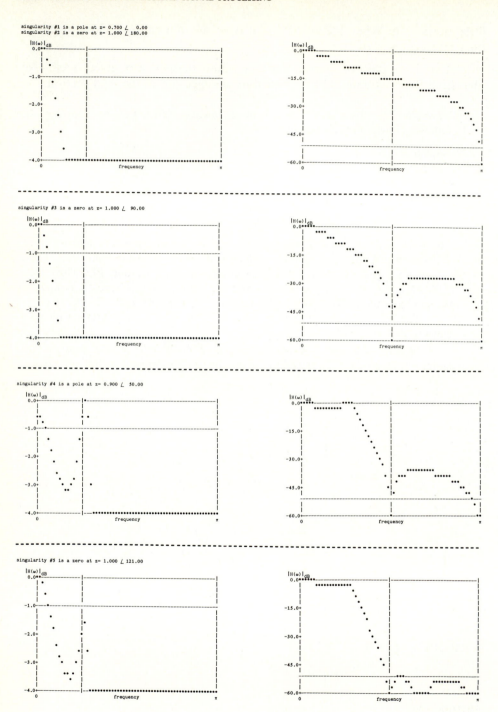

FIGURE 7-10b

singularity #6 is a pole at z= 0.750 ∠ 40.00

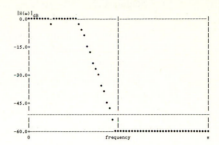

singularity #7 is a pole at z= 0.600 ∠ 38.00

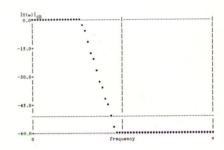

FIGURE 7-10b (*Continued*)

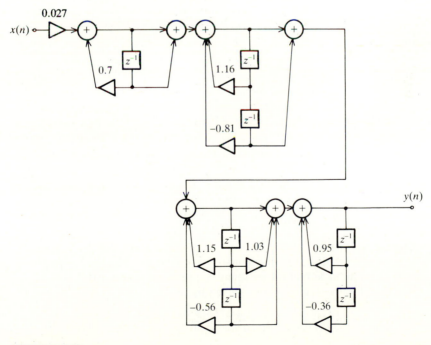

FIGURE 7-10c

Singularity 3: A zero is placed at the edge of the stopband on the unit circle.

Singularity 4: A pole is placed at $z = 0.9e^{\pm j50°}$ to boost the magnitude response in the passband. A maximum is produced slightly past the edge of the passband. Having such a maximum at the edge simplifies the procedure to satisfy the passband requirement.

Singularity 5: A zero is placed on the unit circle at the location of the maximum of the response in the stopband. The stopband specification is then nearly satisfied, but the response at the edge of the passband has been reduced.

Singularity 6: A pole is situated at $z = 0.75e^{\pm j40°}$ to boost the response at the edge of the passband.

Singularity 7: A pole is added at $z = 0.6e^{\pm j38°}$ to boost the response in the middle of the passband. A check of the stopband indicates that it too is satisfied. The final filter system function is given by

$$H(z) = \frac{0.027(1 + z^{-1})(1 + z^{-2})(1 + 1.03z^{-1} + z^{-2})}{(1 - 0.7z^{-1})(1 - 1.16z^{-1} + 0.81z^{-2})(1 - 1.15z^{-1} + 0.56z^{-2})(1 - 0.95z^{-1} + 0.36z^{-2})}$$

The filter structure is shown in Fig. 7.10c. The gain constant 0.0027 was determined by finding the maximum value of the log-magnitude response sequence and applying Eq. (7.16).

Example 7.6. Interactive design of a bandpass filter. Let us consider the following bandpass specification:

passband:

$$-1 < |H(e^{j\omega})|_{dB} \leq 0, \qquad \text{for } 0.42\pi \leq \omega \leq 0.61\pi,$$

stopband:

$$|H(e^{j\omega})|_{dB} < -50, \qquad \text{for } 0 \leq \omega \leq 0.16\pi \text{ and for } 0.87\pi \leq \omega \leq \pi.$$

The pole/zero pattern indicating the placement of singularities is shown in Fig. 7.11a. The sequence of magnitude response curves are shown in Fig. 7.11b.

The following sequence of poles and zeros resulted in a filter design that met the specification:

Singularities 1–3: A pole is placed in the center of the passband and zeros are placed at the centers of the stopbands on the unit circle.

Singularities 4 and 5: Poles are placed at the edges of the passband to increase the gain. The magnitude response indicates maxima at the edges of the passband, in anticipation of the reduction that will occur when zeros are added in the stopbands.

Singularities 6 and 7: Zeros are placed at the edges of the stopbands to satisfy the stopband specification. The magnitude response at the edges of the passband has also been reduced so that the passband requirement is also met.

The final filter system function is given by

$$H(z) = \frac{0.036(1 - z^{-2})(1 + 1.84z^{-1} + z^{-2})(1 - 1.77z^{-1} + z^{-2})}{(1 + 0.07z^{-1} + 0.49z^{-2})(1 - 0.57z^{-1} + 0.77z^{-2})(1 + 0.74z^{-1} + 0.77z^{-2})}$$

The filter structure is shown in Fig. 7.11c.

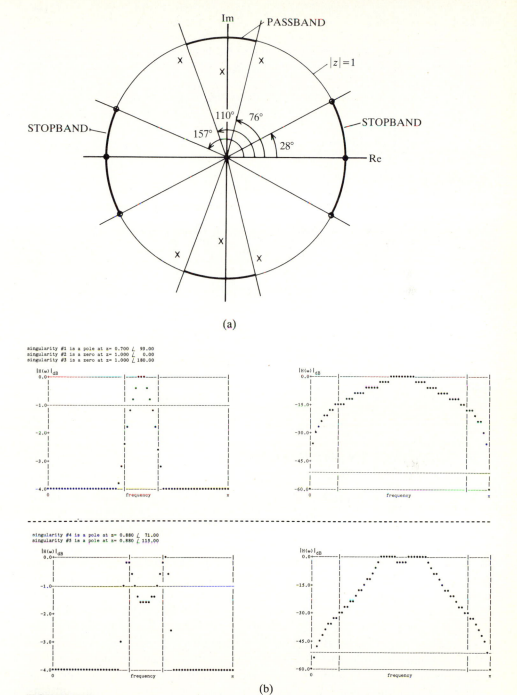

(a)

singularity #1 is a pole at z= 0.700 ∠ 93.00
singularity #2 is a zero at z= 1.000 ∠ 0.00
singularity #3 is a zero at z= 1.000 ∠ 180.00

singularity #4 is a pole at z= 0.880 ∠ 71.00
singularity #5 is a pole at z= 0.880 ∠ 115.00

(b)

FIGURE 7-11
Interactive design of bandpass filter. a) Pole/zero pattern in z plane. b) Sequence of log-magnitude response curves. c) Digital filter structure.

singularity #6 is a zero at z= 1.000 ∠ 157.00
singularity #7 is a zero at z= 1.000 ∠ 28.00

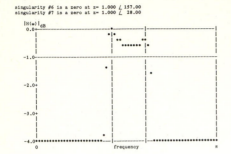
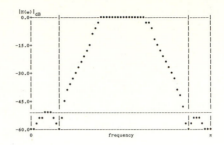

FIGURE 7-11b (*Continued*)

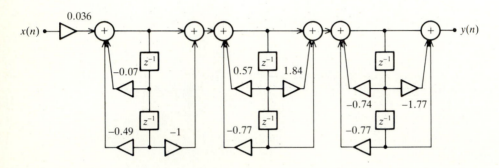

FIGURE 7-11c

7.10 WRITING AN INTERACTIVE DESIGN PROGRAM

To achieve a flexible operation, the interactive filter design program should be composed to include an additional single real or complex-conjugate pair of singularities at each iteration. The log-magnitude response of the filter section generating these singularities is computed and added to the total filter log-magnitude response. The running total log-magnitude response is then compared to the specification as each singularity is added. When the specification is met, the filter design is completed.

The logic diagram of this procedure is shown in Fig. 7.12. The procedure starts with the filter log-magnitude response specification indicating the passband and stopband locations and their respective attenuation levels. These values are inserted into the PLOTLM arguments to indicate these specified frequency and attenuation limits on the log-magnitude response plot. The filter response is monitored with the HTOT(N) array, for $1 \leq N \leq 65$, that contains the current values of the total filter log-magnitude response for $0 \leq \omega \leq \pi$. This total filter log-magnitude response is initialized to 0 dB. The responses of the

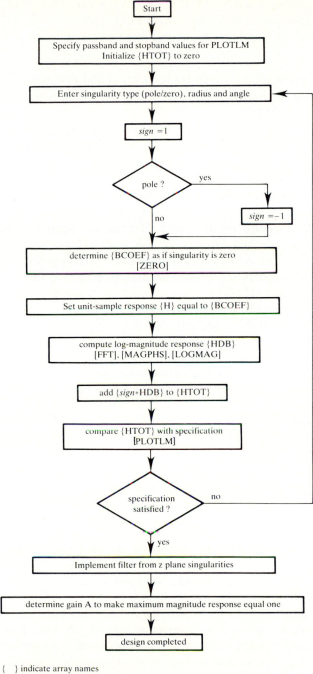

{ } indicate array names
[ ] indicate subroutines

FIGURE 7-12
Logic diagram for interactive filter design program.

sections are computed and accumulated in {HTOT}. The values in the HTOT array are superimposed on the specification curve with PLOTLM.

The interactive filter design procedure should perform the following steps.

Step 1. Input the desired singularity type (pole or zero) and position.

Step 2. Compute the corresponding filter section log-magnitude response and add this section response to {HTOT}.

Step 3. Compare the current values in {HTOT} with the desired specification by plotting with PLOTLM. If the specification is not met, return to Step 1 to include additional singularities. Otherwise, the filter design is done.

The logic diagram illustrates an efficient procedure for determining the log-magnitude response. Further, this procedure does not depend on the duration of the unit-sample response of the poles. It uses the feature that the poles contribute terms to the denominator of the system function. As such, the *inverse* elemental magnitude response contribution is computed by treating the pole as a zero, determining the log-magnitude response of this zero, and then *subtracting* it from {HTOT}. This subtraction is accomplished simply by assigning the correct value to the *sign* variable: if the specified singularity is a zero, $sign = 1$, but if the singularity is a pole, $sign = -1$.

The filter structure should be implemented as a cascade of first and second-order sections which has the minimum number of delay components. To determine the value of the filter gain A, the maximum value HMAX of the final log-magnitude response sequence {HTOT} should be found and substituted into Eq. (7.16).

This same program can be used for evaluating the IIR and FIR filters described in the next two chapters. In those cases, however, the iterative feature should not be necessary.

7.11 SUMMARY

In this chapter, the analytic techniques discussed in the previous chapters were summarized in such a way as to be useful for the synthesis of digital filters. The filter design section of the book began by considering some general issues that should be addressed before a filter design is initiated. A procedure for designing a digital filter by intuitive placement of poles and zeros in the z plane was described. This procedure provided an understanding of the effects of poles and zeros on the transfer function and an appreciation of the conventional filter design techniques to be studied in later chapters.

In the next two chapters, we will present the theory for placing the poles and zeros in the z plane from the magnitude response specification. This theory will help to formalize the intuitive procedure.

PROBLEMS

7.1. Determine the system function $H(z)$ for each of the following magnitude and phase responses (Let $\text{Arg}[H(e^{j0})] = 0$):

a) $|H(e^{j\omega})|^2 = 1/[5 - 4\cos(\omega)]$
   $\text{Arg}[H(e^{j\omega})] = \arctan[-2\sin(\omega)/(2\cos(\omega) - 1)]$

b) $|H(e^{j\omega})|^2 = 17 - 8\cos(2\omega)$
   $\text{Arg}[H(e^{j\omega})] = \arctan[4\sin(2\omega)/(1 - 4\cos(2\omega))]$

c) $|H(e^{j\omega})|^2 = 4/[1 - 2a^2\cos(2\omega) + a^4]$
   $\text{Arg}[H(e^{j\omega})] = \arctan[-a^2\sin(2\omega)/(1 - a^2\cos(2\omega))]$

7.2. Determine the system function $H(z)$ for each of the following magnitude and phase responses:

a) $|H(e^{j\omega})| = 2[1 - \cos(\omega)]$
   $\text{Arg}[H(e^{j\omega})] = -\pi$, for all ω

b) $|H(e^{j\omega})| = 2|\sin(\omega/2)|$
   $\text{Arg}[H(e^{j\omega})] = (\pi - \omega)/2$, for $0 < \omega \leq \pi$
   $= (-\pi - \omega)/2$, for $-\pi < \omega < 0$

c) $|H(e^{j\omega})| = 4|\sin(\omega)|$
   $\text{Arg}[H(e^{j\omega})] = -\pi/2$, for $0 < \omega < \pi$
   $= \pi/2$, for $-\pi < \omega < 0$

7.3. Determine a valid system function $H(z)$ for each of the following magnitude responses:

a) $|H(e^{j\omega})| = |\cos(\omega)|$
b) $|H(e^{j\omega})| = |\sin(2\omega)|$

7.4. Implement three digital filters that have the same structure but different coefficient values and that have the same magnitude response, given by

$$|H(e^{j\omega})| = 1 + \cos(\omega)/2.$$

Hint: assume zero phase and determine the pole/zero pattern.

7.5. Implement the digital filters in Examples 7.5 and 7.6 in the following forms: a) Direct Form II structure, and b) parallel connection of second-order sections.

7.6. Consider a second-order all-pole filter having poles at $z = re^{\pm j\theta}$ and zeros at $z = 0$. Find a general analytic expression for the frequency at which the magnitude response is a maximum. Evaluate this frequency and the maximum magnitude when $\theta = \pi/6$ and $\pi/2$ and $r = 0.5$ and 0.9.

COMPUTER PROJECTS

7.1. Magnitude response curves for poles and zeros
Object. Determine the log-magnitude response curves for single and complex-conjugate singularities as a function of radius and angle in the z plane.

Write the subroutine that computes the log-magnitude response of the magnitude response array given by

LOGMAG(HMAG,NH,HDB).

Analytic results

(a) Determine the log-magnitude response for first and second-order zeros and poles given below.
(b) Determine the maximum value of the magnitude response.
(c) Verify the locations of the half-power frequencies for the singularities below. Convert these frequency values to the nearest index values in the DFT sequence and indicate these on the computer plots below.

Computer results

(a) Compute and plot the log-magnitude responses using PLOTLM with DBVAL = −20., SPEC = −3. and IF1 = IF2 = 0.
(b) For each singularity draw the digital filter block diagram. Determine and include the filter gain A that makes the maximum of $|H(e^{j\omega})|$ over $0 \leq \omega \leq \pi$ equal to 1. Indicate the procedure you used to determine A.

Real-valued singularities (opposite singularity located at $z = 0$):

(i) zero at $z = 1$
(ii) zero at $z = -0.75$
(iii) pole at $z = 0.96$
(iv) pole at $z = -0.6$

Complex-valued singularities (opposite singularities located at $z = 0$):

(i) zeros at $z = e^{j\pi/4}$ and at $z = e^{-j\pi/4}$
(ii) zeros at $z = e^{j\pi/8}$ and at $z = e^{-j\pi/8}$
(iii) poles at $z = 0.8e^{j\pi/4}$ and at $z = 0.8e^{-j\pi/4}$
(iv) poles at $z = 0.8e^{j\pi/8}$ and at $z = 0.8e^{-j\pi/8}$

Instructions for the Filter Design Projects

In the projects below include the following for the final filter:
the filter block diagram
the system function
the pole/zero pattern in the z plane
the log-magnitude response plots
the phase response plot (indicate the causes of the phase jumps).

For these interactive design projects, once a successful filter design has been achieved, include the sequence of log-magnitude response plots obtained as the filter converged to the final design. To economize on paper, show only the log-magnitude plots with each additional two singularities and ignore any singularities that were canceled and not used in the final filter. Below each intermediate log-magnitude response plot, sketch the corresponding z plane pattern of poles and zeros and indicate the two new singularities that were added.

7.2. Intuitive filter design

Object. Design a filter that satisfies a given specification by pole and zero placement in the z plane.

Computer results. Design a digital filter that meets one of the specifications given in Appendix B. The specification is considered satisfied when, in the plot produced by PLOTLM, all the points of the filter log-magnitude response lie completely *inside* the passband and stopband limits.

7.3. Intuitive linear-phase filter design

Object. Design a linear-phase filter that satisfies a given specification by zero placement in the z plane.

A linear-phase filter is an all-zero filter whose zeros in the z plane form complex-conjugate quadruplets. When the zero is located on the unit circle it forms its own reciprocal. When the zero falls on the real axis, it forms its own complex conjugate.

Write an interactive program to design a filter by using only the ZERO and NONREC routines. When a zero is entered that is not on the unit circle, have the program automatically generate the reciprocal zero.

Computer results. Design a digital filter that meets one of the specifications given in Appendix B.

CHAPTER
8

INFINITE-IMPULSE RESPONSE FILTER DESIGN TECHNIQUES

8.1 INTRODUCTION

Infinite-Impulse Response (IIR) filter design procedures are extensions of those originally developed for analog filters and start with the design of the appropriate analog filter in the analog frequency domain. In fact, IIR digital filters are commonly used to replace existing analog filters. The complex-valued Laplace variable, $s = \sigma + j\Omega$, plays the same role for the analysis of analog filters as the variable z for discrete-time filters. The three most commonly used analog filter synthesis techniques for designing a lowpass filter in the s plane are the Butterworth, Chebyshev and elliptic techniques. These three are described and their advantages and disadvantages are compared. The impulse-invariance and the bilinear transformation methods are then presented for translating the s plane singularities of the analog filter into the z plane. A digital filter can then be implemented from these z plane singularities using the methods described in the previous chapters. Frequency transformations are employed to convert lowpass digital filter designs into highpass, bandpass and bandreject digital filters. All-pass filters are employed to alter only the phase response of the IIR digital filter to approximate a linear phase response over the passband.

262

8.2 ANALOG FILTER SYSTEM FUNCTION AND FREQUENCY RESPONSE

In this section, we establish the notation for the analog design techniques. The system function for an analog filter will be denoted by $H(s)$, where s is the Laplace complex variable $s = \sigma + j\Omega$, with Ω the analog frequency. The frequency transfer function of an analog filter, denoted by $H(j\Omega)$, is obtained by evaluating the system function $H(s)$ in the s plane along the frequency axis, or

$$H(j\Omega) = H(s)\big|_{s=j\Omega} \tag{8.1}$$

The squared magnitude response, $|H(j\Omega)|^2$, indicates the *power transfer characteristic* of the filter and is the function commonly employed for analog filter synthesis. The following equivalent forms are useful in the design of analog filters:

$$|H(j\Omega)|^2 = H(j\Omega)\, H^*(j\Omega) = H(s)\, H(-s)\big|_{s=j\Omega} \tag{8.2}$$

The analytic form of the last identity indicates that the singularities of $H(s)\, H(-s)$ in the s plane are symmetric about the $j\Omega$ axis. To obtain a stable system function from $H(s)\, H(-s)$, the poles that lie *in the left half of the s plane* are assigned to $H(s)$. The assignment of the zeros is arbitrary.

Example 8.1. Analog lowpass filter. Let us consider the lowpass power transfer function given by

$$|H(j\Omega)|^2 = (a^2 + \Omega^2)^{-1}$$

To obtain the system function, we replace Ω with $s/j = -js$, to get

$$H(s)\, H(-s) = (a^2 - s^2)^{-1} = [(a+s)(a-s)]^{-1}$$

The poles of $H(s)\, H(-s)$ are located at $s = \pm a$. Assigning the pole in the left half of the s plane to $H(s)$ for stability, we have

$$H(s) = [s + a]^{-1}$$

8.3 ANALOG LOWPASS FILTER DESIGN TECHNIQUES

In this section, we consider the problem of placing poles and zeros in the s plane to satisfy a given magnitude response specification. The Butterworth, Chebyshev and elliptic filter design techniques are described. Each is related to the type of approximation to the desired magnitude response that is employed. Each approach is illustrated by designing a lowpass filter to meet the specification shown in Fig. 8.1. In the next section, two methods are described to change these analog lowpass filters into digital lowpass filters. The following section then describes frequency transformations to convert the digital lowpass filter into one that is either highpass, bandpass or bandreject.

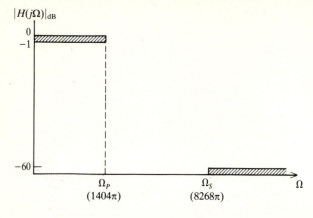

FIGURE 8-1
Analog lowpass filter magnitude specification.

These analog lowpass filters are commonly employed as anti-aliasing filters, that are applied to continuous-time signals before analog-to-digital conversion. They are also suitable as interpolation filters to convert pulses, having amplitudes proportional to the value of the elements in the discrete-time sequence, into signals that are continuous in time. Both these applications were described in Chapter 3.

BUTTERWORTH FILTER DESIGN PROCEDURE. Of the three filter design procedures described in this section, the Butterworth is the easiest to apply. Its simplicity lies in the straightforward pole/zero configuration of the filter transfer function in the s plane: all the poles lie on a circle centered at the origin of the s plane and all the zeros lie at infinity. The Butterworth lowpass filter magnitude response is also the smoothest, decreasing monotonically with frequency.

The Butterworth lowpass filter of order N is defined by

$$H_B(s)\,H_B(-s) = \frac{1}{1 + (s/j\Omega_C)^{2N}} \tag{8.3}$$

where Ω_C is the cutoff frequency. When $s = j\Omega_C$, we have $|H_B(j\Omega_C)|^2 = \frac{1}{2}$, or in logarithmic units, $|H_B(j\Omega_C)|_{dB} = -3$. Since the highest power of s in the denominator of Eq. (8.3) is $2N$, $H_B(s)\,H_B(-s)$ has $2N$ poles. These are located in the s plane at the values of s for which

$$(s/j\Omega_C)^{2N} = -1 \tag{8.4}$$

The roots can be found by multiplying both sides of Eq. (8.4) by $(j\Omega_C)^{2N}$ to give

$$s^{2N} = -(j\Omega_C)^{2N} = e^{j\pi}(\Omega_C\,e^{j\pi/2})^{2N} \tag{8.5}$$

Since the value of a complex number does not change when it is multiplied by $e^{j2\pi m}$, for integer values of m, we have

$$s^{2N} = e^{j\pi}(\Omega_C\,e^{j\pi/2})^{2N}\,e^{j2m\pi}$$
$$= \Omega_C^{2N}\,e^{j\pi(N+2m+1)} \tag{8.6}$$

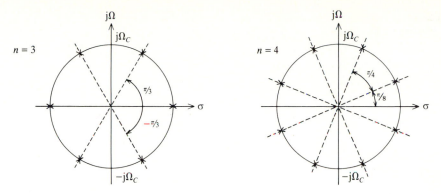

FIGURE 8-2
Pole pattern in the s plane for $H_B(s) H_B(-s)$ for $N = 3$ and $N = 4$.

The *distinct* locations of the poles are found by taking the $2N$th root, to give

$$s_m = \Omega_C \exp\left[j\pi\left(\frac{N + 2m + 1}{2N}\right) \right] \qquad \text{for } 0 \le m \le 2N - 1. \qquad (8.7)$$

The mth pole, for $0 \le m \le 2N - 1$, lies on a circle of radius Ω_C and at an angle with respect to the real axis equal to $(N + 2m + 1)\pi/2N$. The pole configurations in the s plane for $N = 3$ and $N = 4$ are shown in Fig. 8.2. The poles are equally spaced on a circle of radius Ω_C, but no poles lie on the $j\Omega$ axis. The poles of $H_B(s) H_B(-s)$ in the left half-plane are those belonging to the desired system function $H_B(s)$, while those in the right half-plane are from $H_B(-s)$. The $2N$ zeros of $H_B(s) H_B(-s)$ lie at $s = \infty$, and half of these are assigned to $H_B(s)$. Later in this chapter, we present two methods for transferring these s plane singularities into the z plane to obtain a digital Butterworth filter.

The power transfer function of the Butterworth filter is obtained by substituting $j\Omega$ for s into Eq. (8.3), to give

$$|H_B(j\Omega)|^2 = \frac{1}{1 + \left(\dfrac{\Omega}{\Omega_C}\right)^{2N}} \qquad (8.8)$$

For large Ω, the magnitude response decreases as Ω^{-N}, indicating the lowpass nature of this filter. The log-magnitude response is computed as

$$|H_B(j\Omega)|_{dB} = 10 \log_{10} |H_B(j\Omega)|^2$$
$$= -10 \log_{10} [1 + (\Omega/\Omega_C)^{2N}]$$

For large Ω, the log-magnitude response is approximately equal to

$$\lim_{\Omega \to \infty} |H_B(j\Omega)|_{dB} = -20 \log_{10} \Omega^N = -20 N \log_{10} \Omega \qquad (8.9)$$

As Ω increases by a factor of 10, or a *decade*, the log-magnitude response drops 20 dB for every pole, or $-20N$ dB/decade. If Ω increases by a factor of

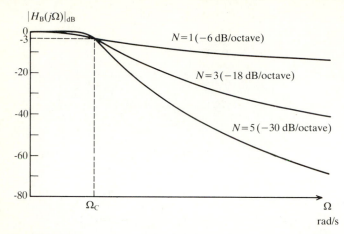

FIGURE 8-3
Analog log-magnitude response curves for Butterworth lowpass filters of order N.

two, or an *octave,* the response drops by $20 \log_{10} 2$, or $-6N$ dB/octave. The log-magnitude responses for different values of N are shown in Fig. 8.3. All the curves pass through -3 dB at $\Omega = \Omega_C$, and the steepness of the log-magnitude curve increases with the value of N. As N is increased, the magnitude response approaches that of the ideal lowpass filter. The value of N required for a particular application is determined by the passband and stopband specifications, as shown in the next example.

> **Example 8.2. Butterworth lowpass analog filter design.** Let us consider the following lowpass specification:
>
> passband:
>
> $$-1 < |H(j\Omega)|_{dB} \le 0 \qquad \text{for } 0 \le \Omega \le 1404\pi \text{ rad/s} \quad (\Omega_P = 1404\pi)$$
>
> stopband:
>
> $$|H(j\Omega)|_{dB} < -60 \qquad \text{for } \Omega \ge 8268\pi \text{ rad/s} \quad (\Omega_S = 8268\pi)$$
>
> **Determining the value of N.** If the cutoff frequency Ω_C were given, then we could solve the following inequality:
>
> $$|H_B(j\Omega_S)|^2 = [1 + (\Omega_S/\Omega_C)^{2N}]^{-1} < 10^{-6} \quad (-60 \text{ dB})$$
>
> for N. The solution is then
>
> $$N > \frac{\log(10^6 - 1)}{2 \log(\Omega_S/\Omega_C)}$$
>
> Since Ω_C is not given, a guess must be made. The Butterworth log-magnitude response approaches a slope of $-6N$ dB/octave at high frequencies. The specification calls for a drop of -59 dB in the frequency range from the edge of the passband (1404π) to the edge of the stopband (8286π). The frequency

difference is equal to $\log_2(8268/1404)$, or 2.56 octaves. Hence, an initial guess is found from

$$-6N = -59/2.56 \quad \text{or} \quad N = 3.8$$

Since the order of the filter must be an integer, we choose the next higher integer value, or $N = 4$.

We must now compute the value for the cutoff frequency Ω_C, which is also the radius of the Butterworth circle. One value for Ω_C is obtained by solving the inequality at the stopband edge with $N = 4$, or

$$|H_B(j\Omega_S)|^2 = [1 + (\Omega_S/\Omega_C)^{2N}]^{-1} < 10^{-6}$$

or

$$\Omega_C < \Omega_S \times 10^{-6/2N} = \Omega_S/5.62 = 1470.3\pi$$

Let us choose $\Omega_C = 1470\pi$. This inequality assures that the stopband specification will be met with these values of N and Ω_C. We need to test whether the passband specification is also satisfied, that is, we must test whether

$$|H_B(j\Omega_P)|^2 = [1 + (\Omega_P/\Omega_C)^{2N}]^{-1} > 0.794 \quad (= -1 \text{ dB})$$

Evaluating, we find

$$[1 + (1404/1470)^8]^{-1} = 1/1.692 = 0.59$$

The result is below the passband specification. Hence, a fourth-order filter is not sufficient. This is not unreasonable, since we employed a large Ω argument to find a first guess for N.

Let us increase N to 5. The stopband specification gives the Ω_c value:

$$\Omega_C < \Omega_S \times 10^{-6/2N} = \Omega_S/3.981 = 2076.8\pi$$

Using $\Omega_C = 2076\pi$ to test the passband specification, we get

$$|H_B(j\Omega_P)|^2 = [1 + (1404/2076)^{10}]^{-1} = 1/1.02 = 0.98$$

Hence, the stopband specification is met and the passband specification is exceeded with $N = 5$ and $\Omega_C = 2076\pi$.

We could also obtain a filter in which the passband specification is met and the stopband specification is exceeded by using $N = 5$ and solving the passband specification for the value of Ω_C. Doing this we get

$$|H_B(j\Omega_P)|^2 = [1 + (\Omega_P/\Omega_C)^{10}]^{-1} > 0.794$$

or

$$\log_{10} \Omega_C > \log_{10} \Omega_P - \frac{1}{10} \log\left(\frac{1}{0.794} - 1\right)$$

to give $\Omega_C > 1607\pi$. We then use this value to find the stopband attenuation

$$|H_B(j\Omega_S)|^2 = [1 + (\Omega_S/\Omega_C)^{2N}]^{-1} = 7.69 \times 10^{-8} \quad (= -71 \text{ dB})$$

Both filters meet or exceed the specification. The particular one to choose depends on other considerations in the design. To continue the example, we choose $N = 5$ and $\Omega_C = 2076\pi$.

Butterworth pole locations. The locations of the N poles on the Butterworth circle of radius Ω_C and in the left half of the s plane are given by

$$s_1 = -2076\pi,$$
$$s_{2,3} = 2076\pi(\cos(4\pi/5) \pm j\sin(4\pi/5)) = -1680\pi \pm j1220\pi = 2076\pi\, e^{\pm j144°}$$
$$s_{4,5} = 2076\pi(\cos(3\pi/5) \pm j\sin(3\pi/5)) = -641.5\pi \pm j1974\pi = 2076\pi\, e^{\pm j108°}$$

These are shown in Fig. 8.4(a). The system function is then given by

$$H_B(s) = \frac{(2076\pi)^5}{[s + 2076\pi][s^2 + 3359\pi s + (2076\pi)^2][s^2 + 1283\pi s + (2076\pi)^2]}$$

The log-magnitude and phase responses for this filter are shown in Fig. 8.4(b).

There are several interesting features that are peculiar to the Butterworth magnitude response:

1. It decreases monotonically as the frequency Ω increases from 0 to ∞. This characteristic is evident from the functional form given in Eq. (8.8). It is due to the configuration of the poles lying on a circle in the s plane. The other two analog filter design procedures described below will not have this smooth magnitude response characteristic.
2. The magnitude response is maximally flat about $\Omega = 0$, in that all its derivatives up to order N are equal to zero at $\Omega = 0$. To illustrate this, let us make $\Omega_C = 1$ and $N = 1$ for simplicity. We then have

$$|H_B(j\Omega)|^2 = (1 + \Omega^2)^{-1} \tag{8.10}$$

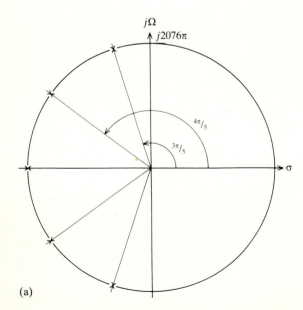

(a)

FIGURE 8-4
Fifth-order Butterworth analog lowpass filter results. a) Pole/zero pattern in the s plane. b) Log-magnitude and phase responses.

PASSBAND:

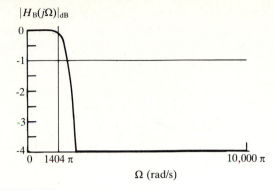

STOPBAND:

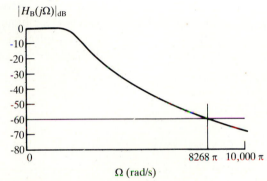

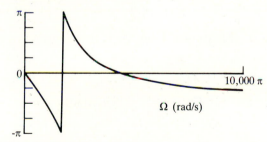

FIGURE 8-4b

The first derivative is then equal to

$$\frac{d\,|H_B(j\Omega)|^2}{d\Omega} = \frac{-2\Omega}{(1+\Omega^2)^2} \tag{8.11}$$

which is zero at $\Omega = 0$. The second derivative is equal to

$$\frac{d^2\,|H_B(j\Omega)|^2}{d\Omega^2} = \frac{-2(1+\Omega^2)^2 + 8\Omega^2(1+\Omega^2)}{(1+\Omega^2)^4}$$

Note that the second derivative is equal to -2 at $\Omega = 0$.

3. The phase response curve approaches $-N\pi/2$ for large Ω, where N is the number of poles on the Butterworth circle in the left half s plane. When the value of N is large, the discontinuities in the principal value of the phase response occasionally cause difficulties in the interpretation of the phase curve.

The Butterworth filter is the easiest one to design. It is commonly used because the smoothness of the magnitude response. This smoothness is bought at a price, however. The Butterworth characteristic requires a relatively large transition range between the passband and stopband. We next consider the Chebyshev filter design that has a smaller transition band, but its magnitude response will not be monotonically decreasing with frequency.

CHEBYSHEV FILTER DESIGN PROCEDURE. The Chebyshev filter is optimum in that for the given passband and stopband levels, it has the smallest transition region, $\Omega_S - \Omega_P$, of any filter that consists *only of finite poles*. Such *all-pole* filters have all their zeros at $s = \infty$. The Chebyshev filter is defined by

$$H_C(s)\, H_C(-s) = [1 + \mu^2 C_N^2(s/j\Omega_P)]^{-1} \qquad (8.12)$$

where μ is a measure of the allowable deviation in the passband, as shown below, and

$$C_N(x) = \cos(N \cos^{-1}(x)) \qquad (8.13)$$

is the Nth-order Chebyshev polynomial. The corresponding power transfer function is given by

$$|H_C(j\Omega)|^2 = [1 + \mu^2 C_N^2(\Omega/\Omega_P)]^{-1} \qquad (8.14)$$

The Chebyshev polynomials can be determined from the following recursion formula:

$$C_{N+1}(x) = 2xC_N(x) - C_{N-1}(x) \qquad \text{for } N \geq 1 \qquad (8.15)$$

with $C_0(x) = 1$ and $C_1(x) = x$. The first five Chebyshev polynomials are given in Table 8.1.

TABLE 8.1
Chebyshev polynomials $C_N(x)$

N	$C_N(x)$
0	1
1	x
2	$2x^2 - 1$
3	$4x^3 - 3x$
4	$8x^4 - 8x^2 + 1$
5	$16x^5 - 20x^3 + 5x$

The following two features of Chebyshev polynomials are important for filter design.

1. $|C_N(x)| \leq 1$ for $|x| \leq 1$. Inserting this result into Eq. (8.14) indicates that, in the passband, the power transfer function lies in the range

$$(1 + \mu^2)^{-1} \leq |H_C(j\Omega)|^2 \leq 1 \qquad \text{for } 0 \leq \Omega \leq \Omega_P \qquad (8.16)$$

Whereas the frequency value important for the design of the Butterworth filter was the cutoff frequency Ω_C, the relevant frequency for the Chebyshev filter is the *edge of the passband* Ω_P. This allows the Chebyshev filter to be specified directly in terms of the two important design parameters consisting of the passband frequency and the allowable deviation in the passband.

2. For $|x| \gg 1$, $|C_N(x)|$ increases as the Nth power of x. This indicates that for $\Omega \gg \Omega_P$ the magnitude response decreases as Ω^{-N}, or the log-magnitude response decreases by $6N$ dB/octave. This large-frequency result is identical to that for the Butterworth filter.

Where the Butterworth filter poles lie on a circle in the s plane, it can be shown that the poles of $H_C(s)$ lie on an ellipse. The ellipse is defined by two circles determining the major and minor axes, as shown in Fig. 8.5a. If we define the parameter ρ as

$$\rho = \mu^{-1} + \sqrt{1 + \mu^{-2}} \qquad (8.17)$$

then r, the radius of the circle defining the minor axis, is equal to

$$r = \Omega_P(\rho^{1/N} - \rho^{-1/N})/2 \qquad (8.18)$$

where N is the order of the filter. The radius R of the circle defining the major axis, is equal to

$$R = \Omega_P(\rho^{1/N} + \rho^{-1/N})/2 \qquad (8.19)$$

The locations of the poles on the ellipse are found by first finding the Butterworth pole locations. These are then projected onto the ellipse by keeping the frequency constant, as shown in Fig. 8.5b.

The locations of the Chebyshev poles can be determined without actually constructing the ellipse. As shown in Fig. 8.5b, the ordinate ($j\Omega$) locations of the poles are defined by the intersections of the N lines with the larger circle. The abscissa (σ) locations are defined by the intersections of the lines with the smaller circle. For example, let us consider the line that makes an angle θ with respect to the positive real axis. The corresponding pole of the Chebyshev filter is located in the left half-plane at $s = s_P$, where

$$s_P = r \cos(\theta) + jR \sin(\theta) \qquad (8.20)$$

As with the Butterworth filter, the poles of $H_C(s) H_C(-s)$ that lie in the left half of the s plane are assigned to the system function $H_C(s)$, and all the zeros of the Chebyshev filter lie at $s = \infty$.

(a)

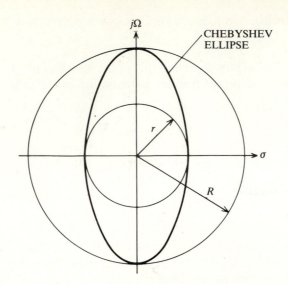

(b)

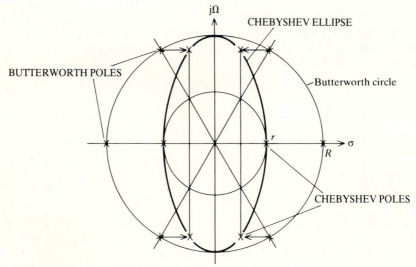

FIGURE 8-5
Chebyshev ellipse. a) Construction of ellipse. b) Pole pattern in the s plane for $H_C(s) H_C(-s)$ for $N = 3$.

Example 8.3. Chebyshev lowpass analog filter design. Consider the same lowpass specification given in Example 8.2:

passband:

$$-1 < |H(j\Omega)|_{dB} \leq 0 \qquad \text{for } 0 \leq \Omega \leq 1404\pi \text{ rad/s} \quad (\Omega_P = 1404\pi)$$

stopband:

$$|H(j\Omega)|_{dB} < -60 \qquad \text{for } \Omega \geq 8268\pi \text{ rad/s} \quad (\Omega_S = 8268\pi).$$

Value of μ. The value of μ is determined from the passband ripple, or

$$10 \log(1 + \mu^2)^{-1} > -1 \text{ dB}$$

This gives the value as

$$\mu < [10^{0.1} - 1]^{1/2} = 0.508$$

We will take $\mu = 0.508$.

Value of N. The order of the Chebyshev filter is determined from the stopband inequality

$$|H_C(j\Omega_S)|^2 = [1 + \mu^2 C_N^2(\Omega_S/\Omega_P)]^{-1} < 10^{-6} \quad (-60 \text{ dB})$$

Since $\Omega_S/\Omega_P = 5.9$, we evaluate $C_N(5.9)$ for increasing N until the stopband inequality is satisfied, or

$$C_N(5.9) > [(10^6 - 1)/\mu^2]^{1/2} = 1969.$$

Evaluating, we find $C_3(5.9) = 804$, $C_4(5.9) = 9416$. Hence, $N = 4$ is sufficient. Since this last inequality is easily satisfied with $N = 4$ the value of μ can be reduced to as small as 0.11, to decrease passband ripple while satisfying the stopband. The value $\mu = 0.4$ provides a margin in both the passband and stopband. We proceed with the design with $\mu = 0.508$ to show the 1 dB ripple in the passband.

Axes of ellipse. The radii of the circles defining the major and minor axes of the ellipse are determined from the constant ρ, which is equal to

$$\rho = 0.508^{-1} + (1 + 0.508^{-2})^{1/2} = 4.17$$

The radii are equal to

$$R = \frac{1404\pi}{2}(4.17^{1/4} + 4.17^{-1/4}) = 702\pi(1.43 + 0.67) = 1473\pi$$

and

$$r = \frac{1404\pi}{2}(4.17^{1/4} - 4.17^{-1/4}) = 702\pi(1.43 - 0.67) = 533\pi.$$

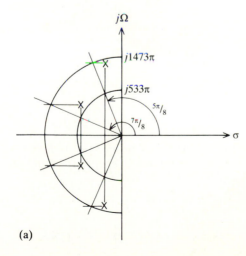

(a)

FIGURE 8-6
Fourth-order Chebyshev analog lowpass filter results. a) Pole/zero pattern in the s plane. b) Log-magnitude and phase responses.

PASSBAND:

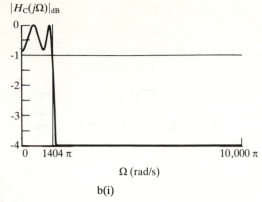

b(i)

STOPBAND:

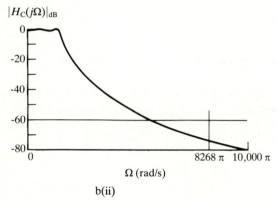

b(ii)

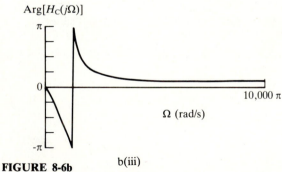

b(iii)

FIGURE 8-6b

Pole locations. Since there are four poles, the four lines dividing the s-plane have angles equal to $\pm 7\pi/8$ and $\pm 5\pi/8$ as shown in Fig. 8.6(a). The four poles of the Chebyshev filter are then at

$$s_{1,2} = 533\pi \cos(7\pi/8) \pm j1473\pi \sin(7\pi/8) = -492\pi \pm j564\pi = 748\pi \, e^{\pm j131°}$$

and

$$s_{3,4} = 533\pi \cos(5\pi/8) \pm j1473\pi \sin(5\pi/8) = -204\pi \pm j1361\pi = 1376\pi \, e^{\pm j98.5°}$$

The system function is then given by

$$H_C(s) = \frac{1.06\pi^4 \times 10^{12}}{[s^2 + 981\pi s + (748\pi)^2][s^2 + 407\pi s + (1376\pi)^2]}$$

The log-magnitude and phase responses for this Chebyshev filter are shown in Fig. 8.6(b).

Owing to the closer proximity of the Chebyshev filter poles to the $j\Omega$ axis than those in the Butterworth filter, the magnitude response of the Chebyshev filter exhibits a ripple in the passband. There is a peak in the passband for each pole in the filter, located approximately at the ordinate value of the pole. Even though the passband contains ripples, the magnitude response beyond the passband is a smooth function that decreases monotonically with frequency. But, of importance of implementing the filter having the smallest number of components, the Chebyshev magnitude response exhibits a smaller transition region to reach the desired attenuation in the stopband, when compared to the Butterworth filter.

The Chebyshev phase response is similar to that of the Butterworth filter, in that it decreases monotonically to $-N\pi/2$. In the example, this asymptote is equal to -2π.

Because of the proximity of the Chebyshev filter poles to the $j\Omega$ axis, small errors in their locations, caused by numerical round-off in the computations, can result in significant changes in the magnitude response. This is especially true in the passband. Choosing a smaller value of μ will provide some margin for keeping the ripples within the passband specification. However, too small a value for μ may require an increase in the filter order.

In the Butterworth and Chebyshev filters, we were constrained to use only finite poles to achieve the magnitude specification. It is reasonable to expect that if relevant (finite) zeros were included in the system function, a lower-order filter can be found to satisfy the specification. These relevant zeros could serve to achieve additional attenuation in the stopband. The elliptic filter, described next, does exactly this.

ELLIPTIC FILTER DESIGN PROCEDURE. The elliptic filter has the smallest transition band of the three approaches described here, but is the most complicated analytically. The elliptic filter is defined by

$$H_E(s)\, H_E(-s) = [1 + \varepsilon^2 E_N^2(s/j\Omega_P)]^{-1} \tag{8.21}$$

where ε is a measure of the allowable deviation in the passband, Ω_P is the edge of the passband, and $E_N(x)$ is the Jacobian elliptic function of order N. The analysis of this filter is rather complicated and beyond the scope of this book, but can be found in the references. It is sufficient to state here that, unlike the Chebyshev *polynomial,* $E_N(s)$ can be considered to be a *rational function,* consisting of both a numerator polynomial and a denominator polynomial in s. The denominator polynomial of $E_N^2(s/j\Omega_P)$ would then appear in the numerator of $H_E(s)\, H_E(-s)$ and its roots would then become the zeros of the system function. For example, although simpler than the first Jacobian elliptic function, let us consider the following rational function,

$$E(s) = \frac{(s^2 - 1)}{(s^2 + 1)} \tag{8.22}$$

Then

$$E^2(s) = \frac{(s^2 - 1)^2}{(s^2 + 1)^2} \tag{8.23}$$

The system function is then given by

$$H(s)\, H(-s) = \cfrac{1}{1 + \varepsilon^2 \cfrac{(s^2 - 1)^2}{(s^2 + 1)^2}}$$

$$= \frac{(s^2 + 1)^2}{(s^2 + 1)^2 + \varepsilon(s^2 - 1)^2} \tag{8.24}$$

Note that $H(s)$ has zeros at $s = \pm j$, and that these zeros came from the denominator of $E(s)$.

The elliptic filter follows an obvious extension of the pole/zero placement procedure of the Butterworth and Chebyshev filters. The Butterworth transfer function decreases monotonically over the entire frequency range. By allowing ripples to occur in the passband with the Chebyshev filter, the transition region is reduced and the attenuation decreases monotonically in the stopband. Following this logic, the transition region can be reduced still further by also allowing a ripple to occur in the stopband. This ripple is accomplished in the elliptic filter by placing the zeros on the $j\Omega$ axis in the stopband. For the Butterworth and Chebyshev filters, all the zeros were located at $s = \infty$.

The zeros of $H_E(s)\, H_E(-s)$ appear in sets of pairs on the finite $j\Omega$ axis. One of each pair is assigned to $H_E(s)$, when $H_E(s)\, H_E(-s)$ is factored and the singularities are distributed between $H_E(s)$ and $H_E(-s)$. In addition, for the coefficients to be real-valued, the zeros must also occur in complex-conjugate pairs. Hence, if there are two zeros at $s = j\Omega_0$, then there are also two at $s = -j\Omega_0$. If the order of the system function N is odd, $H_E(s)$ has one zero at $s = \infty$. For a lowpass filter, there are no zeros at $s = 0$.

This additional complication with the placement of the zeros makes the design of the elliptic filter very complicated analytically. To circumvent the mathematics, but yet achieve an approximate elliptic filter magnitude response, we use the insights gained by the intuitive pole/zero placement described in the previous chapter. The elliptic filter design will be approximated by initially designing a Chebyshev filter, to determine the approximate location of the poles. Then the pole/zero pattern is modified by placing zeros on the $j\Omega$ axis to achieve additional stopband attenuation. These additional zeros, however, also reduce the magnitude response in the passband. If the passband specification is not met after these additional zeros are included, the pole locations must be slightly modified to satisfy the specification. This procedure is illustrated in the following example.

Example 8.4. Approximate elliptic analog filter design. Let us consider the previous lowpass specification:

passband:

$$-1 < |H(j\Omega)|_{dB} \leq 0 \qquad \text{for } 0 \leq \Omega \leq 1404\pi \text{ rad/s} \quad (\Omega_P = 1404\pi)$$

stopband:

$$|H(j\Omega)|_{dB} < -60 \qquad \text{for } \Omega \geq 8268\pi \text{ rad/s} \quad (\Omega_s = 8268\pi).$$

To approximate the elliptic filter, we design a Chebyshev filter having a lower order than that in the previous example, and then add zeros to satisfy the specification.

Value of μ. The value of μ was determined from the passband ripple in Example 8.3, to give $\mu = 0.508$.

Value of N. The order of the Chebyshev filter in the previous example was 4. Let us consider a third-order Chebyshev filter.

Axes of the ellipse. The radii of the circles defining the major and minor axis of the ellipse are determined from the constant ρ. In Example 8.3, $\rho = 4.17$. The radii are equal to

$$R = \frac{1404\pi}{2}(4.17^{1/3} + 4.17^{-1/3}) = 702\pi(1.61 + 0.62) = 1566\pi$$

and

$$r = \frac{1404\pi}{2}(4.17^{1/3} - 4.17^{-1/3}) = 702\pi(1.61 - 0.62) = 694\pi$$

Pole locations. Since there are three poles, the three lines dividing the s plane have angles equal to π and $\pm 2\pi/3$, as shown in Fig. 8.7(*a*). The poles are then located at

$$s_1 = -694\pi$$

$$s_{2,3} = 694\pi \cos(2\pi/3) \pm j1566\pi \sin(2\pi/3) = -347\pi \pm j1356\pi = 1400\pi \, e^{\pm j104.5}$$

The log-magnitude response for this Chebyshev filter is shown in Fig. 8.7(*b*). As expected, the stopband specification is not met with this third-order filter, although the passband specification is satisfied.

Zero locations. To achieve the stopband specification, we move two of the three zeros of the Chebyshev filter, which are all located at $\Omega = \infty$, to finite values on the $j\Omega$ axis. An interactive pole/zero placement procedure, similar to that described in the previous chapter, was employed to place the zeros at

$$s_{Z1,Z2} = \pm j9000\pi$$

These zeros also reduced the magnitude response in the passband, such that the passband specification is not met. The locations of the complex-conjugate Chebyshev poles were changed, until both the passband and stopband specifica-

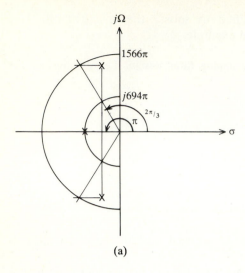

(a)

PASSBAND:

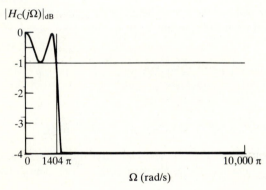

STOPBAND:

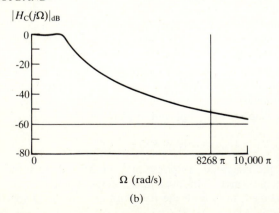

(b)

FIGURE 8-7
Approximate elliptic filter design results. a) Third-order Chebyshev analog lowpass filter pole/zero pattern in the s plane. b) Log-magnitude response of third-order filter. c) Log-magnitude and phase responses after zeros are included at $s = \pm j9000\pi$.

PASSBAND:

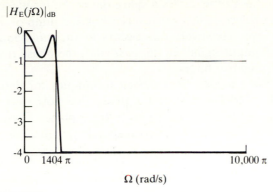

STOPBAND:

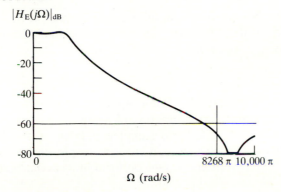

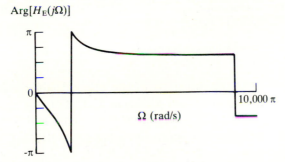

FIGURE 8-7c

tions were met. The final pole locations are given by

$$s_{P1} = -750\pi$$

$$s_{P2,P3} = 1450\, e^{\pm j105}$$

The system function is then given by

$$H_E(s) = \frac{61[s^2 + (9000\pi)^2]}{[s + 750\pi][s^2 - 751\pi s + (1450\pi)^2]}$$

The log-magnitude and phase responses are shown in Fig. 8.7(c).

In the previous example, the stopband magnitude response is no longer monotonically decreasing, but rather shows a sharp dip at the frequency where the zero is located. The presence of this zero causes a phase jump of π radians in the phase response curve. However, this occurs in the stopband, where there is little effect on the output signal. In the passband, the phase response is similar to that of the previous two filters.

Having considered three analog filter design techniques for determining the locations of the poles and zeros in the s plane, we now describe two procedures for transferring these poles and zeros to the z plane. These z plane singularities are then used to implement the corresponding digital filter.

8.4 METHODS TO CONVERT ANALOG FILTERS INTO DIGITAL FILTERS

To convert an analog filter into a digital filter, the singularities in the s plane must be mapped into equivalent singularities in the z plane. This section describes two such methods: the impulse-invariance method and the bilinear transform method.

THE IMPULSE-INVARIANCE METHOD. When replacing an analog filter with its digital counterpart, it would be intuitively pleasing if the unit-sample response of the digital filter resembled the sampled version of the impulse response of the analog filter. If such a resemblance were present in the time domain, then, intuitively, the two filters should perform in a similar manner in the frequency domain. The impulse-invariance method does exactly this. It produces the digital filter whose unit-sample response is equal to samples of the analog filter impulse response, or

$$h(nT_s) = h_A(t)\big|_{t=nT_s} \tag{8.25}$$

where T_s is the sampling period.

We now compare the transfer functions of the original analog filter and the resulting digital filter obtained with this method. Let us recall the relationship between the spectrum of a continuous-time signal and that of the sampled signal, discussed in Chapter 3. Here the analog signal is the impulse response of the analog filter. Its spectrum is the filter transfer function, $H_A(j\Omega)$, which is found by taking the continuous-time Fourier transform of the impulse response, given by

$$H_A(j\Omega) = \int_0^\infty h_A(t)\,e^{-j\Omega t}\,dt \tag{8.26}$$

For sampled-data applications, we must scale the discrete-time values to the analog frequencies by dividing ω by T_s. Hence, the digital filter transfer function is denoted by $H(e^{j\omega/T_s})$. By sampling the impulse response to obtain the unit-sample response, the transfer function of the digital filter becomes the periodic extension of that of the analog filter:

$$H(e^{j\omega/T_s}) = \sum_{k=-\infty}^{\infty} H_A(j\omega/T_s + j2\pi k/T_s) \qquad (8.27)$$

If $H_A(j\Omega)$ is limited to the frequency range $-\Omega_M \le \Omega \le \Omega_M$, then no aliasing would occur if the sampling period T_s is chosen to be less than π/Ω_M. However, since practical analog filters are composed of a finite number of R, L and C components, $H_A(j\Omega)$ can be zero at no more than N frequency points, where N is the order of the filter. These points correspond to the locations of the zeros that may be located on the finite $j\Omega$ axis. Hence, $H_A(j\Omega)$ cannot be truly band-limited, and some degree of aliasing will *always* occur. For filters having a $|H_A(j\Omega)|$ that approaches zero as Ω approaches infinity, such as those of lowpass and bandpass filters, this aliasing error decreases as the sampling period T_s approaches 0.

Example 8.5. Impulse-invariance method applied to a first-order filter. Let us consider the RC lowpass filter shown in Fig. 8.8. If we let $a = 1/RC$, the impulse response is equal to

$$h_A(t) = \begin{cases} e^{-at} & \text{for } t \ge 0 \\ 0 & \text{otherwise} \end{cases}$$

The analog system function is defined as the Laplace transform of $h_A(t)$, or

$$H_A(s) = \int_0^{\infty} h_A(t)\, e^{-st}\, dt = (s+a)^{-1}$$

The analog transfer function is the system function evaluated at $s = j\Omega$, or

$$H_A(j\Omega) = (a + j\Omega)^{-1}$$

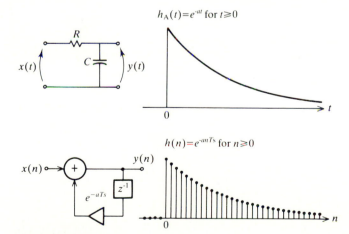

FIGURE 8-8
Comparison of analog and digital filters using the impulse-invariance method. Filter structures and time-domain responses are shown.

MAGNITUDE:

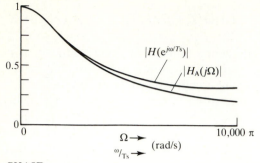

PHASE:

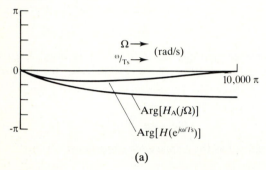

(a)

MAGNITUDE:

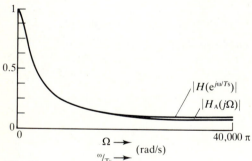

PHASE:

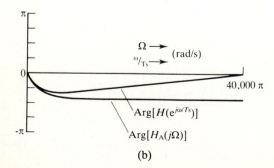

(b)

FIGURE 8-9

Comparison of frequency responses of digital and analog filters. a) $T_s = 10^{-4}$ seconds. b) $T_s = 0.25 \times 10^{-4}$ seconds.

The unit-sample response derived by sampling $h_A(t)$ every T_s seconds is given by

$$h(nT_s) = \begin{cases} e^{-anT_s} & \text{for } n \geq 0 \\ 0 & \text{otherwise} \end{cases}$$

The digital frequency response is given by

$$H(e^{j\omega/T_s}) = \sum_{n=0}^{\infty} e^{-anT_s} e^{-jn\omega/T_s} = (1 - e^{-aT_s} e^{-j\omega/T_s})^{-1}$$

The digital filter system function is determined by substituting z for $e^{j\omega/T_s}$, or

$$H(z) = \sum_{n=0}^{\infty} h(nT_s)z^{-n} = (1 - e^{-aT_s}z^{-1})^{-1}$$

The digital filter that replaces the RC lowpass filter is shown in Fig. 8.8.

Comparing the values of the analog and digital frequency responses at zero frequency, we find that $H_A(0)$ is $1/a$ and $H(e^{j0}) = 1/(1 - e^{-aT_s})$. For small values of T_s, $H(e^{j0}) \cong 1/(aT_s)$.

A comparison of $|H(e^{j\omega/T_s})|$ and $|H_A(j\Omega)|$ is shown in Fig. 8.9 for different values of T_s. As T_s approaches zero, $\{h(n)\}$ appears more similar to $h_A(t)$, indicating that the aliasing is less severe, which, in turn, implies that $H(e^{j\omega/T_s})$ approaches $H_A(j\Omega)$ over the range $-\pi/T_s \leq \omega/T_s < \pi/T_s$.

Relationship of the s plane and the z plane. We now relate locations of the singularities in the s plane to those in the z plane when the impulse-invariance method is employed. In the z plane, the frequency "axis" corresponds to the unit circle, with $\omega = 0$ located at $z = 1$. In the s plane, the frequencies are located on the vertical axis with $\Omega = 0$ at the origin. These zero-frequency points in the two planes should coincide. By equating individual frequency points in the range $-\pi/T_s \leq \Omega \leq \pi/T_s$ with $-\pi/T_s \leq \omega/T_s \leq \pi/T_s$, as shown in Fig. 8.10, we find that the $j\Omega$-axis *wraps around* upon the unit circle.

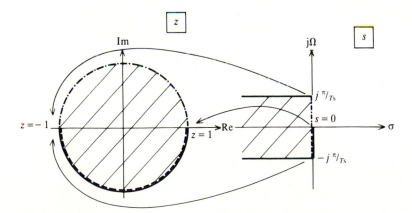

FIGURE 8-10
Relationship of the s plane to the z plane for the impulse-invariance method.

Continuing this point-by-point mapping, we find that the interval $(2k - 1)\pi/T_s \leq \Omega < (2k + 1)\pi/T_s$, for any integer k, maps onto $-\pi/T_s \leq \omega T_s < \pi/T_s$. With this mapping, the left half of the s plane becomes the inside of the unit circle and the right half of the s plane becomes the outside of the unit circle. This relationship between the s plane and the z plane is reasonable, since, for stability, the poles of the analog filter must lie in the left half of the s plane, while those for stable digital filters must lie within the unit circle.

Pole-mapping procedure. We now derive the mapping rule that transfers the poles from the s plane to the z plane. Let us factor $H_A(s)$ into the sum of first-order system functions given by

$$H_A(s) = \sum_{k=1}^{N} \frac{\alpha_k}{s + a_k} \tag{8.28}$$

where both α_k and a_k may be complex numbers. The total impulse response is then equal to the sum of the responses of the individual first-order systems, or

$$h_A(t) = \sum_{k=1}^{N} \alpha_k e^{-a_k t} \tag{8.29}$$

Sampling this impulse response, we obtain the discrete-time sequence given by

$$h(nT_s) = \sum_{k=1}^{N} \alpha_k e^{-a_k nT_s} \tag{8.30}$$

When viewed as an exponentially decreasing unit-sample response, the kth sequence could be generated by a first-order recursive digital filter having its pole located at $z = e^{-a_k T_s}$. The linearity of the z-transform allows us to sum these components to obtain the digital filter system function given by

$$H(z) = \sum_{k=1}^{N} \frac{\alpha_k}{1 - e^{-a_k T_s} z^{-1}} \tag{8.31}$$

By comparing the pole locations in this last equation with those in (8.28), we arrive at the pole-mapping rule of the impulse-invariance method. This rule can be stated as

a pole located at $s = s_P$ in the s plane is transformed into a pole in the z plane located at $z = e^{s_P T_s}$.

In the above formulation, $s_P = -a_k$. Examining the magnitude of the pole in z plane, we find that poles lying in the left half of the s plane are mapped into the inside of the unit circle. Hence, a stable analog filter converts into a stable digital filter after the mapping.

In the derivation of the rule, $H_A(s)$ was factored into the sum of terms in (8.28), corresponding to a parallel connection of filters. As discussed in Chapter 5, the location of the finite *zeros* of a parallel connection of filters does not relate in such a simple manner, but is rather complicated. It can be shown, however, that the zeros at $s = \infty$ map to $z = 0$. We now consider an example.

Example 8.6. Impulse-invariance method applied to a second-order filter. Let

$$H_A(s) = \frac{s + a}{(s + a)^2 + b^2}$$

$$= \frac{s + a}{(s + a + jb)(s + a - jb)}$$

The inverse Laplace transform yields

$$h_A(t) = \begin{cases} e^{-at} \cos(bt) & \text{for } t \geq 0 \\ 0 & \text{otherwise} \end{cases}$$

Sampling this function produces

$$h(nT_s) = \begin{cases} e^{-anT_s} \cos(bnT_s) & \text{for } n \geq 0 \\ 0 & \text{otherwise} \end{cases}$$

The z-transform of $h(nT_s)$ is equal to

$$H(z) = \frac{1 - e^{-aT_s} \cos(bT_s)\, z^{-1}}{1 - 2e^{-aT_s} \cos(bT_s)z^{-1} + z^{-2}}$$

$$= \frac{1 - e^{-aT_s} \cos(bT_s)\, z^{-1}}{(1 - e^{-(a+jb)T_s}z^{-1})(1 - e^{-(a-jb)T_s}z^{-1})}$$

Comparing the two system functions, we see that the pole located in the s plane at $s = p$ is transformed into a pole in the z plane at $z = e^{pT_s}$, as before. However, the finite zero located in the s plane at $s = -a$ was not converted into a zero in the z plane at $z = e^{-aT_s}$, although the zero at $s = \infty$ was placed at $z = 0$. The digital filter structure is shown in Fig. 8.11.

We illustrate the application of the impulse invariance method to the design of a Chebyshev lowpass filter in the next example.

Example 8.7. Chebyshev lowpass digital filter using impulse-invariance method. From Example 8.3, the system function poles of the Chebyshev filter

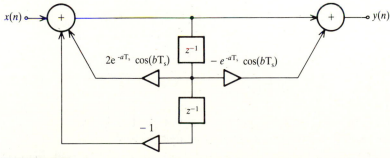

FIGURE 8-11
Second-order digital filter implementation using impulse-invariance method.

were found to be

$$s_{1,2} = -492\pi \pm j564\pi \qquad s_{3,4} = -204\pi \pm j1361\pi$$

We must now choose a value for T_s. The frequency responses for the analog filters were plotted over the frequency range from 0 to $10{,}000\pi$ rad/s. To set the discrete-time frequency range $[0, \pi/T_s]$ equal to this analog range, we choose $T_s = 10^{-4}$ seconds. With this value for T_s, the poles are mapped to

$$z_{1,2} = e^{s_{1,2}T_s} = e^{-0.155 \pm j0.177} = 0.857e^{\pm j10.2°}$$

$$z_{3,4} = e^{s_{3,4}T_s} = e^{-0.064 \pm j0.428} = 0.938e^{\pm j24.5°}$$

and the four zeros at $s = \infty$ are all mapped to $z = 0$. The system function of the digital filter is given by

$$H(z) = \frac{7.23 \times 10^{-3}}{(1 - 1.687z^{-1} + 0.734z^{-2})(1 - 1.707z^{-1} + 0.88z^{-2})}$$

The pole/zero pattern, the digital filter structure, the log-magnitude and phase responses are shown in Fig. 8.12. The gain of the digital filter is set so that that maximum value in the passband is equal to 1 (0 dB).

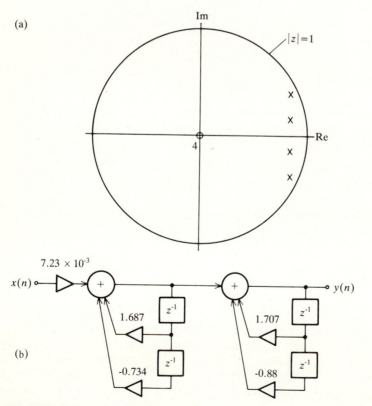

FIGURE 8-12
Chebyshev digital lowpass filter results when using impulse-invariance method. a) Pole/zero pattern for $H_C(z)$. b) Digital filter structure. c) Log-magnitude and phase responses.

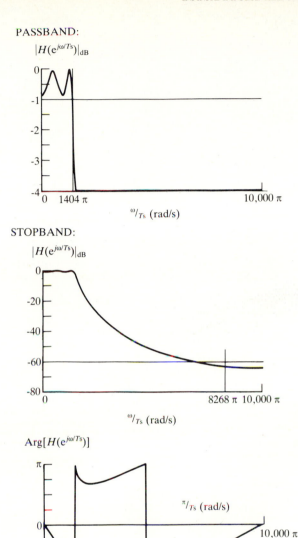

PASSBAND:

STOPBAND:

FIGURE 8-12c

Although useful for implementing lowpass and bandpass filters, the impulse-invariance method is unsuccessful for implementing digital filters for which $|H_A(j\Omega)|$ does not approach zero for large Ω, such as a highpass filter. In these cases, $|H_A(j\Omega)|$ does not have small values for high frequencies and an appropriate sampling period cannot be found to prevent catastrophic aliasing. Hence, the aliasing error terms will dominate the transfer function, as shown in Fig. 8.13. Because of these aliasing error terms, a highpass digital filter does not result when the impulse-invariance method is employed. To overcome this problem, we need a method in which the entire $j\Omega$ axis, for $-\infty < \Omega < \infty$, maps

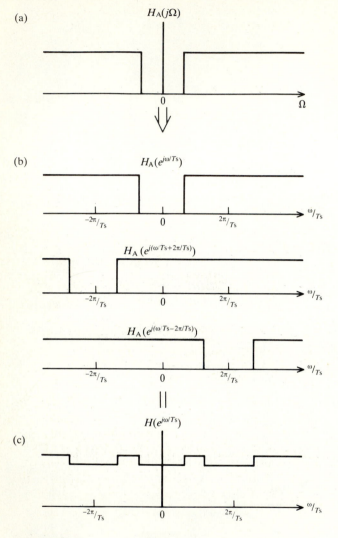

FIGURE 8-13
Aliasing effects prevents the application of the impulse-invariance method to implement highpass filters. a) Original analog highpass filter frequency response. b) Components in the periodic extension after sampling the impulse response. c) Resulting digital filter magnitude response.

uniquely onto the unit circle, for $-\pi/T_s < \omega/T_s < \pi/T_s$. This is accomplished by the bilinear transform method described in the next section.

BILINEAR TRANSFORM METHOD. The bilinear transform method provides a nonlinear one-to-one mapping of the frequency points on the $j\Omega$ axis in the s plane to those on the unit circle in the z plane. This procedure also allows us to implement digital highpass filters from their analog counterparts. To illustrate the method, let us consider the following simple bilinear transformation:

$$s = \frac{2}{T_s}\frac{z-1}{z+1} \tag{8.32}$$

To find $H(z)$, each occurrence of s in $H_A(s)$ is replaced by $(2/T_s)(z-1)/(z+1)$. The inverse transformation is given by

$$z = \frac{sT_s/2 + 1}{sT_s/2 - 1} \tag{8.33}$$

We now explore the properties of this bilinear transformation that are useful for digital filter synthesis.

Relationship of the s plane and z plane. To find the mapping of the frequencies from Ω to ω, we set $s = j\Omega$ and $z = e^{j\omega}$ in Eq. (8.33) to get

$$e^{j\omega} = \frac{j\Omega T_s/2 + 1}{j\Omega T_s/2 - 1} = \frac{(\Omega^2(T_s/2)^2 + 1)^{1/2} \, e^{j \, \arctan[\Omega T_s/2]}}{(\Omega^2(T_s/2)^2 + 1)^{1/2} \, e^{j \, \arctan[-\Omega T_s/2]}}$$
$$= e^{j2 \, \arctan[\Omega T_s/2]} \tag{8.34}$$

Equating the exponents, we can relate the values of Ω to ω by

$$\omega = 2 \arctan[\Omega T_s/2] \tag{8.35}$$

This transformation is shown in Fig. 8.14. The entire $j\Omega$ axis in the s plane, $-\infty < j\Omega \leq \infty$ maps *exactly once* onto the unit circle $-\pi < \omega \leq \pi$, such that there is a one-to-one correspondence between the continuous-time and discrete-time frequency points. It is this one-to-one mapping that allows analog highpass filters to be implemented in digital filter form. As in the impulse-invariance method, the left half of the s plane maps onto the inside of the unit circle in the z plane, and the right half of the s plane maps onto the outside.

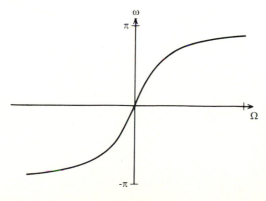

FIGURE 8-14
The arctangent relationship between continuous-time and discrete-time frequencies that occurs in applying the bilinear transform.

The inverse relationship, relating ω to Ω, is given by

$$\Omega = \frac{2}{T_s} \tan[\omega/2] \tag{8.36}$$

For small values of frequency, this relationship reduces to

$$\Omega = \frac{2\sin(\omega/2)}{T_s \cos(\omega/2)} = \frac{2(\omega/2 - \omega^3/8 + \cdots)}{T_s(1 - \omega^2/4 + \cdots)} = \omega/T_s \quad \text{for small } \omega$$

This result indicates that the mapping is approximately linear for small Ω and ω. For larger frequency values, the nonlinear compression that occurs in the mapping of Ω to ω is more apparent. This compression causes the transfer function at the high Ω frequencies to be highly distorted when it is translated to the ω-domain.

Prewarping procedure. For some special types of transfer functions, the effect of this nonlinear compression of Ω to ω values can be compensated. When the desired magnitude response is *piece-wise constant* over frequency, this compression can be compensated by introducing a suitable prescaling, or *prewarping* to the Ω frequency scale. For the bilinear transform, the Ω scale is converted into the Ω^* scale, where

$$\Omega^* = \frac{2}{T_s} \tan\left(\frac{\Omega T_s}{2}\right) \tag{8.37}$$

This scale conversion is shown in Fig. 8.15.

An analog filter design is then performed to determine $|H_A(j\Omega^*)|$ that satisfies the specifications on the Ω^* scale. When the bilinear transformation is applied, the digital filter magnitude response will satisfy the original specifications. This can be seen by substituting Ω^* into Eq. (8.35): $\omega = 2\arctan[\Omega^* T_s/2] = \Omega T_s$.

For example, if the original specification calls for a passband frequency of Ω_P, then the prewarped frequency value, to which the analog filter is designed, occurs at $\Omega_P^* = (2/T_s)\tan(\Omega_P/T_s)$. When this prewarped value is employed in the bilinear transform method, the passband of the resulting digital filter design is located at $\omega_P/T_s = \Omega_P$, or the original desired value.

Pole mapping procedure. We now derive the rule by which the poles are mapped from the s plane to the z plane. Let us consider the analog system function having a pole at $s = s_p$, where $s_p < 0$, given by

$$H_A(s) = \frac{1}{s - s_p} \tag{8.38}$$

Applying the bilinear transform, we get

$$H(z) = 1 \Big/ \left(\frac{2}{T_s}\frac{1 - z^{-1}}{1 + z^{-1}} - s_p\right) = \frac{T_s}{2 - s_p T_s}(1 + z^{-1}) \Big/ \left(1 - \frac{2 + s_p T_s}{2 - s_p T_s}z^{-1}\right) \tag{8.39}$$

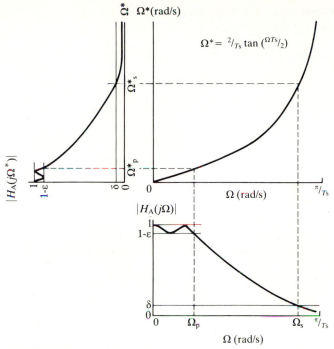

FIGURE 8-15
Frequency prewarping procedure. Original specification in Ω scale is modified by prewarping to produce the specification in the Ω^* scale.

Comparing the pole location in the s plane with that in the z plane, we can state the pole mapping rule for the bilinear transform as

A pole at $s = s_p$ in the s plane gets mapped into a zero at $z = -1$ and a pole at $z = (2 + s_p T_s)/(2 - s_p T_s)$ in the z plane when the bilinear transform is applied.

By examining the magnitude of the pole in the z plane, we find that poles lying in the left half of the s plane are mapped into the inside of the unit circle. Hence, a stable analog filter produces a stable digital filter after the mapping.

The following examples illustrate the application of the bilinear transform.

Example 8.8. Chebyshev lowpass filter design using the bilinear transformation. Consider the previous lowpass specification given by

passband:

$$-1 < |H(j\Omega)|_{\mathrm{dB}} \leq 0 \qquad \text{for } 0 \leq \Omega \leq 1404\pi \text{ rad/s} = 4411 \text{ rad/s}$$

stopband:

$$|H(j\Omega)|_{\mathrm{dB}} < -60 \qquad \text{for } \Omega \geq 8268\pi \text{ rad/s} = 25\,975 \text{ rad/s}$$

and let the sampling period $T_s = 10^{-4}$ seconds.

Prewarped frequency values. Since we intend to employ the bilinear transform method, we must prewarp these frequencies. The prewarped values are given by

$$\Omega_P^* = \frac{2}{T_s} \tan[\Omega_P T_s/2] = 2 \times 10^4 \tan[0.0702\pi] = 4484 \text{ rad/s}$$

and

$$\Omega_S^* = \frac{2}{T_s} \tan[\Omega_S T_s/2] = 2 \times 10^4 \tan[0.4134\pi] = 71\,690 \text{ rad/s}$$

The modified specifications to which the analog filter is to be designed are given by

passband:

$$-1 < |H(j\Omega^*)|_{dB} \le 0 \qquad \text{for } 0 \le \Omega^* \le 4484 \text{ rad/s}$$

stopband:

$$|H(j\Omega^*)|_{dB} < -60 \qquad \text{for } \Omega^* \ge 71\,690 \text{ rad/s}$$

Value of μ. The value of μ is determined from the passband ripple: $\mu = 0.508$.

Value of N. The order of the Chebyshev filter N is determined from

$$|H_C(j\Omega_S^*)|^2 = [1 + \mu^2 C_N^2(\Omega_S^*/\Omega_P^*)]^{-1} < 10^{-6} \quad (-60 \text{ dB})$$

Since $\Omega_S^*/\Omega_P^* = 16$, we evaluate $C_N(16)$ for increasing N until

$$C_N(16) > [(10^6 - 1)/\sigma^2]^{1/2} = 1969.$$

Doing this, we find $C_3(16) = 16\,301$. Hence, $N = 3$ is sufficient. When using the impulse-invariance method a value of $N = 4$ was required. The prewarping of the frequencies has resulted in a reduction in the filter order.

Axes of the ellipse. The radii of the circles defining the major and minor axes of the ellipse are determined from the constant defined in Eq. (8.17) to give $\rho = 4.17$. The radii are equal to

$$R = \frac{4484}{2} (4.17^{1/4} + 4.17^{-1/4}) = 5001$$

and

$$r = \frac{4484}{2} (4.17^{1/4} - 4.17^{-1/4}) = 2216$$

Since there are three poles, the three lines dividing the s plane have angles equal to π and $\pm 2\pi/3$. The three poles of the Chebyshev filter are then at

$$s_1 = -2216$$

$$s_{2,3} = 2216 \cos(2\pi/3) \pm j5001 \sin(2\pi/3) = -1108 \pm j4331 = 4470e^{\pm j104.4}$$

The analog system function is then given by

$$H_C(s) = \frac{4.43 \times 10^{10}}{[s + 2216][s^2 + 2223s + (4470)^2]}$$

Pole mapping. We now apply the pole mapping dictated by Eq. (8.39) to each pole in the s plane. For $s = s_1$ in the s plane, in the z plane there is a zero at

$z = -1$ and a pole at

$$z = \frac{2 + (-2216 \times 10^{-4})}{2 - (-2216 \times 10^{-4})} = 0.801$$

For $s = s_{2,3}$, there are two zeros at $z = -1$ and poles at

$$z = \frac{2 + (-1108 \pm j4331) \times 10^{-4}}{2 - (-1108 \pm j4331) \times 10^{-4}} = 0.819 \pm j0.373 = 0.90 e^{\pm j24.5°}$$

The digital filter system function is then given by

$$H(z) = 4.29 \times 10^{-3} \frac{1 + z^{-1}}{1 - 0.801 z^{-1}} \frac{1 + 2z^{-1} + z^{-2}}{1 - 1.638 z^{-1} + 0.81 z^{-2}}$$

The pole/zero pattern, filter block diagram, the log-magnitude response and the phase response are shown in Fig. 8.16. Only four multipliers and three delays are needed to implement this filter. This is a reduction over the requirements for the filter designed with the impulse-invariance method.

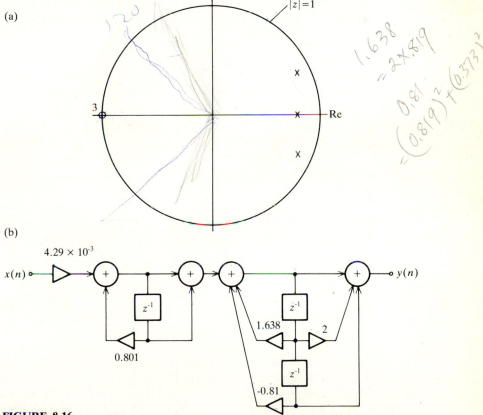

FIGURE 8-16
Chebyshev digital lowpass filter results when using the bilinear transformation method. a) Pole/zero pattern for $H_C(z)$. b) Digital filter structure. c) Log-magnitude and phase responses.

PASSBAND:

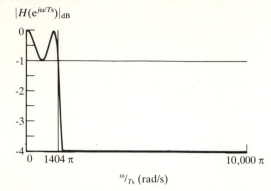

STOPBAND:

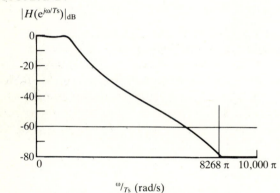

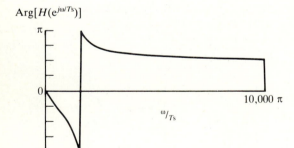

FIGURE 8-16c

The bilinear transform for the lowpass filter maps the zeros at $\Omega = \infty$ to the zeros at $\omega = \pi$, as illustrated in the previous example. These zeros provide additional attenuation in the stopband at the high frequencies. In contrast, by the impulse invariance method these zeros are located at $z = 0$ and have no effect on the magnitude response. Hence, the bilinear transform is a more effective, albeit more complicated, procedure for obtaining digital filters from analog filters.

In designing highpass digital filters, it is more common to start with an analog lowpass filter, determine the digital lowpass filter and then employ a lowpass-to-highpass frequency transformation that then results in a highpass filter. This and other common transformations are described in the next section.

8.5 FREQUENCY TRANSFORMATIONS FOR CONVERTING LOWPASS FILTERS INTO OTHER TYPES

Once we have designed the digital lowpass filter, we can employ frequency transformations to produce lowpass filters with different cutoff frequency values, highpass filters, bandpass filters, or bandreject filters. The transformations that generate these filters are given in Table 8.2. These transformations are applied by replacing each occurrence of z^{-1} in the prototype lowpass filter with the entry in the table.

The lowpass and highpass transformations have analytic forms that are similar to the bilinear transformation. In fact, the transformations act by stretching and relocating the frequency points on the unit circle onto itself to produce the desired filter. The characteristic frequency values in the table serve as reference points, where the magnitude response has the same value in both the prototype and transformed filter. Typically, the edge of the passband is the appropriate characteristic frequency, to assure that the passband characteristics match.

TABLE 8.2

Frequency transformations to convert a lowpass filter of characteristic frequency β to a filter having characteristic frequency θ

Lowpass	$\dfrac{z^{-1} - \alpha}{1 - \alpha z^{-1}}$	$\alpha = \dfrac{\sin[\frac{1}{2}(\beta - \theta)]}{\sin[\frac{1}{2}(\beta + \theta)]}$
Highpass	$-\dfrac{z^{-1} + \alpha}{1 + \alpha z^{-1}}$	$\alpha = -\dfrac{\cos[\frac{1}{2}(\beta + \theta)]}{\cos[\frac{1}{2}(\beta - \theta)]}$
Bandpass	$-\dfrac{z^{-2} - \dfrac{2\alpha k}{k+1}z^{-1} + \dfrac{k-1}{k+1}}{\dfrac{k-1}{k+1}z^{-2} - \dfrac{2\alpha k}{k+1}z^{-1} + 1}$	$\alpha = \dfrac{\cos[\frac{1}{2}(\sigma_U + \sigma_L)]}{\cos[\frac{1}{2}(\sigma_U - \sigma_L)]}$
(σ_L is lower characteristic frequency σ_U is upper characteristic frequency)		$k = \cot[\frac{1}{2}(\sigma_U - \sigma_L)]\tan(\beta/2)$
Bandreject	$\dfrac{z^{-2} - \dfrac{2\alpha}{1+k}z^{-1} + \dfrac{1-k}{1+k}}{\dfrac{1-k}{1+k}z^{-2} - \dfrac{2\alpha}{1+k}z^{-1} + 1}$	$\alpha = \dfrac{\cos[\frac{1}{2}(\sigma_U + \sigma_L)]}{\cos[\frac{1}{2}(\sigma_U - \sigma_L)]}$
		$k = \tan[\frac{1}{2}(\sigma_U - \sigma_L)]\tan(\beta/2)$

POLE-MAPPING RULES. We now consider the pole-mapping rules for the various transformations. In the discussions below, we consider two elemental system functions. The first generates a zero at $z = c$ and a pole at $z = 0$ and is given by

$$H_Z(z) = 1 - cz^{-1} \tag{8.40}$$

The second elemental system function generates a pole at $z = d$ and a zero at $z = 0$ and is given by

$$H_P(z) = \frac{1}{1 - dz^{-1}} \tag{8.41}$$

Both c and d can be complex-valued numbers.

POLE-MAPPING FOR THE LOWPASS-TO-LOWPASS FILTER. Applying the lowpass-to-lowpass transformation to $H_Z(z)$, we get $H_{LZ}(z)$ given by

$$\begin{aligned}
H_{LZ}(z) &= 1 - c \frac{z^{-1} - \alpha}{1 - \alpha z^{-1}} \\
&= (1 + c\alpha) \frac{1 - [(\alpha + c)/(1 + c\alpha)]z^{-1}}{1 - \alpha z^{-1}}
\end{aligned} \tag{8.42}$$

This relationship indicates that the lowpass zero at $z = c$ is transformed into a zero at $z = c_1$, where

$$c_1 = \frac{\alpha + c}{1 + c\alpha} \tag{8.43}$$

The pole at $z = 0$ is transformed into a pole at $z = \alpha$. The location of the zero is related to c, the location of the lowpass zero, while the location of the pole is not. If we do the same to $H_P(z)$, then we get the transformed system function that is similar to the inverse of $H_{LZ}(z)$ above, given by

$$H_{LP}(z) = \frac{1 - \alpha z^{-1}}{(1 + d\alpha)(1 - [(\alpha + d)/(1 + \alpha d)] z^{-1})} \tag{8.44}$$

This relationship indicates that the pole at $z = d$ is transformed into a pole at $z = (\alpha + d)/(1 + \alpha d)$ and the zero at $z = 0$ is transformed into a zero at $z = \alpha$. The location of the pole is related to d, the location of the lowpass pole, while the location of the zero is not.

If the original lowpass system function contains both relevant poles and zeros, a cancellation of the singularities that are produced at $z = \alpha$ will occur after the transformation is accomplished.

Since the poles for practical filters are typically close to the unit circle, the coefficient values determined by these transformations are very sensitive to numerical round-off. This coefficient error can result in unexpected distortions in the magnitude response that cause the passband specification not to be satisfied. To reduce this sensitivity in applying the frequency transformations, one pole should be transferred at a time. Even though a pole may be complex, the analytic form for a first-order complex pole is still much simpler than that for the second-order complex-conjugate pair. After the transformation is accomplished, the complex-conjugate poles can then be recombined into a second-order section with real-valued coefficients. This procedure is illustrated in Example 8.9.

Example 8.9. Application of frequency transformations—lowpass-to-lowpass. Consider the transformation of the Chebyshev lowpass digital filter, given in Example 8.8, into a Chebyshev lowpass digital filter having the passband specification, given by

$$-1 < |H(e^{j\omega})|_{dB} \leq 0 \qquad \text{for } 0 \leq \omega \leq 0.2808\pi \quad (\theta = 0.2808\pi)$$

For the prototype lowpass filter, the passband edge is at $\omega = 0.1404\pi$ ($\beta = 0.1404\pi$). First, we must evaluate the constant α:

$$\alpha = \frac{\sin[\frac{1}{2}(\beta - \theta)]}{\sin[\frac{1}{2}(\beta + \theta)]} = \frac{\sin(-0.0702\pi)}{\sin(0.2106\pi)} = -0.356$$

The system function of the prototype Chebyshev filter can be written as

$$H_C(z) = A \frac{1 + z^{-1}}{1 - z_1 z^{-1}} \frac{1 + z^{-1}}{1 - z_2 z^{-1}} \frac{1 + z^{-1}}{1 - z_3 z^{-1}}$$

which contains three zeros at $z = -1$ and three relevant poles, at $z = z_1$, $z = z_2$ and $z = z_3$. Each of the zeros at $z = -1$ is transformed into a zero at

$$z = \frac{\alpha - 1}{1 - \alpha} = -1$$

and each of the corresponding poles at $z = 0$ is transformed into a pole at

$$z = \alpha = -0.356$$

The pole of the prototype lowpass filter at $z = z_i$ is transformed into a pole located at

$$c_i = \frac{\alpha + z_i}{1 + \alpha z_i}$$

and the corresponding zero at $z = 0$ is transformed into a zero at $z = \alpha = -0.3561$. These latter zeros cancel the poles at $z = \alpha$ that resulted from the zeros of the prototype lowpass filter.

The pole locations are given by:

for $z_1 = 0.801$, $c_1 = 0.622$

for $z_2 = 0.819 + j0.373$ $\quad c_2 = (0.462 + j0.373)/(0.709 - j0.133)$
$$= 0.824e^{j49.5°}$$

for $z_3 = 0.819 - j0.373$ $\quad c_3 = 0.824e^{-j49.5°}$

The system function of the new lowpass filter is given by

$$H(z) = 0.029 \frac{1 + z^{-1}}{1 - 0.622z^{-1}} \frac{1 + 2z^{-1} + z^{-2}}{1 - 1.07z^{-1} + 0.674z^{-2}}$$

The pole/zero pattern in the z plane, the filter block diagram and the log-magnitude and phase responses are shown in Fig. 8.17. For the prototype lowpass filter, the log-magnitude response fell below -60 dB for $\omega > 0.68\pi$ (see Fig. 8.16). For the new filter, this occurs for $\omega > 0.83\pi$. The stopband performance of the new filter must be checked, since there is no control over it when the passband frequency is used as the characteristic frequency in the transformation. If the stopband is important, the stopband frequency should be used in the transformation.

POLE-MAPPING FOR THE LOWPASS-TO-HIGHPASS FILTER. Applying the lowpass to highpass transformation to $H_Z(z)$, we get $H_{HZ}(z)$ given by

$$H_{HZ}(z) = 1 + c\frac{z^{-1} + \alpha}{1 + \alpha z^{-1}}$$

$$= (1 + c\alpha)\frac{1 + [(c + \alpha)/(1 + c\alpha)]\,z^{-1}}{1 + \alpha z^{-1}} \tag{8.45}$$

(a)

(b)

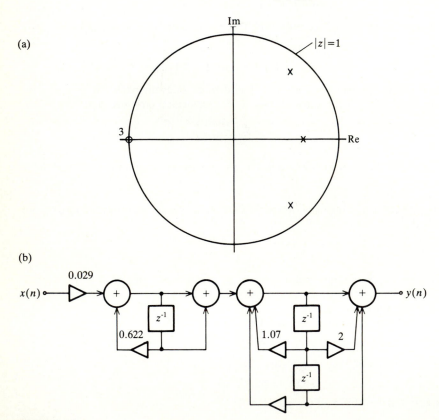

FIGURE 8-17
Results of applying lowpass-to-lowpass frequency transformation to the digital filter shown in Fig. 8.16. a) Pole/zero pattern for $H(z)$. b) Digital filter structure. c) Log-magnitude and phase responses.

PASSBAND:

STOPBAND:

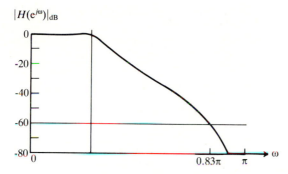

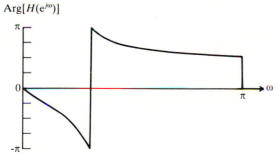

FIGURE 8-17c

This relationship indicates that the lowpass zero at $z = c$ is transformed into a zero at $z = -(c + \alpha)/(1 + c\alpha)$ and the pole at $z = 0$ is transformed into a pole at $z = -\alpha$. The location of the zero is related to c, the location of the lowpass zero, while the location of the pole is not. If we do the same to $H_P(z)$, then we get the transformed system function that is similar to the inverse of $H_{HZ}(z)$ above, given by

$$H_{HP}(z) = \frac{1}{1 + d\alpha} \frac{1 + \alpha z^{-1}}{(1 + [(d + \alpha)/(1 + \alpha d)]z^{-1})} \qquad (8.46)$$

This relationship indicates that the pole at $z = d$ is transformed into a pole at $z = -(d + \alpha)/(1 + \alpha d)$ and the zero at $z = 0$ is transformed into a zero at $z = -\alpha$. The location of the pole is related to d, the location of the lowpass pole, while the location of the zero is not.

If the original lowpass system function contains both relevant poles and zeros, a cancellation of the singularities that are produced at $z = -\alpha$ will occur after the highpass transformation is accomplished.

Example 8.10.   Application of frequency transformations—lowpass-to-highpass.
Consider the transformation of the Chebyshev lowpass digital filter, given in Example 8.8, into a Chebyshev highpass digital filter having the passband specification given by

$$-1 < |H(e^{j\omega})|_{dB} \leq 0 \qquad \text{for } \pi/2 \leq \omega \leq \pi \quad (\theta = 0.5\pi)$$

For the prototype lowpass filter, the passband edge is at $\omega = 0.1404\pi$ ($\beta = 0.1404\pi$). First, we must evaluate the constant α:

$$\alpha = -\frac{\cos[\frac{1}{2}(\beta + \theta)]}{\cos[\frac{1}{2}(\beta - \theta)]} = -\frac{\cos(0.3202\pi)}{\cos(-0.1798\pi)} = -0.634$$

The system function of the prototype Chebyshev filter can be written as

$$H_C(z) = A \frac{1 + z^{-1}}{1 - z_1 z^{-1}} \frac{1 + z^{-1}}{1 - z_2 z^{-1}} \frac{1 + z^{-1}}{1 - z_3 z^{-1}}$$

$H_C(z)$ contains three relevant zeros at $z = -1$ and three relevant poles, at $z = z_1$, $z = z_2$ and $z = z_3$. Each of the zeros at $z = -1$ is transformed into a zero at

$$z = -\frac{-1 + \alpha}{1 - \alpha} = 1$$

and each of the corresponding poles at $z = 0$ are transformed into a pole at $z = -\alpha = 0.6337$. The pole of the prototype lowpass filter at $z = z_i$ is transformed into a pole located at

$$c_i = -\frac{z_i + \alpha}{1 + \alpha z_i}$$

and the corresponding zero at $z = 0$ is transformed into a zero at $z = -\alpha = 0.6337$. These latter zeros cancel the poles at $z = -\alpha$ that resulted from the zeros of the prototype lowpass filter.

The pole locations are given by:

for $z_1 = 0.801$ $c_1 = -0.339$

for $z_2 = 0.819 + j0.373$ $c_2 = -(0.1848) + j0.373)/(0.4813 - j0.2364)$

$$= 0.777e^{-j90.2°}$$

for $z_3 = 0.819 - j0.373$ $c_3 = 0.777e^{j90.2°}$

The system function of the highpass filter is given by

$$H(z) = 0.132 \frac{1 - z^{-1}}{1 + 0.339z^{-1}} \frac{1 - 2z^{-1} + z^{-2}}{1 + 0.603z^{-2}}$$

The pole/zero pattern in the z plane, the filter block diagram and the log-magnitude and phase responses are shown in Fig. 8.18.

POLE-MAPPING FOR THE LOWPASS-TO-BANDPASS FILTER. Applying the lowpass-to-bandpass transformation to $H_Z(z)$, we get $H_{BZ}(z)$ given by

$$
\begin{aligned}
H_{BZ}(z) &= 1 + c \frac{z^{-2} - [2\alpha k/(k+1)]z^{-1} + [(k-1)/(k+1)]}{[(k-1)/(k+1)]z^{-2} - [2\alpha k/(k+1)]z^{-1} + 1} \\
&= \frac{k+1 + c(k-1) - 2\alpha k(c+1)z^{-1} + [k-1+c(k+1)]z^{-2}}{k+1 - 2\alpha kz^{-1} + (k-1)z^{-2}} \\
&= \frac{[k+1+c(k-1)](1 - c_1 z^{-1})(1 - c_2 z^{-1})}{(k+1)(1 - d_1 z^{-1})(1 - d_2 z^{-1})}
\end{aligned} \tag{8.47}
$$

This relationship indicates that the lowpass zero at $z = c$ is transformed into

(a)

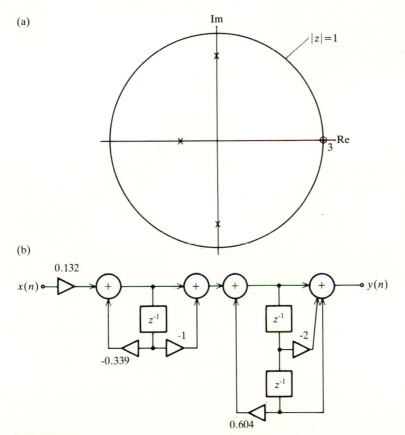

(b)

FIGURE 8-18
Results of applying lowpass-to-highpass frequency transformation to the digital filter shown in Fig. 8.16. a) Pole/zero pattern for $H(z)$. b) Digital filter structure. c) Log-magnitude and phase responses.

PASSBAND:

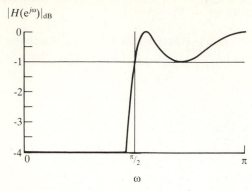

STOPBAND:

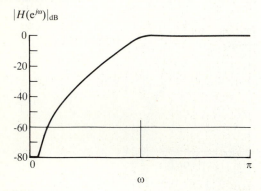

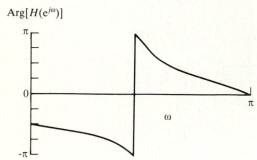

FIGURE 8-18c

two zeros, at $z = c_1$ and $z = c_2$, where

$$c_1 = \frac{\alpha k(c+1) + \{\alpha^2 k^2(c+1)^2 - [k+1+c(k-1)][k-1+c(k+1)]\}^{1/2}}{k+1+c(k-1)}$$

and
$$(8.48)$$

$$c_2 = \frac{\alpha k(c+1) - \{\alpha^2 k^2(c+1)^2 - [k+1+c(k-1)][k-1+c(k+1)]\}^{1/2}}{k+1+c(k-1)}$$

The pole at $z = 0$ is transformed into two poles, at $z = d_1$ and $z = d_2$, where

$$d_1 = \frac{\alpha k + [\alpha^2 k^2 - (k + 1)(k - 1)]^{1/2}}{k + 1}$$

and

$$d_2 = \frac{\alpha k - [\alpha^2 k^2 - (k + 1)(k - 1)]^{1/2}}{k + 1}$$

(8.49)

The locations of the zeros are related to c, the location of the lowpass zero, while the locations of the poles are not.

If we do the same to the elemental system function $H_P(z)$ then we get the transformed system function that is similar to the inverse of $H_{BZ}(z)$ above, given by

$$H_{BP}(z) = \frac{(k + 1)(1 - d_1 z^{-1})(1 - d_2 z^{-1})}{[k + 1 + c(k - 1)](1 - c_1 z^{-1})(1 - c_2 z^{-1})}.$$

(8.50)

The above relationship indicates that the lowpass pole at $z = d$ is transformed into two poles at $z = c_1$ and $z = c_2$, where

$$c_1 = \frac{\alpha k(d + 1) + \{\alpha^2 k^2 (d + 1)^2 - [k + 1 + d(k - 1)][k - 1 + d(k + 1)]\}^{1/2}}{k + 1 + d(k - 1)},$$

(8.51)

and

$$c_2 = \frac{\alpha k(d + 1) - \{\alpha^2 k^2 (d + 1)^2 - [k + 1 + d(k - 1)][k - 1 + d(k + 1)]\}^{1/2}}{k + 1 + d(k - 1)}.$$

The zero at $z = 0$ is transformed into two zeros at $z = d_1$ and $z = d_2$, where

$$d_1 = \frac{\alpha k + [\alpha^2 k^2 - (k + 1)(k - 1)]^{1/2}}{k + 1}$$

(8.52)

and

$$d_2 = \frac{\alpha k - [\alpha^2 k^2 - (k + 1)(k - 1)]^{1/2}}{k + 1}$$

The locations of the poles are related to d, the location of the lowpass pole, while the locations of the zeros are not. If the original lowpass system function contains both relevant poles and zeros, a cancellation of the singularities that are produced at $z = d_1$ and $z = d_2$ will occur after the transformation is accomplished.

Example 8.11. Application of frequency transformations—lowpass-to-bandpass. Consider the transformation of the Chebyshev lowpass digital filter, given in Example 8.8, into a Chebyshev bandpass digital filter having the passband specification given by

bandpass:

$$-1 < |H(e^{j\omega})|_{dB} \leq 0 \qquad \text{for } 0.421\pi \leq \omega \leq 0.608\pi$$

($\sigma_L = 0.421\pi$ and $\sigma_U = 0.608\pi$). For the prototype lowpass filter, the passband edge is at $\omega = 0.1404\pi$ ($\beta = 0.1404\pi$).

First, we must evaluate the constants α and k:

$$\alpha = \frac{\cos[\frac{1}{2}(\sigma_U + \sigma_L)]}{\cos[\frac{1}{2}(\sigma_U - \sigma_L)]} = \frac{\cos(0.515\pi)}{\cos(0.094\pi)} = -0.0486$$

$$k = \cot[\tfrac{1}{2}(\sigma_U - \sigma_L)]\tan(\beta/2) = \frac{\tan(0.070\pi)}{\tan(0.094\pi)} = 0.7403$$

The product αk is equal to -0.0360. The system function of the prototype Chebyshev filter can be written as

$$H_C(z) = A\,\frac{1 + z^{-1}}{1 - z_1 z^{-1}}\frac{1 + z^{-1}}{1 - z_2 z^{-1}}\frac{1 + z^{-1}}{1 - z_3 z^{-1}}$$

$H_C(z)$ contains three relevant zeros at $z = -1$ and three relevant poles, at $z = z_1$, $z = z_2$ and $z = z_3$. Each of the zeros at $z = -1$ is transformed into two zeros, at $z = c_1$ and $z = c_2$, where

$$c_{1,2} = \frac{\pm\{-[k+1-(k-1)][k-1-(k+1)]\}^{1/2}}{k+1-(k-1)} = \pm 1$$

Each of the corresponding poles at $z = 0$ is transformed into two poles at $z = d_1$ and $z = d_2$, where

$$d_{1,2} = \frac{\alpha k \pm [\alpha^2 k^2 - (k+1)(k-1)]^{1/2}}{k+1}$$

$$= \frac{-0.036 \pm [0.0129 + 0.4519]^{1/2}}{1.7403} = -0.021 \pm 0.392$$

The pole of the prototype lowpass filter at $z = z_i$ is transformed into two poles located at

$$c_{i,1,2} = \frac{-0.036(z_i + 1) \pm \{0.013(z_i + 1)^2 - [1.740 - 0.260z_i][-0.260 + 1.740z_i]\}^{1/2}}{1.740 - 0.260z_i}$$

The zero at $z = 0$ is transformed into two zeros at $z = d_1$ and $z = d_2$, where

$$d_{1,2} = \frac{\alpha k \pm [\alpha^2 k^2 - (k+1)(k-1)]^{1/2}}{k+1} = -0.021 \pm 0.392$$

These latter zeros cancel the poles that resulted from the zeros of the prototype lowpass filter.

The pole locations are given by:

for $z_1 = 0.801$ $c_{1,1,2} = -0.0422 \pm j0.859 = 0.860e^{\pm j92.8°}$

for $z_2 = 0.819 + j0.373$, $c_{2,1} = 0.221 - j0.906 = 0.932e^{-j76.3°}$

                         $c_{2,2} = -0.305 + j0.883 = 0.934e^{j109.1°}$

for $z_3 = 0.819 - j0.373$ $c_{3,1} = 0.221 + j0.906 = 0.932e^{j76.3°}$

                         $c_{3,2} = -0.305 - j0.883 = 0.934e^{-j109.1°}$

The system function of the bandpass filter is given by

$$H(z) = 9.58 \times 10^{-3} \frac{1 - z^{-2}}{1 - 0.441z^{-1} + 0.869z^{-2}}$$

$$\times \frac{1 - z^{-2}}{1 + 0.084z^{-1} + 0.74z^{-2}} \frac{1 - z^{-2}}{1 + 0.611z^{-1} + 0.872z^{-2}}$$

The pole/zero pattern in the z plane, the filter block diagram and the log-magnitude and phase responses are shown in Fig. 8.19.

(a)

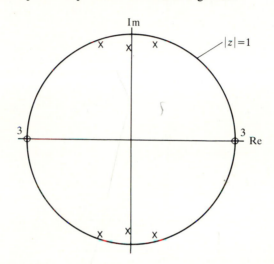

(b)

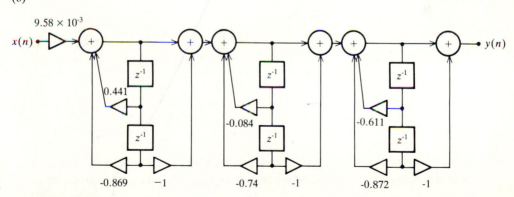

FIGURE 8-19

Results of applying lowpass-to-bandpass frequency transformation to the digital filter shown in Fig. 8.16. a) Pole/zero pattern for $H(z)$. b) Digital filter structure. c) Log-magnitude and phase responses.

PASSBAND:

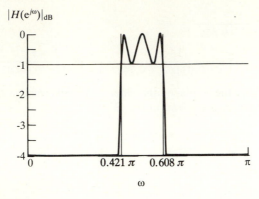

STOPBAND:

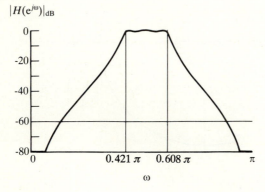

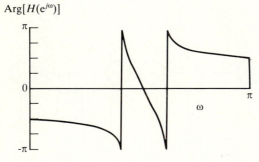

FIGURE 8-19c

POLE-MAPPING FOR THE LOWPASS-TO-BANDREJECT FILTER. Applying the lowpass-to-bandreject transformation to $H_Z(z)$, we get $H_{RZ}(z)$ given by

$$
\begin{aligned}
H_{RZ}(z) &= 1 - c\,\frac{z^{-2} - [2\alpha/(k+1)]z^{-1} + [(1-k)/(k+1)]}{[(1-k)z^{-2}/(k+1)] - [2\alpha/(k+1)]z^{-1} + 1} \\
&= \frac{[k+1+c(k-1)](1 - c_1 z^{-1})(1 - c_2 z^{-1})}{(k+1)(1 - d_1 z^{-1})(1 - d_2 z^{-1})}
\end{aligned}
\tag{8.53}
$$

This relationship indicates that the lowpass zero at $z = c$ is transformed into two zeros, at $z = c_1$ and $z = c_2$, where

$$c_1 = \frac{-\alpha(c-1) + \{\alpha^2(c-1)^2 - [k+1+c(k-1)][1-k-c(k+1)]\}^{1/2}}{k+1+c(k-1)}$$

and

$$c_2 = \frac{-\alpha(c-1) - \{\alpha^2(c-1)^2 - [k+1+c(k-1)][1-k-c(k+1)]\}^{1/2}}{k+1+c(k-1)}$$

(8.54)

The pole at $z = 0$ is transformed into two poles, at $z = d_1$ and $z = d_2$, where

$$d_1 = \frac{\alpha + [\alpha^2 - (k+1)(1-k)]^{1/2}}{k+1}$$

(8.55)

and

$$d_2 = \frac{\alpha - [\alpha^2 - (k+1)(1-k)]^{1/2}}{k+1}$$

The locations of the zeros are related to c, the location of the lowpass zero, while the locations of the poles are not.

If we do the same to the elemental system function $H_P(z)$ that generates a pole at $z = d$ and a zero at $z = 0$, then we get the transformed system function that is similar to the inverse of $H_{RZ}(z)$ above, given by

$$H_{RP}(z) = \frac{(k+1)(1-d_1 z^{-1})(1-d_2 z^{-1})}{[k+1+d(k-1)](1-c_1 z^{-1})(1-c_2 z^{-1})}$$

(8.56)

This relationship indicates that the lowpass pole at $z = d$ is transformed into two poles at $z = c_1$ and $z = c_2$, where

$$c_1 = \frac{-\alpha(d-1) + \{\alpha^2(d-1)^2 - [k+1+d(k-1)][1-k-d(k+1)]\}^{1/2}}{k+1+d(k-1)}$$

and

$$c_2 = \frac{-\alpha(d-1) - \{\alpha^2(d-1)^2 - [k+1+d(k-1)][1-k-d(k+1)]\}^{1/2}}{k+1+d(k-1)}$$

(8.57)

The zero at $z = 0$ is transformed into two zeros at $z = d_1$ and $z = d_2$, where

$$d_1 = \frac{\alpha + [\alpha^2 - (k+1)(1-k)]^{1/2}}{k+1}$$

(8.58)

and

$$d_2 = \frac{\alpha - [\alpha^2 - (k+1)(1-k)]^{1/2}}{k+1}$$

The locations of the poles are related to d, the location of the lowpass pole, while the locations of the zeros are not.

If the original lowpass system function contains both relevant poles and zeros, a cancellation of the singularities that are produced at $z = d_1$ and $z = d_2$ will occur after the transformation is accomplished.

Example 8.12. Application of frequency transformations—lowpass-to-bandreject. Consider the transformation of the Chebyshev lowpass digital filter, given in Example 8.8, into a Chebyshev bandreject digital filter having the passband specification given by

$$-1 < |H(e^{j\omega})|_{\text{dB}} \leq 0 \qquad \text{for } 0 \leq \omega \leq 0.421\pi \text{ and } 0.608\pi \leq \omega \leq \pi.$$

($\sigma_L = 0.421\pi$ and $\sigma_U = 0.608\pi$). For the prototype lowpass filter, the passband edge is at $\omega = 0.1404\pi$ ($\beta = 0.1404\pi$).

First, we must evaluate the constants α and k:

$$\alpha = \frac{\cos[\frac{1}{2}(\sigma_U + \sigma_L)]}{\cos[\frac{1}{2}(\sigma_U - \sigma_L)]} = \frac{\cos(0.515\pi)}{\cos(0.094\pi)} = -0.0486$$

$$k = \tan[\tfrac{1}{2}(\sigma_U - \sigma_L)]\tan(\beta/2) = \tan(0.094\pi)\tan(0.070\pi) = 0.068$$

The system function of the prototype Chebyshev filter can be written as

$$H_C(z) = A\frac{1 + z^{-1}}{1 - z_1 z^{-1}}\frac{1 + z^{-1}}{1 - z_2 z^{-1}}\frac{1 + z^{-1}}{1 - z_3 z^{-1}}$$

$H_C(z)$ contains three relevant zeros at $z = -1$ and three relevant poles, at $z = z_1$, $z = z_2$ and $z = z_3$. Each of the zeros at $z = -1$ is transformed into two zeros, at $z = c_1$ and $z = c_2$, where

$$c_{1,2} = \frac{2\alpha \pm \{4\alpha^2 - [k + 1 - (k - 1)][1 - k + k + 1)]\}^{1/2}}{k + 1 - (k - 1)} = \alpha \pm (\alpha^2 - 1)^{1/2}$$

$$= -0.0486 \pm j0.999 = e^{\pm j92.8°}$$

The corresponding pole at $z = 0$ is transformed into two poles at $z = d_1$ and $z = d_2$, where

$$d_{1,2} = \frac{\alpha \pm [\alpha^2 - (k + 1)(1 - k)]^{1/2}}{k + 1}$$

$$= \frac{-0.0486 \pm [0.0024 + 0.995]^{1/2}}{1.068} = -0.0455 \pm 0.934$$

The pole of the prototype lowpass filter at $z = z_i$ is transformed into two poles located at

$$c_{1,2} = \frac{0.0486(z_i - 1) \pm \{0.0024(z_i - 1)^2 - [1.068 - 0.932z_i][0.932 - 1.068z_i]\}^{1/2}}{1.068 - 0.932z_i}$$

The corresponding zero at $z = 0$ is transformed into two zeros at $z = d_1$ and $z = d_2$, where

$$d_{1,2} = \frac{\alpha \pm [\alpha^2 - (k + 1)(k - 1)]^{1/2}}{k + 1} = -0.0455 \pm 0.934$$

These latter zeros cancel the poles that resulted from the zeros of the prototype lowpass filter. The pole locations are given by:

for $z_1 = 0.801$ $c_{1,1,2} = 0.489e^{\pm j93.5°}$

for $z_2 = 0.819 + j0.373$ $c_{2,1} = 0.932e^{-j109.2°}$

$$c_{2,2} = 0.931e^{j76.2°}$$

for $z_3 = 0.819 - j0.373 = z_2^*$ $c_{3,1} = 0.932e^{j109.2°}$

$$c_{3,2} = 0.931e^{-j76.2°}$$

The system function of the bandpass filter is given by

$$H(z) = 0.497 \frac{1 - 0.1z^{-1} + z^{-2}}{1 + 0.06z^{-1} + 0.239z^{-2}}$$

$$\times \frac{1 - 0.1z^{-1} + z^{-2}}{1 - 0.444z^{-1} + 0.867z^{-2}} \frac{1 - 0.1z^{-1} + z^{-2}}{1 + 0.613z^{-1} + 0.869z^{-2}}$$

The pole/zero pattern in the z plane, the filter block diagram and the log-magnitude and phase responses are shown in Fig. 8.20.

(a)

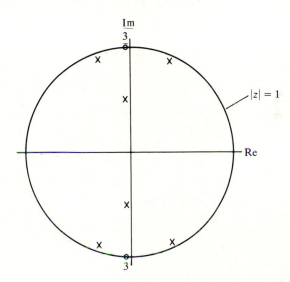

(b)

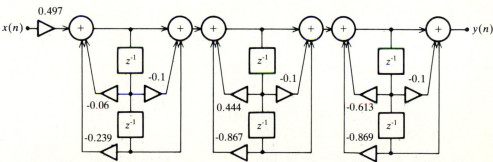

FIGURE 8-20
Results of applying lowpass-to-bandreject frequency transformation to the digital filter shown in Fig. 8.16. a) Pole/zero pattern for $H(z)$. b) Digital filter structure. c) Log-magnitude and phase responses.

PASSBAND:

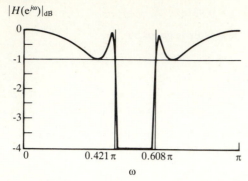

STOPBAND:

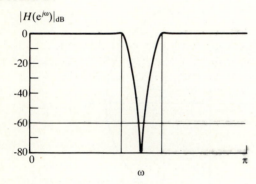

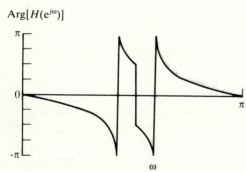

FIGURE 8-20c

8.6 ALL-PASS FILTERS FOR PHASE RESPONSE COMPENSATION

The phase response curves of the IIR filters above are not linear with frequency. Since the unit-sample response has an infinite duration, a point of symmetry cannot exist. Since this is a necessary condition for linear phase, IIR filters cannot have exactly linear phase. In practice, however, it is usually desirable to have approximately linear phase over the passband. In the stopband, the phase response is less important since these components get eliminated anyway. In this section, we investigate the use of all-pass filters to modify the phase response in the passband. These modify the phase by being connected in cascade with the existing filter. Since these filters have a constant magnitude response, that of the existing filter is left unchanged.

An *all-pass filter* is one whose magnitude response is constant for all frequencies, but whose phase response is not identically zero. We consider both first and second-order filter sections that can be connected in cascade with the IIR filters discussed previously in this chapter to modify the overall phase characteristic.

First-order all-pass filter. The system function of a first-order all-pass filter is given by

$$H_{\text{AP}}(z) = \frac{1 - r^{-1}z^{-1}}{1 - rz^{-1}} \qquad \text{for } |r| < 1 \tag{8.59}$$

This system function has a pole at $z = r$ and a zero at $z = r^{-1}$. For stability, r must be less than one. We can easily show that the magnitude response is constant by computing the squared magnitude response, given by

$$
\begin{aligned}
H_{\text{AP}}(e^{j\omega}) \, H_{\text{AP}}(e^{-j\omega}) &= \frac{(1 - r^{-1}e^{-j\omega})(1 - r^{-1}e^{j\omega})}{(1 - re^{-j\omega})(1 - re^{j\omega})} \\
&= \frac{1 - 2r^{-1}\cos(\omega) + r^{-2}}{1 - 2r\cos(\omega) + r^2} = r^{-2}
\end{aligned}
\tag{8.60}
$$

To compute the phase response, we write the Fourier transform as

$$
\begin{aligned}
H_{\text{AP}}(e^{j\omega}) &= \frac{(1 - r^{-1}e^{-j\omega})(1 - re^{j\omega})}{(1 - re^{-j\omega})(1 - re^{j\omega})} \\
&= \frac{2 - (r + r^{-1})\cos(\omega) - j(r - r^{-1})\sin(\omega)}{1 - 2r\cos(\omega) + r^2}
\end{aligned}
\tag{8.61}
$$

The phase response is given by

$$\text{Arg}[H_{\text{AP}}(z)] = \arctan\left[\frac{-(r - r^{-1})\sin(\omega)}{2 - (r + r^{-1})\cos(\omega)}\right] \tag{8.62}$$

The phase responses for different values of r are shown in Fig. 8.21. When $0 < r < 1$, the zero lies on the positive real axis. The phase over $0 < \omega < \pi$ is positive, at $\omega = 0$ it is equal to π and decreases until $\omega = \pi$, where it is zero. When $-1 < r < 0$, the zero lies on the negative real axis. The phase over $0 < \omega < \pi$ is negative, starting at 0 for $\omega = 0$ and decreasing to $-\pi$ at $\omega = \pi$. Since the zero is outside the unit circle, the phase will always decrease by π as ω goes from 0 to π. The shape of the phase response is related to the radius r.

Second-order all-pass filter. Second-order all-pass filters are obtained by allowing r to take on complex values and including the complex-conjugate

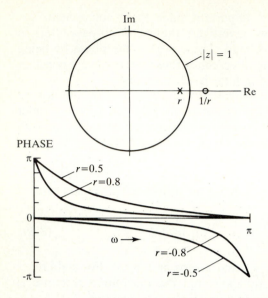

FIGURE 8.21
Pole/zero pattern and phase response for first-order all-pass filter.

singularity locations. Doing this we have the second-order all-pass filter system function given by

$$H_{\text{AP}}(z) = \frac{(1 - r^{-1}z^{-1})(1 - r^{-1*}z^{-1})}{(1 - rz^{-1})(1 - r^*z^{-1})}$$

$$= \frac{1 - 2\,\text{Re}(r^{-1})z^{-1} + |r^{-1}|^2\,z^{-2}}{1 - 2\,\text{Re}(r)z^{-1} + |r|^2\,z^{-2}} \qquad (8.63)$$

The phase responses for different values of r are shown in Fig. 8.22. Since there are two zeros outside the unit circle, the phase at $\omega = 0$ is 2π, or equivalently, equal to zero, when $r > 0$. When $r < 0$, the phase at $\omega = 0$ is zero again. Since the two zeros lie outside the unit-circle, the phase will decrease by 2π as ω goes from 0 to π.

The phase response of a given filter can be tailored in a narrow band of frequencies by varying the value of r and including additional all-pass filters. However, the type of tailoring that can be done is limited, since the phase responses of all-pass filters all show a negative trend with frequency.

8.7 SUMMARY

The Butterworth, Chebyshev and elliptic filter synthesis techniques for designing analog lowpass filters in the s plane have been presented. For a given lowpass specification, it was found that the Butterworth filter was the simplest, but required the highest order. The elliptic filter provided the lowest order, but was the most difficult to design analytically. The Chebyshev filter provides a reasonable tradeoff between filter order (hardware cost) versus design effort (design cost).

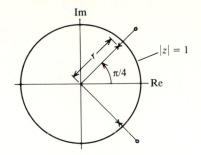

PHASE:

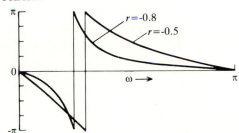

FIGURE 8.22
Pole/zero pattern and phase response for second-order all-pass filter.

The impulse-invariance and the bilinear transformation methods were described for translating the s plane singularities into the z plane. The digital filter could then be implemented from these z plane singularities as described in the previous chapters. For the lowpass filter specification, the bilinear transform provided better stopband results because the zeros were mapped to $z = -1$, rather than $z = 0$ with the impulse-invariance method.

Frequency transformations were presented to convert the prototype lowpass digital filter designs into highpass, bandpass and bandreject digital filters. All-pass filters were used to compensate the phase response of IIR filters. An exactly linear phase is impossible to obtain with a filter having an infinite duration unit-sample response. In practice, an approximately linear phase response can be achieved over the passband of the original filter by including all-pass filters.

FURTHER READING

Antoniou, Andreas: *Digital Filters*: *Analysis and Design*, McGraw-Hill, New York, 1979.

Gold, Bernard, and Charles M. Rader: *Digital Processing of Signals*, McGraw-Hill, New York, 1969.

Hamming, R. W.: *Digital Filters*, 2nd ed., Prentice-Hall, Englewood Cliffs, NJ, 1983.

Oppenheim, Alan V., and Ronald W. Schafer: *Digital Signal Processing*, Prentice-Hall, Englewood Cliffs, NJ, 1975.

Rabiner, Lawrence R., and Bernard Gold: *Theory and Application of Digital Signal Processing*, Prentice-Hall, Englewood Cliffs, NJ, 1975.

REFERENCE TO TOPIC

Elliptic filter reference

Rader, C. M., and B. Gold: "Digital filter design techniques in the frequency domain," *Proc. IEEE,* **55:** 149–171 (1967).

PROBLEMS

8.1. In replacing an analog filter with a digital filter, aliasing effects must be eliminated by including an anti-aliasing analog lowpass filter prior to the analog-to-digital converter. The Butterworth and Chebyshev filters are suitable candidates in practice. A possible specification for an anti-aliasing filter is to have a negligible attenuation (between 0 and 0.1 dB) in the passband and to have a large attenuation (greater than 60 dB) at the analog frequency corresponding to the near edge of the passband of the next period of $|H(e^{j\omega/T_s})|$, that is, the replica of the magnitude response centered at $\omega = 2\pi/T_s$.

Consider the analog lowpass filter specification that is to be satisfied by the digital filter to be given by

passband:

$$-0.1 < |H(j\Omega)|_{dB} \le 0 \qquad \text{for } 0 \le \Omega \le 10 \text{ rad/s}$$

stopband:

$$|H(j\Omega)|_{dB} < -60 \qquad \text{for } \Omega \ge 50 \text{ rad/s}$$

Let the sampling period $T_s = 10^{-2}$ s.

(a) Sketch the digital filter magnitude response specification over the frequency range $-2\pi/T_s \le \omega/T_s \le 2\pi/T_s$.

(b) Determine the order of the Butterworth analog filter that acts as a suitable anti-aliasing filter.

(c) Determine the order of the Chebyshev analog filter that acts as a suitable anti-aliasing filter.

8.2. Analog filter design. Design the Butterworth analog lowpass filter that satisfies the analog specification given in Table Proj8.1. Use one set of the frequency values given in the table. Show the steps in the design procedure. Sketch the pole/zero pattern in the s plane and give the system function $H_A(s)$.

Digital filter design. Design a digital lowpass filter to replace the analog filter above. Let the sampling period be $T_s = 10^{-4}$ seconds.

Impulse invariance method. Determine the pole/zero pattern obtained by performing the impulse-invariance mapping. Write the digital filter system function $H(z)$ and sketch the block diagram of the filter.

Bilinear transform method. Determine the pole/zero pattern obtained by performing the bilinear mapping. Write the digital filter system function $H(z)$ and sketch the block diagram of the filter.

8.3. Repeat the steps in Problem 8.2 with a Chebyshev filter design.

8.4. Consider the analog lowpass filter specification given by

passband:
$$-1 < |H(j\Omega)|_{dB} \le 0 \qquad \text{for } 0 \le \Omega \le 10 \text{ rad/s}$$

stopband:
$$|H(j\Omega)|_{dB} < -60 \qquad \text{for } \Omega \ge 50 \text{ rad/s}$$

Let the sampling period $T_s = 10^{-2}$ s.

(a) Determine the order of the Butterworth filter that satisfies this specification and draw the s plane pole/zero pattern. Transfer the poles to the z plane using the impulse invariance method and implement the digital filter.
(b) Determine the order of the Chebyshev filter that satisfies this specification and draw the s plane pole/zero pattern. Transfer the poles to the z plane using the impulse invariance method and implement the digital filter.
(c) Compare the order of the two filters.

8.5. Bilinear-transform method. Consider the analog lowpass filter specification given by

passband:
$$-1 < |H(j\Omega)|_{dB} \le 0 \qquad \text{for } 0 \le \Omega \le 10 \text{ rad/s}$$

stopband:
$$|H(j\Omega)|_{dB} < -60 \qquad \text{for } \Omega \ge 200 \text{ rad/s}$$

Let the sampling period $T_s = 10^{-2}$ s.

(a) Determine the specification in terms of the prewarped frequency values.
(b) Determine the order of the Butterworth filter that satisfies this specification and draw the s plane pole/zero pattern.
(c) Determine the order of the Chebyshev filter that satisfies this specification and draw the s plane pole/zero pattern.

COMPUTER PROJECTS

8.1. Comparison of impulse-invariance and bilinear transform methods for a Butterworth lowpass filter

Object. Design a Butterworth analog lowpass filter and compare the magnitude and phase responses of the digital filters produced by using the impulse-invariance and bilinear transform methods.

Analytic results

Analog filter design. Design the Butterworth analog lowpass filter that satisfies the analog specification given in Table Proj8.1. Use one set of the frequency values given in the table. Show the steps in the design procedure. Sketch the pole/zero pattern in the s plane and give the system function $H_A(s)$.

Digital filter design. Design a digital lowpass filter to replace the analog filter above. Let the sampling period be $T_s = 10^{-4}$ s.

Impulse invariance method. Determine the pole/zero pattern obtained by performing the impulse-invariance mapping. Write the digital filter system function $H(z)$ and sketch the block diagram of the filter.

TABLE Proj8.1

Frequency values for prototype analog lowpass specification (values are given in radians/second)

				Analog lowpass specification					

Possible values:

Ω_1	Ω_2	Ω_1	Ω_2	Ω_1	Ω_2	Ω_1	Ω_2	Ω_1	Ω_2
1094π	5938π	1250π	5938π	1406π	5938π	1563π	6094π	1719π	6094π
1094π	6094π	1250π	6094π	1406π	6094π	1563π	6250π	1719π	6250π
1094π	6250π	1250π	6250π	1406π	6250π	1563π	6406π	1719π	6406π
1094π	6406π	1250π	6406π	1406π	6406π	1563π	6563π	1719π	6563π
1094π	6563π	1250π	6563π	1406π	6563π	1563π	6720π	1719π	6720π

Bilinear transform method. Determine the pole/zero pattern obtained by performing the bilinear mapping. Write the digital filter system function $H(z)$ and sketch the block diagram of the filter.

Computer results

Implement both filters and plot the following

(a) the unit-sample response.
(b) the log-magnitude responses using PLOTLM with DBVAL $= -4$ and -60. The specification limits of -1 and -50 dB should be shown.
(c) the phase response.

Comparison

Compare the stopband performance of the two methods. Explain the phase response in terms of the pole/zero pattern in the z plane.

8.2. Comparison of impulse-invariance and bilinear transform methods for a Chebyshev lowpass filter.

Object. Design a Chebyshev analog lowpass filter and compare the magnitude and phase responses of the digital filters produced by using the impulse-invariance and bilinear transform methods.

Perform the steps listed in Project 8.1 for the Chebyshev filter design.

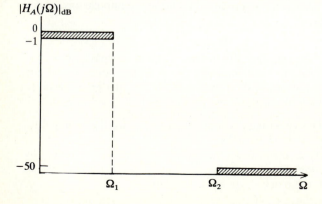

FIGURE Proj8.1
Lowpass, highpass, bandpass and bandreject specifications for the projects.

8.3. Butterworth IIR filter design

Object. Design an analog Butterworth filter that satisfies the lowpass specification; use the bilinear transform to obtain the digital filter; and use frequency transformations to obtain the digital filter to meet a bandpass, highpass or bandreject specification.

Do the design by performing the following steps (let $T_s = 10^{-4}$ s):

1. Design the analog Butterworth filter using the values for the analog lowpass filter found in Table Proj8.1. These values have not been prewarped. Show the analog Butterworth lowpass filter design procedure. Include sketches showing the frequency prewarping transformation.

2. Design the digital lowpass filter by performing a bilinear transformation. For the digital lowpass filter determine the system function obtained by the bilinear transform and sketch the pole/zero pattern. Sketch the block diagram and plot the unit-sample response. Plot the log-magnitude response with DVBAL = −4 and −60 with the specification limits shown. Plot the phase response.

3. Design the digital filter that satisfies one of the highpass, bandpass or band reject specifications given in Appendix B. Use the frequency transformations with the edge of the passband as the characteristic frequency. Sketch the pole/zero pattern, and the block diagram. Plot the log-magnitude response of the digital bandpass filter with DBVAL = −4 and −60, and with the specification limits indicated. Also plot the phase response.

   (a) If your stopband specification is not met after performing the frequency transformation, satisfy the specification by adding zeros to the filter system function to achieve additional attenuation.
   (b) If your specification is met after performing the frequency transformation, add zeros so that either of the following is achieved:
   (1) a reduction in the transition band by a factor of 2, or
   (2) a reduction in the stopband attenuation by 6 dB.

Provide the following for the final filter: the system function, the filter block diagram and the z plane pole/zero pattern.

8.4. Chebyshev IIR filter design

Object. Design an analog Chebyshev filter that satisfies the lowpass specification; use the bilinear transform to obtain the digital filter; and use frequency transformations to obtain the digital filter to meet a bandpass, highpass or bandreject specification.

Perform the steps listed in Project 8.3 for a Chebyshev filter. Numerical round-off errors in your computations may shift the implemented pole position so that the specification is not satisfied. If this occurs, manually shift the pole positions in your final filter until the specification is satisfied. Do this interactively by examining the magnitude response after a modification is made. List the steps that you took.

CHAPTER
9

FINITE-IMPULSE RESPONSE FILTER DESIGN TECHNIQUES

9.1 INTRODUCTION

In this chapter, we describe procedures for implementing *finite-impulse response (FIR)* filters. In contrast to infinite-impulse response (IIR) filters, the FIR filter is characterized by a unit-sample response that has a finite duration. Understanding the performance of FIR filters is important because the FIR filter is the discrete-time-domain equivalent to processing data in the frequency domain with the fast Fourier transform (FFT) algorithm.

FIR filters have two advantages over IIR filters. First, the design procedures for IIR filters are well established only for magnitude responses that are piece-wise constant over frequency, such as lowpass, highpass, bandpass and bandreject filters. In contrast, the magnitude response of an FIR filter can be designed to have an *arbitrary* shape. Hence, if the magnitude response specification is other than piece-wise constant, the FIR filter is the filter of choice. The second advantage of FIR filters is that they can be designed to have a phase response that is exactly linear with frequency. In Chapter 6, it was shown that a special FIR linear-phase filter structure can be used to reduce the number of multiplications by approximately one-half. The disadvantage of FIR filters is that, *if* an IIR filter can be found that satisfies the same specification as an FIR filter, the IIR filter will typically be faster and require less memory.

318

Three FIR design procedures are described in this chapter. The one to be used in a particular case depends on whether the *desired unit-sample response* $\{h_D(n)\}$ is known and whether it has a finite duration. If $\{h_D(n)\}$ is known and has a finite duration, the FIR design is straightforward and can be implemented directly by using the structures considered in Chapter 6. If the $\{h_D(n)\}$ is known, but has an infinite duration, the response must be truncated to form the finite-duration response $\{h_N(n)\}$ having N samples, where N is a finite integer. To do this, the *windowing method* for designing FIR filters is described. It employs an N-point tapered window sequence to perform the truncation. When the form of the unit-sample response is not known, $\{h_N(n)\}$ must be determined from the desired magnitude specification $|H_D(e^{j\omega})|$ and an appropriate phase response. With the *DFT method*, $\{h_N(n)\}$ is determined from the inverse discrete Fourier transform of N samples of the desired magnitude response $\{|H_D(k)|\}$ and an appropriate linear phase response. The windowing and DFT methods are combined to find the filter having the shortest-duration unit-sample response that still satisfies the specification. The third FIR filter design procedure is the *frequency-sampling method*. It produces an interesting filter structure by using $\{|H_D(k)|\}$ and an appropriate phase response. With this third method any continuous magnitude response having an arbitrary shape can be achieved. The resulting filter structure contains a parallel combination of second-order resonators, which allows the processing to be performed in parallel. This structure is most useful for applications in which processing speed must be reduced to a minimum. The chapter concludes with a discussion on how to use the FFT algorithm to perform filtering tasks on long data sequences. A comparison of multiplication counts for performing the filtering operation with a digital filter and with the FFT algorithm indicates when one is to be preferred over the other.

9.2 DESIGNING FIR FILTERS WITH THE WINDOWING METHOD

As described in Chapter 6, the system function of an FIR filter can be written as

$$H_{\text{FIR}}(z) = \sum_{k=-N_F}^{N_P} b_k z^{-k} \tag{9.1}$$

Using the nonrecursive structure, this filter can be implemented directly from the coefficient values b_k, for $-N_F \le k \le N_P$. Further, the unit-sample response of the FIR filter is equal to this coefficient sequence, or

$$h(n) = b_n \qquad \text{for } -N_F \le n \le N_P \tag{9.2}$$

Hence, the FIR filter can be implemented directly from the unit-sample response. We first consider implementing the FIR filter to approximate a desired unit-sample response sequence. Then we consider the case of requiring the FIR filter to satisfy a desired magnitude specification.

To take advantage of the efficient nonrecursive structure for linear-phase FIR filters, described in Chapter 6, we will design FIR filters to a have a linear-phase response. Let $\{h_D(n)\}$ denote the desired unit-sample response sequence of a linear-phase filter. Further, let $\{h_D(n)\}$ be a finite-duration symmetric sequence extending over $-N_F \leq n \leq N_P$, where $N_F = N_P$ by symmetry. This desired sequence can be achieved exactly by an FIR filter by setting the coefficients equal to $h_D(n)$, or

$$b_n = h_D(n) \qquad \text{for } -N_F \leq n \leq N_P \tag{9.3}$$

A problem arises for filters, such as the ideal lowpass filter, which have a unit-sample response $\{h_D(n)\}$ that has an infinite duration. The sequence $\{h_D(n)\}$ has an infinite duration whenever the desired magnitude response $|H_D(e^{j\omega})|$, or any of its derivatives exhibit a discontinuity. For such cases, we must truncate the infinite-duration $\{h_D(n)\}$ to produce one that has a finite duration of N samples, denoted by $\{h_N(n)\}$.

As shown in Chapter 3, to have a linear-phase response, the unit-sample response must be symmetric or antisymmetric about some point in time. To simplify the analysis below, we consider $\{h_D(n)\}$ to be symmetric about $n = 0$, or

$$h_D(n) = h_D(-n) \qquad \text{for all } n \tag{9.4}$$

To retain the linear-phase response in the FIR filter having the response $\{h_N(n)\}$, the truncation must be performed symmetrically around $n = 0$ to maintain the symmetry in $\{h_N(n)\}$, as shown in Fig. 9.1. The N-point

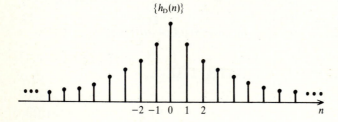

for $N = 7$

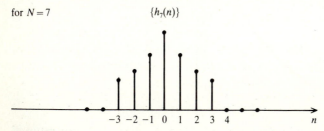

FIGURE 9-1
Comparison of desired unit-sample response $\{h_D(n)\}$ having an infinite duration and the 7-point finite-duration approximation $\{h_7(n)\}$.

unit-sample response, where N is an odd-valued integer, is then given by

$$h_N(n) = \begin{cases} h_D(n) & \text{for } -(N-1)/2 \le n \le (N-1)/2 \\ 0 & \text{otherwise} \end{cases} \tag{9.5}$$

If a causal linear-phase filter is desired, $\{h_N(n)\}$ can be delayed by $(N-1)/2$, to start at $n = 0$. In the truncation above, the value of N is taken to be an odd-valued integer. In this case, the point of symmetry falls on a sample point. When the point of symmetry in $\{h_D(n)\}$ falls between two sample points, N must be an even-valued integer.

To analyze the truncation process, we model it as a multiplication of the desired sequence by a finite-duration *window sequence*, denoted by $\{w(n)\}$. The general truncation of an even sequence can be written as

$$h_N(n) = h_D(n)\, w(n) \qquad \text{for all } n \tag{9.6}$$

where

$$w(n) = w(-n) \ne 0 \qquad \text{for } |n| \le (N-1)/2$$
$$= 0 \qquad\qquad \text{for } |n| > (N-1)/2 \tag{9.7}$$

The sequence $\{w(n)\}$ has a duration of N samples and is symmetric about $n = 0$ to maintain the symmetry that is present in $\{h_D(n)\}$. The goal of the windowing synthesis approach is, given the desired filter unit-sample response $\{h_D(n)\}$, to determine the values of $\{w(n)\}$, such that the finite-duration $\{h_N(n)\}$ results in the FIR filter that satisfies the desired magnitude response specification. Since the number of multiplications in the FIR filter increases with N, the window having the smallest duration is desired.

It was shown in Chapter 3 that the multiplication of two discrete-time sequences corresponds to a convolution of their Fourier transforms, or spectra. If $H_D(e^{j\omega})$ is the spectrum of $\{h_D(n)\}$, and $W(e^{j\omega})$ is the spectrum of $\{w(n)\}$, then the transfer function of the FIR filter $H_N(e^{j\omega})$ is given by

$$H_N(e^{j\omega}) = H_D(e^{j\omega}) \circledast W(e^{j\omega})$$

$$= \frac{1}{2\pi} \int_{-\pi}^{\pi} H_D(e^{j\theta})\, W(e^{j[\omega-\theta]})\, d\theta \tag{9.8}$$

Since the two functions in the integral are periodic, a *circular convolution* results and the limits are taken over one period. Trying to evaluate this convolution in general is difficult because the spectra are complex-valued and tend to have complicated analytic forms involving sine and cosine functions. However, if all the discrete-time sequences are symmetric about $n = 0$, then all the spectra are real-valued functions of ω. This property helps us to visualize the effects of convolution in the following discussion.

To illustrate the effects of this convolution, let us consider the problem of approximating an ideal lowpass filter by using the windowing method. For the

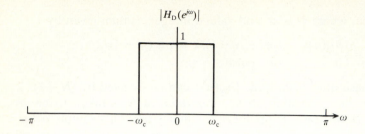

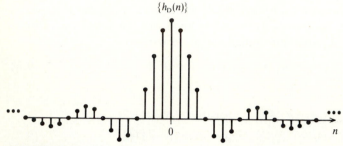

FIGURE 9-2
Ideal lowpass filter magnitude response $|H_D(e^{j\omega})|$ and the corresponding infinite-duration desired unit-sample response $\{h_D(n)\}$. Case for $\omega_c = \pi/4$ is shown.

cutoff frequency specified as ω_c, the desired transfer function is equal to

$$H_D(e^{j\omega}) = \begin{cases} 1 & \text{for } -\omega_c \le \omega \le \omega_c \\ 0 & \text{for } -\pi \le \omega < -\omega_c \text{ and } \omega_c < \omega < \pi. \end{cases} \tag{9.9}$$

This is shown in Fig. 9.2 for $\omega_c = \pi/4$. The unit-sample response of the ideal lowpass filter is given by the inverse Fourier transform, or

$$h_D(n) = \frac{1}{2\pi} \int_{-\pi}^{\pi} H_D(e^{j\omega}) \, e^{j\omega n} \, d\omega$$

$$= \frac{1}{2\pi} \int_{-\omega_c}^{\omega_c} e^{j\omega n} \, d\omega = \frac{e^{j\omega_c n} - e^{-j\omega_c n}}{j2\pi n}$$

$$= \frac{\sin(\omega_c n)}{\pi n} \qquad \text{for all } n \tag{9.10}$$

The desired unit-sample response is an even sequence that extends to both plus and minus infinity. We will consider several different finite-duration approximations to this ideal response by employing several common window sequences: the rectangular, triangular, and cosine windows.

THE RECTANGULAR WINDOW. The rectangular window sequence is defined by

$$w_R(n) = \begin{cases} 1 & \text{for } -(N-1)/2 \le n \le (N-1)/2 \\ 0 & \text{otherwise} \end{cases} \tag{9.11}$$

An example is shown in Fig. 9.3 for $N = 31$. The infinite-duration unit-sample response of the ideal lowpass filter is approximated by a finite-duration sequence containing N elements by multiplying it with the rectangular window, or

$$h_N(n) = w_R(n)\, h_D(n) \qquad \text{for all } n$$

$$= \frac{\sin(\omega_c n)}{\pi n} \qquad \text{for } -(N-1)/2 \le n \le (N-1)/2 \qquad (9.12)$$

The real-valued transfer function $H_N(e^{j\omega})$ of the FIR approximation is shown in Fig. 9.4. The shape of $H_N(e^{j\omega})$ can be explained in terms of the convolution

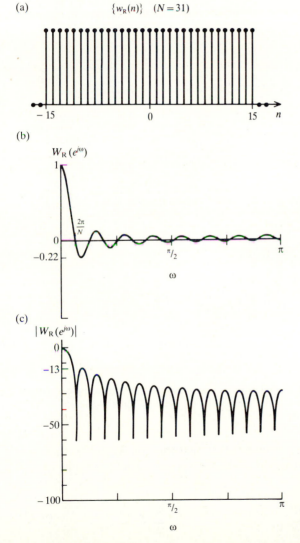

(a) $\{w_R(n)\}$ $(N = 31)$

-15 0 15 n

(b)

$W_R(e^{j\omega})$

1

$\frac{2\pi}{N}$

0

-0.22

$\pi/2$

π

ω

(c)

$|W_R(e^{j\omega})|$

0

-13

-50

-100

$\pi/2$

π

ω

FIGURE 9-3
Rectangular data window ($N = 31$).
(a) Window sequence; (b) spectrum;
(c) log-magnitude spectrum.

(a)

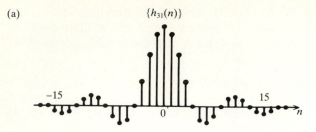

(b) $H_{31}(e^{j\omega})$

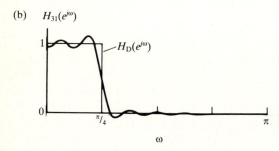

(c) $|H_{31}(e^{j\omega})|_{\text{dB}}$

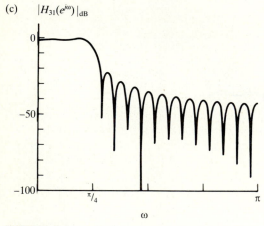

FIGURE 9-4
Results of applying a 31-point rectangular window to the ideal lowpass filter unit-sample response. (*a*) Finite-duration unit-sample response; (*b*) magnitude response of approximation to ideal lowpass filter; (*c*) log-magnitude response.

of $W_R(e^{j\omega})$ and $H_D(e^{j\omega})$ by examining the features of $W_R(e^{j\omega})$ that are responsible for the deviations from the ideal response. The spectrum of the rectangular window is equal to

$$W_R(e^{j\omega}) = \sum_{n=-(N-1)/2}^{(N-1)/2} e^{-j\omega n} = e^{j\omega(N-1)/2} \sum_{n=0}^{N-1} e^{-j\omega n} \qquad (9.13)$$

The finite geometric sum formula can be applied to evaluate the sum, to give

$$W_R(e^{j\omega}) = e^{j\omega(N-1)/2} \frac{1 - e^{-j\omega N}}{1 - e^{-j\omega}} \tag{9.14}$$

To simplify this analytic form, we express the numerator and denominator as sine functions, or

$$W_R(e^{j\omega}) = e^{j\omega(N-1)/2} \frac{e^{-j\omega N/2}}{e^{-j\omega/2}} \frac{e^{j\omega N/2} - e^{-j\omega N/2}}{e^{j\omega/2} - e^{-j\omega/2}}$$

$$= \frac{\sin(N\omega^2)}{\sin(\omega/2)} \tag{9.15}$$

The spectrum $W_R(e^{j\omega})$ for $N = 31$ is shown in Fig. 9.3. The spectrum $W_R(e^{j\omega})$ has two features that are important for this application, the width of the main lobe and the sidelobe amplitude. The main-lobe width is defined as the distance between the two points closest to $\omega = 0$ where $W_R(e^{j\omega})$ is zero. For the rectangular window, the main-lobe width is equal to $4\pi/N$. The maximum sidelobe magnitude for $W_R(e^{j\omega})$ occurs for the first sidelobe and is equal to approximately 22% of the main-lobe amplitude, or -13 dB relative to the maximum value at $\omega = 0$.

We now consider the effect of these two features in the convolution to produce the lowpass FIR filter transfer function $H_N(c^{j\omega})$. The transfer function $H_N(e^{j\omega})$ and the log-magnitude response $|H_N(e^{j\omega})|_{dB}$ are shown in Fig. 9.4 for the 31-point rectangular window. The filter response differs from the desired response in several ways. The sharp transition in the ideal response at $\omega = \omega_c$ has been converted into a gradual transition. In the passband, a series of *overshoots* and *undershoots* occur, and in the stopband, where the desired response is zero, the FIR filter has a nonzero response, called *leakage*. These effects can be explained in terms of the features of the window spectrum.

The main lobe of $W_R(e^{j\omega})$ causes the smearing of the desired transfer function features, especially noticeable at frequencies where the desired magnitude response is discontinuous. The discontinuity in $H_D(e^{j\omega})$ is converted into a gradual transition in $H_N(e^{j\omega})$, whose width is related to the width of the main lobe of $W_R(e^{j\omega})$. Since the main-lobe width of $W_R(e^{j\omega})$ is equal to $4\pi/N$, the size of this transition region can be reduced to any desired size by increasing the size of the window sequence. The forms of $H_N(e^{j\omega})$ for different values of N are shown in Figs. 9.5 and 9.6. However, this increase in N also increases the number of computations necessary to implement the FIR filter.

The sidelobes of $W_R(e^{j\omega})$ cause a different distortion to the filter magnitude response. Since the sidelobes of $W_R(e^{j\omega})$ extend over a wide frequency range, large amplitude components in $H_D(e^{j\omega})$ become smeared over a wide range of frequencies in $H_N(e^{j\omega})$. This effect is most evident in the frequency ranges in which the $H_D(e^{j\omega})$ is either constant (in the passband) or has a small magnitude (in the stopband). In the passband, these sidelobe effects appear both as overshoots and undershoots to the desired response. In

(a)

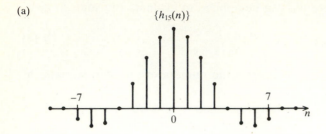

(b)

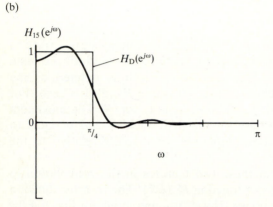

(c)

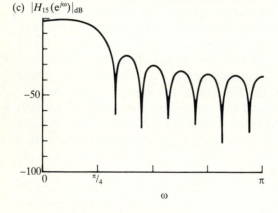

FIGURE 9-5
Results of applying a 15-point rectangular window to the ideal lowpass filter unit-sample response. (*a*) Finite-duration unit-sample response; (*b*) transfer function of approximation to ideal lowpass filter; (*c*) log-magnitude response.

the stopband, these effects appear as a nonzero response. Unfortunately, and unexpectedly, these sidelobe effects do not diminish significantly, but remain almost constant, as the duration of the rectangular window is increased, as shown in Figs. 9.4 through 9.6. These results indicate that no matter how many elements of the $\{h_D(n)\}$ are included in $\{h_N(n)\}$, the magnitudes of the overshoot and leakage will not change significantly when the rectangular window is used. This result is known as the *Gibbs phenomenon,* after the American mathematician Josiah Willard Gibbs of Yale, who first noted this effect.

(a)

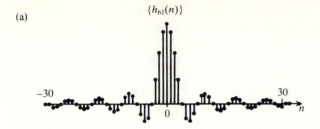

$\{h_{61}(n)\}$

(b)

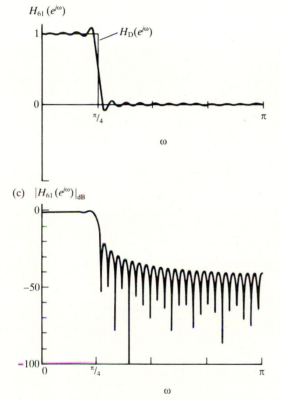

$H_{61}(e^{j\omega})$

$H_D(e^{j\omega})$

(c) $|H_{61}(e^{j\omega})|_{dB}$

FIGURE 9-6
Results of applying a 61-point rectangular window to the ideal lowpass filter unit-sample response. (*a*) Finite-duration unit-sample response; (*b*) transfer function of approximation to ideal lowpass filter; (*c*) log-magnitude response.

To reduce these sidelobe effects, we must consider alternate window sequences having spectra exhibiting smaller sidelobes. To indicate how the form of the window should change, we observe that the sidelobes of the window spectrum $W(e^{j\omega})$ represent the contribution of the high-frequency components in the window sequence. For the rectangular window, these high-frequency components are due to the sharp transitions from zero to one at the edges of the window sequence. Hence, the amplitudes of these high-frequency components, the sidelobe levels, can be reduced by replacing these sharp transitions by more gradual ones. This is the motivation for considering the triangular, or Bartlett, window described next.

THE TRIANGULAR, OR BARTLETT, WINDOW. Let us consider the effect of tapering the rectangular window sequence linearly from the middle to the ends. We then obtain the N-point triangular window given by

$$w_T(n) = 1 - \frac{2|n|}{N-1} \quad \text{for } -(N-1)/2 \le n \le (N-1)/2 \tag{9.16}$$

This window and its spectrum $W_T(e^{j\omega})$ are shown in Fig. 9.7. As expected, the

(a)

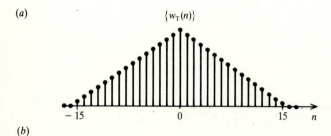

(b)

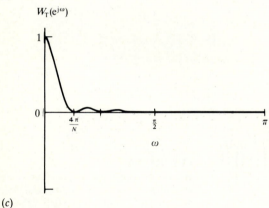

(c)

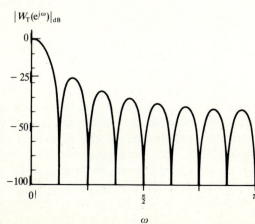

FIGURE 9-7
Triangular data window results ($N = 31$). (a) Window sequence; (b) spectrum; (c) log-magnitude spectrum.

sidelobe level is smaller than that of the rectangular window, being reduced from -13 to -25 dB relative to the maximum. However, the main-lobe width is now $8\pi/N$, or twice that of the rectangular window having the same duration. This result illustrates that there is a tradeoff between main-lobe width and sidelobe level.

The FIR lowpass magnitude response obtained by applying the triangular window to the ideal lowpass filter unit-sample response is shown in Fig. 9.8. Comparing it with that obtained by using the rectangular window, we find that the triangular window produces a smoother magnitude response. The transition from passband to stopband is not as steep as that for the rectangular

(a)

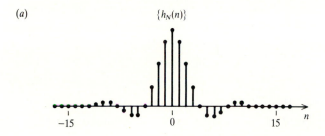

(b)

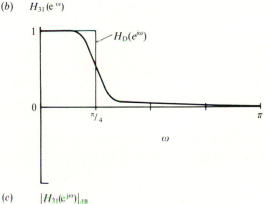

(c)

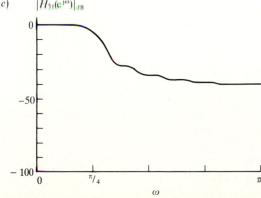

FIGURE 9-8

Results of applying a 31-point triangular window to the ideal lowpass filter unit-sample response. (a) Finite-duration unit-sample response; (b) magnitude response of approximation to ideal lowpass filter; (c) log-magnitude response.

window. In the stopband, the response is smoother, but the attenuation is less than that produced by the rectangular window. Because of these characteristics, the triangular window is not usually a good choice.

We next consider a class of raised-cosine windows that, compared with the triangular window, are smoother at the ends, but closer to one at the middle. The smoother taper at the ends should reduce the sidelobe level, while the broader middle section produces less distortion of $\{h_D(n)\}$ around $n = 0$.

RAISED-COSINE WINDOWS. To reduce the sidelobe level further, we can consider an even more gradual taper at the ends of the window sequence by using the raised-cosine sequence. We now discuss three such windows. The first is the *Hanning*, or simply *raised-cosine window*, denoted by $\{w_C(n)\}$, given by

$$w_C(n) = \begin{cases} 0.5 + 0.5 \cos(2\pi n/(N-1)) & \text{for } -(N-1)/2 \le n \le (N-1)/2 \\ 0 & \text{otherwise} \end{cases}$$

(9.17)

This window and its spectrum $W_C(e^{j\omega})$ are shown in Fig. 9.9 for $N = 31$. The magnitude of the first sidelobe level is -31 dB relative to the maximum value, an improvement of 6 dB over that for the triangular window. Since the main-lobe width is the same, the Hanning window is preferable to the triangular.

This reduction in the sidelobe level can be explained by considering $\{w_C(n)\}$ itself to be an infinite-duration raised-cosine sequence that is multiplied by the rectangular window, or

$$w_C(n) = \left[0.5 + 0.5 \cos\left(\frac{2\pi n}{N-1}\right) \right] w_R(n) \qquad \text{for all } n$$

$$= c(n) w_R(n) \qquad \text{for all } n, \qquad (9.18)$$

where $\{c(n)\}$ is the infinite-duration raised-cosine sequence as shown in Fig. 9.10(a). This multiplication in discrete-time is equivalent to the circular convolution of the two Fourier transforms. Since $\{c(n)\}$ is composed of a constant term and a cosine term at $\omega = 2\pi/(N-1)$, its Fourier transform can be expressed in terms of three Dirac delta functions, given by

$$C(e^{j\omega}) = \pi \, \delta(\omega) + 0.5\pi \, \delta\left(\omega - \frac{2\pi}{N-1}\right) + 0.5\pi \, \delta\left(\omega + \frac{2\pi}{N-1}\right)$$

$$\text{for } -\pi \le \omega \le \pi \quad (9.19)$$

When convolving $W_R(e^{j\omega})$ with $C(e^{j\omega})$, versions of $W_R(e^{j\omega})$ are shifted to the locations of these impulses, the shifted spectra are scaled by the weight of the impulse, and the results are added to produce the final curve, as shown in Fig. 9.10(b). The shifts caused by the impulses from the cosine function locate the sidelobes of $W_R(e^{j\omega})$ so that they tend to cancel, thus reducing the sidelobe magnitudes of the resulting spectrum $W_C(e^{j\omega})$.

(a)

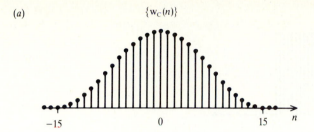

(b) $W_C(e^{j\omega})$

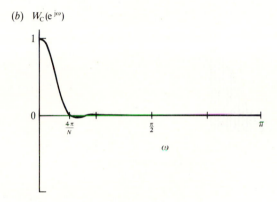

(c) $|W_C(e^{j\omega})|_{dB}$

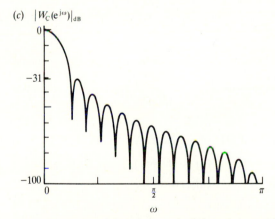

FIGURE 9-9
Raised-cosine (Hanning) window results ($N = 31$). (a) Window sequence; (b) spectrum; (c) log-magnitude spectrum.

The discrete-time and frequency approximations to the ideal lowpass filter produced by applying the Hanning window are shown in Fig. 9.11. Most notable is the improved stopband attenuation characteristic. The largest peak is approximately −44 dB relative to the passband level. At higher frequencies the stopband attenuation is even greater.

Hamming noted that a reduction in the first sidelobe level can be achieved by adding a small constant value to the cosine window. This produces

(a) $\{w_R(n)\}$

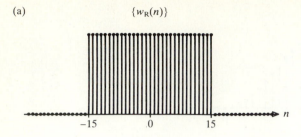

$\{c(n)\} = \{1 + \cos(^{\pi n}/_{15})\}$

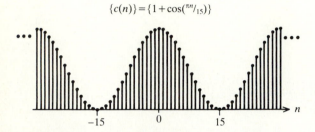

$\{w_C(n)\}$

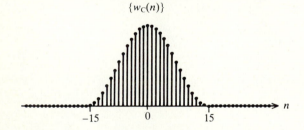

FIGURE 9-10
Interpreting the raised-cosine (Hanning) window results. (a) Time domain; (b) frequency domain.

the *Hamming window*, denoted by $\{w_H(n)\}$, and given by

$$w_H(n) = \begin{cases} 0.54 + 0.46\cos(2\pi n/(N-1)) & \text{for } -(N-1)/2 \le n \le (N-1)/2 \\ 0 & \text{otherwise} \end{cases}$$

$$(9.20)$$

This reduction in the first sidelobe level can be explained in terms of the addition of the shifted and scaled window spectra $W_R(e^{j\omega})$, shown in Fig. 9.10(b). For the raised-cosine window, the Dirac delta functions at $\omega = 0$ and at $\omega = \pi/15$ formed a ratio of two to one. This ratio resulted in a negative amplitude for the first sidelobe in $W_C(e^{j\omega})$. By changing this ratio of delta functions to become 2.35 to one (0.54 to 0.46/2), the magnitude of the first sidelobe in $W_C(e^{j\omega})$ can be reduced. The Hamming window and its spectrum

(b)

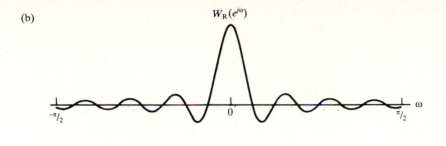

$W_R(e^{j\omega})$

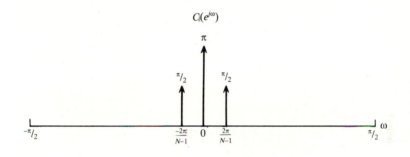

$C(e^{j\omega})$

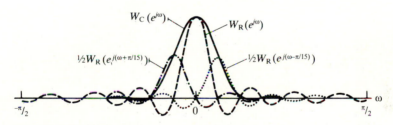

$W_C(e^{j\omega})$ $W_R(e^{j\omega})$

$\tfrac{1}{2}W_R(e^{j(\omega+\pi/15)})$ $\tfrac{1}{2}W_R(e^{j(\omega-\pi/15)})$

FIGURE 9-10b

$W_H(e^{j\omega})$ are shown in Fig. 9.12 for $N = 31$. The magnitude of the first sidelobe has been reduced to -41 dB, a reduction of 10 dB relative to the Hanning window. But this reduction is achieved at the expense of the sidelobe magnitudes at the higher frequencies, which are almost constant with frequency. With the Hanning window, the sidelobe amplitudes decrease with frequency.

The lowpass filter magnitude and log-magnitude responses when using the Hamming window are shown in Fig. 9.13. It is noted that the first sidelobe peak is -51 dB, -7 dB with respect to the Hanning window filter. However, as the frequency increases, the stopband attenuation does not increase as much as with the filter produced by the Hanning window.

The stopband attenuation in the lowpass filter magnitude response is

(a)

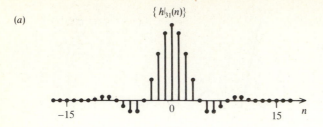

(b)

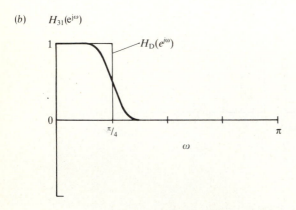

(c)

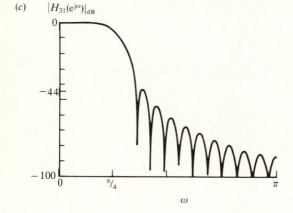

FIGURE 9-11
Results of applying a 31-point raised-cosine (Hanning) window to the ideal lowpass filter unit-sample response. (a) Finite-duration unit-sample response; (b) magnitude response of approximation to ideal lowpass filter; (c) log-magnitude response.

limited by the sidelobe level of the window function. Even though the Hamming window achieved an attenuation of 51 dB for our lowpass filter, it may not be sufficiently small for some applications. A window having smaller magnitude sidelobes is needed. Hamming reduced the sidelobe magnitude while maintaining the main-lobe width, equal to $8\pi/N$. The following window reduces the sidelobe level at the expense of the main-lobe width.

(a)

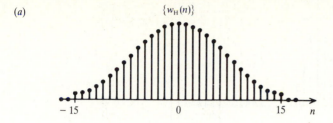

(b) $W_H(e^{j\omega})$

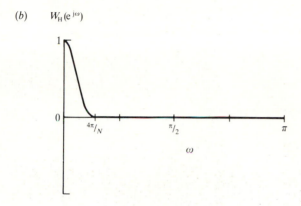

(c) $|W_H(e^{j\omega})|_{dB}$

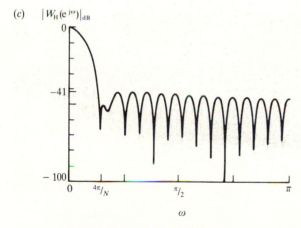

FIGURE 9-12
Hamming window results ($N =$ 31). (a) Window sequence; (b) spectrum; (c) log-magnitude spectrum.

Blackman noted that the sidelobes can be reduced further by including an additional cosine term in the window sequence. The Blackman window is given by

$$w_B(n) = 0.42 + 0.5\cos(2\pi n/(N-1)) + 0.08\cos(4\pi n/(N-1))$$

$$\text{for } -(N-1)/2 \leq n \leq (N-1)/2$$

$$= 0 \quad \text{otherwise.} \tag{9.21}$$

(a)

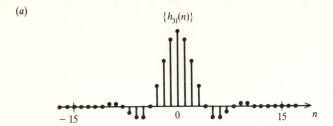

$\{h_{31}(n)\}$

(b) $H_{31}(e^{j\omega})$

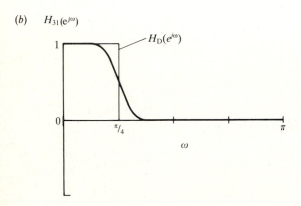

$H_D(e^{j\omega})$

(c) $|H_{31}(e^{j\omega})|_{dB}$

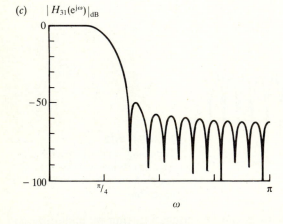

FIGURE 9-13
Results of applying a 31-point Hamming window to the ideal lowpass filter unit-sample response. (a) Finite-duration unit-sample response; (b) magnitude response of approximation to ideal low-pass filter; (c) log-magnitude response.

This window and its spectrum $W_B(e^{j\omega})$ are shown in Fig. 9.14. The magnitude of the first sidelobe level has been reduced to -57 dB. However, the main-lobe width has increased to $12\pi/N$. This is another illustration of the tradeoff between the sidelobe magnitude and the main-lobe width.

This reduction in the sidelobe level can be explained in a manner similar to the raised-cosine window. The Blackman window can be written as the

(a)

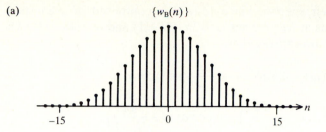

$\{w_B(n)\}$

(b) $W_B(e^{j\omega})$

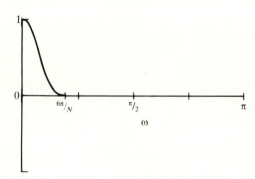

(c) $|W_B(e^{j\omega})|_{dB}$

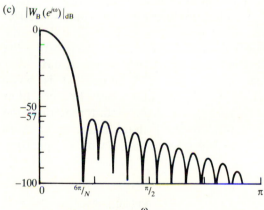

FIGURE 9-14
Blackman window results $(N = 31)$.
(a) Window sequence; (b) spectrum;
(c) log-magnitude spectrum.

product of an infinite-duration sequence and a rectangular window, or

$$w_B(n) = [0.42 + 0.5 \cos(2\pi n/(N - 1)) + 0.08 \cos(4\pi n/(N - 1))]w_R(n)$$

for all n

$$= c_2(n)w_R(n) \qquad \text{for all } n \tag{9.22}$$

where $\{w_R(n)\}$ is the N-point rectangular window. As before, the multiplication of two sequences in the time-domain corresponds to a convolution in the frequency domain of their spectra. Since $\{c_2(n)\}$ is composed of a constant term and two cosine terms at frequencies $\omega = 2\pi/(N-1)$ and $\omega = 4\pi/(N-1)$, its Fourier transform can be expressed as

$$C_2(e^{j\omega}) = 0.84\pi\delta(\omega) + 0.5\pi\delta\left(\omega - \frac{2\pi}{N-1}\right) + 0.5\pi\delta\left(\omega + \frac{2\pi}{N-1}\right)$$

$$+ 0.04\pi\delta\left(\omega - \frac{4\pi}{N-1}\right) + 0.04\pi\delta\left(\omega + \frac{4\pi}{N-1}\right)$$

$$\text{for} \quad -\pi \leq \omega \leq \pi. \tag{9.23}$$

This spectrum is shown in Fig. 9.15(a). The individual shifted and scaled versions of $W_R(e^{j\omega})$ produced by the convolution with $C_2(e^{j\omega})$ are shown in Fig. 9.15(b). Their sum produces the spectrum of the Blackman window $W_B(e^{j\omega})$ shown in Fig. 9.15(c). The main-lobe width is extended by the components centered at $\omega = \pm 4\pi/(N-1)$, but these components serve to reduce the sidelobe levels as well.

The digital lowpass filter results obtained when applying the Blackman window are shown in Fig. 9.16. The minimum attenuation in the stopband is an impressive -74 dB, but it occurs for $\omega > \pi/2$. Such a large transition region is necessitated by the wide main-lobe width of the Blackman window spectrum.

The windows discussed above are compared in terms of their main-lobe width and the maximum sidelobe level in Table 9.1.

Although other more complicated windows have been proposed, their performance is only slightly better than that produced by the windows above. These other windows are not discussed here, but can be found in the references. Another reason for discussing only these simple windows is that the shape of the window spectrum itself is not the important parameter, but rather its effect on the transfer function of the final FIR filter that is being approximated, so that different windows may be better for different filter types. Hence, it is impossible to make a general statement regarding the appropriateness of a particular window for the general filter. It is more important to be aware of the general effects produced by the window spectrum main-lobe and sidelobe structures.

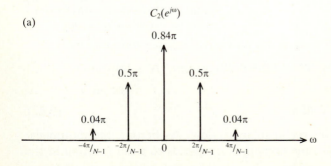

(a)

FIGURE 9-15
Frequency-domain interpretation of the Blackman window results. (*a*) Spectrum of infinite-duration sequence; (*b*) shifted and scaled versions of the rectangular window spectrum generated by the convolution; (*c*) spectrum of the Blackman window is the sum of the components in (*b*).

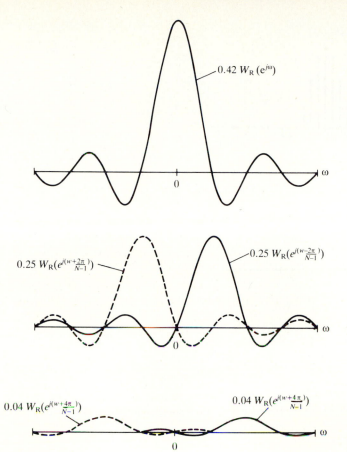

$0.42\ W_R\ (e^{j\omega})$

$0.25\ W_R(e^{j(w+\frac{2\pi}{N-1})})$

$0.25\ W_R(e^{j(w-\frac{2\pi}{N-1})})$

$0.04\ W_R(e^{j(w+\frac{4\pi}{N-1})})$

$0.04\ W_R(e^{j(w+\frac{4\pi}{N-1})})$

FIGURE 9-15b

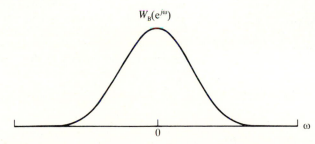

$W_B(e^{j\omega})$

FIGURE 9-15c

The main problem with applying the windowing method is that the desired unit-sample response $\{h_D(n)\}$ must be known. For a particular desired magnitude response, this unit-sample response may not be known. In the next section, we describe the DFT method that can be used to obtain a finite-duration approximation of the desired unit-sample response. In the following section, we will apply the windowing method to this approximation.

(a)|

$\{h_{31}(n)\}$

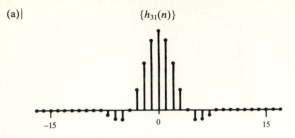

(b) $H_{31}(e^{j\omega})|$

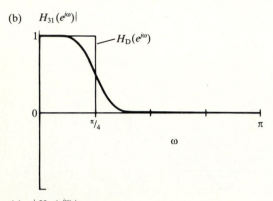

(c) $|H_{31}(e^{j\omega})|_{dB}$

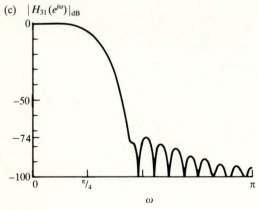

FIGURE 9-16
Results of applying a 31-point Blackman window to the ideal lowpass filter unit-sample response. (*a*) Finite-duration unit-sample response; (*b*) magnitude response of approximation to ideal lowpass filter; (*c*) log-magnitude response.

TABLE 9.1
Spectral properties of N-point windows

Window	Main-lobe width	Maximum sidelobe level (relative to maximum), dB
Rectangular	$4\pi/N$	-13
Bartlett	$8\pi/N$	-25
Hanning	$8\pi/N$	-31
Hamming	$8\pi/N$	-41
Blackman	$12\pi/N$	-57

9.3 THE DFT METHOD FOR APPROXIMATING THE DESIRED UNIT-SAMPLE RESPONSE

In this section, we describe a procedure to obtain a finite-duration approximation, denoted by $\{h_d(n)\}$, to the desired unit-sample response $\{h_D(n)\}$ by sampling the desired magnitude response $|H_D(e^{j\omega})|$ and performing an inverse DFT. To compute the inverse DFT, we must also specify a phase response sequence, for which we will consider a linear-with-frequency response. Since the inverse DFT is employed, the approximation to $\{h_D(n)\}$ is finite-duration, having N_d samples when an N_d-point DFT is employed. In this case, N_d should be much larger than N, the duration of the unit-sample response of the final FIR filter. Having determined an approximation to $\{h_D(n)\}$, we can then apply the windowing method for designing the final FIR filter.

The DFT method is important because it is the discrete-time equivalent of performing signal processing in the frequency domain by operating directly on the spectrum of the input signal, most often by using the FFT algorithm. By examining this frequency-domain operation as the equivalent time-domain filtering operation, we obtain a better understanding of the effects associated with the frequency-domain approach. The DFT method also serves as an introduction to the frequency-sampling filter design considered later in this chapter.

To find $\{h_d(n)\}$, we need to compute the inverse of an N_d-point DFT sequence. The inverse DFT produces a sequence defined over $0 \leq n \leq N_d - 1$. In Chapter 4, it was shown that the N_d-point sequence is one period of the periodic sequence $\{\bar{h}_d(n)\}$, which is the periodic extension of the desired sequence $\{h_D(n)\}$, or

$$\bar{h}_d(n) = \sum_{k=-\infty}^{\infty} h_D(n + kN_d) \tag{9.24}$$

The relationships of $\{h_D(n)\}$, $\{h_d(n)\}$ and $\{\bar{h}_d(n)\}$ for the ideal lowpass filter are shown in Fig. 9.17. If $\{h_D(n)\}$ has an infinite duration, some degree of *time-aliasing* will occur. For a stable filter, $\{h_D(n)\}$ is equal to or approaches zero for both positive and negative values of n. Hence, the time-aliasing effect can be made negligible by choosing N_d sufficiently large. In the design procedures below, we will consider $h_d(n) = h_D(n)$ for the N_d points over which $\{h_d(n)\}$ is defined.

To illustrate the DFT approach, let us consider the ideal lowpass filter. To obtain this N_d-point DFT sequence, the magnitude response is sampled at frequency points $\omega_k = 2\pi k/N_d$, for $0 \leq k \leq N_d - 1$. Let us define the sequence $\{H_D(k)\}$ to be

$$H_D(k) = H_D(e^{j\omega})\big|_{\omega = 2\pi k/N_d} \qquad \text{for } 0 \leq k \leq N_d - 1 \tag{9.25}$$

Applied to the ideal lowpass magnitude response, we have

$$|H_D(k)| = \begin{cases} 1 & \text{for } 0 \leq k \leq N_L \text{ and for } N_d - N_L \leq k \leq N_d - 1, \text{ and} \\ 0 & \text{otherwise} \end{cases}$$

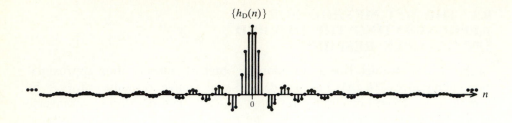

$\{h_D(n)\}$

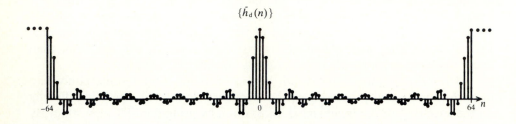

$\{\bar{h}_d(n)\}$

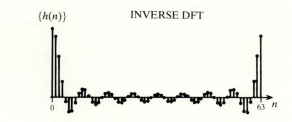

$\{h(n)\}$ INVERSE DFT

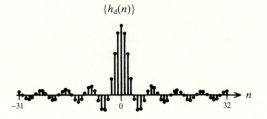

$\{h_d(n)\}$

FIGURE 9-17
Relationships of desired unit-sample response $\{h_D(n)\}$, its periodic extension $\{\bar{h}_d(n)\}$ ($N_d = 64$), the sequence produced by the inverse DFT assuming a zero phase sequence, and the approximate unit-sample response $\{h_d(n)\}$ for the non-causal FIR filter.

(a) $\{|H_D(k)|\}$

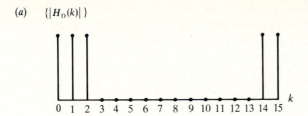

$\{\text{Arg}[H_D(k)]\}$

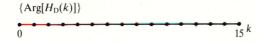

(b) $\{h(n)\}$

(c) $\{h_d(n)\}$

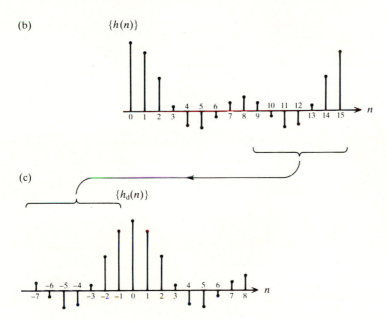

FIGURE 9-18
(a) Magnitude and phase DFT sequences for approximating the ideal lowpass filter ($N = 16$); (b) resulting 16-point inverse DFT sequence; (c) approximate unit-sample response for the non-causal FIR filter. Parts (b) and (c) indicate the shift of samples to be performed.

where N_L defines the edge of the passband, as shown in Fig. 9.18 for $N_d = 16$ and $N_L = 2$. To determine the unit-sample response from the inverse DFT, we must also specify a phase response sequence. Let us first choose the simplest, the zero-phase response sequence, given by

$$\mathrm{Arg}[H_D(k)] = 0 \qquad \text{for all } k$$

For this zero-phase response sequence, the real and imaginary sequences are equal to

$$H_R(k) = |H_D(k)| \qquad \text{and} \qquad H_I(k) = 0$$

Hence, the DFT element values are given by

$$H_D(k) = H_R(k) + jH_I(k) = H_R(k) \qquad \text{for } 0 \le k \le N_d - 1 \qquad (9.26)$$

The N_d-point unit-sample response $\{h_d(n)\}$ is then related to the inverse DFT sequence $\{h(n)\}$ computed as

$$h(n) = \frac{1}{N_d} \sum_{k=0}^{N_d-1} H_D(k)\, e^{j2\pi kn/N_d} \qquad \text{for } 0 \le n \le N_d - 1 \qquad (9.27)$$

This sequence is shown in Fig.9.18(b). The analytic evaluation of the inverse DFT can be simplified by noting that, for the zero-phase response sequence, $\{H_D(k)\}$ is an even sequence of k. The limits of the sum can then be taken from $-N_L$ to N_L, over which $H_D(k) = 1$. Hence,

$$
\begin{aligned}
h(n) &= \frac{1}{N_d} \sum_{k=-N_L}^{N_L} e^{j2\pi kn/N_d} \\
&= \frac{e^{-j2\pi N_L n/N_d}}{N_d} \sum_{k=0}^{2N_L} e^{j2\pi kn/N_d} \\
&= \frac{e^{-j2\pi N_L n/N_d}}{N_d} \frac{1 - e^{j2\pi(2N_L+1)n/N_d}}{1 - e^{j2\pi n/N_d}} \\
&= \frac{\sin[2\pi(N_L + \tfrac{1}{2})n/N_d]}{N_d \sin[\pi n/N_d]}
\end{aligned}
\qquad (9.28)
$$

Since a zero-phase response sequence has been specified, the unit-sample response should be an even sequence of n. The N_d-point noncausal sequence, corresponding to the zero-phase response sequence, can be obtained from the inverse DFT results by shifting the second half of the sequence, the values for $(N_d/2) + 1 \le n \le N_d - 1$, into negative values of n, as shown in Fig. 9.18(c). This then defines $\{h_d(n)\}$, the FIR approximation to the desired zero-phase lowpass filter. However, even after this shift is accomplished, the N_d-point sequence may not be exactly symmetric about $n = 0$ because N_d is an even integer and there may be one additional nonzero element without a partner at $n = N_d/2$. Since this symmetry is a necessary condition for a linear-phase response, even though the *sequence* $\{\mathrm{Arg}[H_D(k)]\}$ was specified to be zero-valued, the phase response *function* of the filter $\mathrm{Arg}[H_D(e^{j\omega})]$ cannot be equal to zero, except at the sampling points.

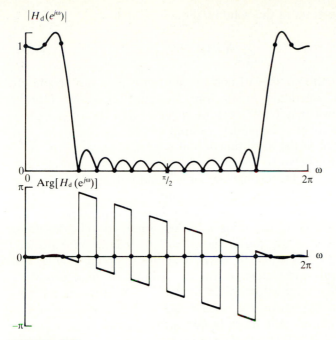

FIGURE 9-19
Comparison of the Fourier transform of $\{h_d(n)\}$ shown in Fig. 9-18 and the DFT sequence values.
The DFT samples are shown as heavy dots.

To resolve this dilemma, let us define $H_d(e^{j\omega})$ to be the filter transfer function determined by computing the Fourier transform of $\{h_d(n)\}$, or

$$H_d(e^{j\omega}) = \sum_{n=-N_d/2+1}^{N_d/2} h_d(n)e^{-j\omega n} \tag{9.29}$$

This transfer function is plotted in Fig. 9.19. Note that the function passes through the sample values of $\{H_D(k)\}$, but is not zero-phase.

We now want to relate $H_d(e^{j\omega})$ to the desired transfer function $H_D(e^{j\omega})$. The connection between the two is that $\{h_d(n)\}$ is one period of the periodic extension of $\{h_D(n)\}$. Expressing $\{h_d(n)\}$ as the rectangular window truncation of this periodic extension, we have

$$H_d(e^{j\omega}) = \sum_{n=-\infty}^{\infty} w_R(n)\,\bar{h}_d(n)\,e^{-j\omega n} \tag{9.30}$$

where $\{w_R(n)\}$ is the N_d-point rectangular window and $\{\bar{h}_d(n)\}$ is the periodic extension of $\{h_D(n)\}$. As shown in Chapter 3, the multiplication of two discrete-time sequences corresponds to the circular convolution of the two corresponding Fourier transforms. We then have

$$H_d(e^{j\omega}) = \frac{1}{2\pi} \int_{-\pi}^{\pi} \bar{H}(e^{j\theta})\,W_R(e^{j(\omega-\theta)})\,d\theta \tag{9.31}$$

where $W_R(e^{j\omega})$ is the spectrum of the window sequence, and

$$\tilde{H}(e^{j\omega}) = \sum_{n=-\infty}^{\infty} \tilde{h}_d(n) \, e^{-j\omega n} \tag{9.32}$$

Since $\{\tilde{h}_d(n)\}$ has an infinite number of nonzero elements, it is not absolutely summable and a true functional expression does not exist for $\tilde{H}(e^{j\omega})$. However, it can be shown that $\tilde{H}(e^{j\omega})$ can be expressed as the set of Dirac delta functions located at the frequency sample points $\omega_k = 2\pi k/N_d$, for $0 \le k \le N_d - 1$. The weight of the delta function at $\omega = \omega_k$ is equal to the DFT sequence value $H_D(k)$, as shown in Fig. 9.20. When convolving any function, in this case $W_R(e^{j\omega})$, with a Dirac impulse, only a shift is produced in the function to the location of the impulse. Because integration is a linear operation, each delta function in $\tilde{H}(e^{j\omega})$ can be considered separately and the contributions from all the delta functions are then summed. The resultant continuous-frequency transfer function is shown in Fig. 9.21b. As expected, the DFT sequence $\{H_D(k)\}$ is equal to $H_d(e^{j\omega})$ at the sampling points, but $H_d(e^{j\omega})$ exhibits overshoots and leakage terms very similar to those produced by the rectangular window method described in the previous section. In fact, the windowing operation was performed when $\{w_R(n)\}$ extracted the N_d-point unit-sample response from the periodic extension in Eq. (9.30).

It is important to realize that the relevant function to be specified for a digital filter is the transfer function $H_d(e^{j\omega})$, and not the DFT sequence $\{H_D(k)\}$. Even though the N_d-point DFT sequence $H_D(k)$, for $0 \le k \le N_d - 1$, is a sampled version of the desired transfer function $H_D(e^{j\omega})$ over $0 \le \omega < 2\pi$,

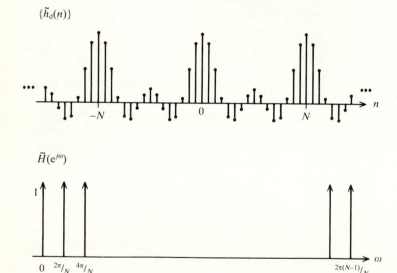

FIGURE 9-20
Periodic extension of $\{h_D(n)\}$ forms the infinite-duration sequence $\{\tilde{h}_d(n)\}$. The spectrum of the infinite-duration sequence $\tilde{H}(e^{j\omega})$ is composed of Dirac delta functions.

the values of the FIR filter transfer function $H_d(e^{j\omega})$ between the sample points may deviate significantly from $H_D(e^{j\omega})$. The differences are most significant around frequencies for which $|H_D(e^{j\omega})|$ is discontinuous, as illustrated by the edge of the passband in the ideal lowpass filter. When $|H_D(e^{j\omega})|$ is a smooth function of frequency, the sampled sequence $\{|H_D(k)|\}$ is a good representation of the form of $|H_D(e^{j\omega})|$, provided the sampling is dense enough, which occurs for large values of N_d. In this case, the filter transfer function $H_d(e^{j\omega})$ will be essentially equal to $H_D(e^{j\omega})$.

The above analysis also illustrates the problems that can occur when filtering is performed by taking the DFT of the input data and merely eliminating the undesired frequency components by setting them equal to zero. In effect, a zero-phase filter is assumed having a symmetric finite-duration unit-sample response $\{h_d(n)\}$.

In the next section, we combine the windowing and DFT methods for the design of practical FIR filters.

TIME-DOMAIN OPERATIONS

$\{\tilde{h}_d(n)\}$

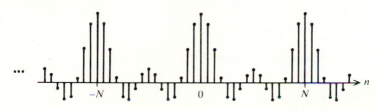

$\{w_R(n)\}$

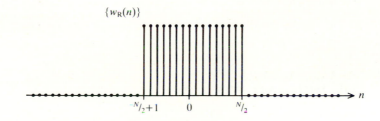

$\{h_d(n)\}$

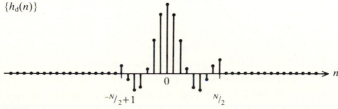

FIGURE 9-21
Interpreting the overshoots and leakage effects in $H_d(e^{j\omega})$. (a) Time domain; (b) frequency domain.

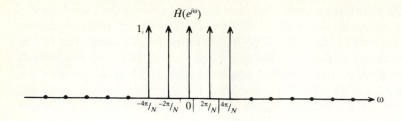

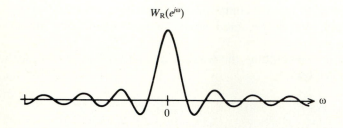

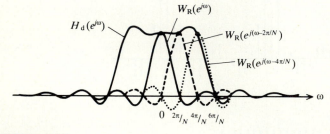

FIGURE 9-21b

9.4 DESIGNING FIR FILTERS BY COMBINING THE DFT AND WINDOW METHODS

In this section, we combine the windowing and DFT methods to design FIR filters. This combination eliminates the disadvantages of each method. For the windowing method, we needed to know the form of the desired unit-sample response $\{h_D(n)\}$. The DFT method produces $\{h_d(n)\}$, the finite-duration approximation to $\{h_D(n)\}$ containing N_d-points, where N_d is a large number. When applying the DFT method directly, overshoots and leakage effects are observed in the FIR filter transfer function $H_d(e^{j\omega})$. The windowing method can be employed to reduce these effects. The windowing procedure allows the duration of $\{h_d(n)\}$ to be reduced, to take advantage of the tolerance limits in the magnitude specification. This cannot be done easily by using the DFT method directly. We first consider the method of obtaining a desired magnitude response sequence $\{|H_D(k)|\}$ from the magnitude response specification, followed by the procedure for specifying an appropriate phase response. An inverse DFT then produces the N_d-point unit-sample response $\{h_d(n)\}$. The application of an N-point window, where $N < N_d$, to obtain $\{h_N(n)\}$ from $\{h_d(n)\}$ is then discussed. Finally, a step-by-step procedure for designing an FIR filter is presented.

In this section, we will describe the DFT/windowing procedure for implementing lowpass, highpass, bandpass and bandreject filters. When discussing the application of windows earlier in this chapter, we observed that the resulting filter magnitude response was a distorted version of the desired magnitude response $|H_D(e^{j\omega})|$. This distortion was caused by the convolution of $H_D(e^{j\omega})$ with the spectrum of the window. This distortion prevents the application of this procedure to designing filters having an arbitrary magnitude response shape, such as linear-with-frequency, for example. A magnitude response having a specific shape is more easily achieved with the frequency-sampling method described in the next section.

SPECIFYING THE APPROPRIATE MAGNITUDE RESPONSE SEQUENCE. In the design of lowpass filters by using windows, the resulting magnitude response exhibited a reduction in the passband. This reduction is one consequence of the convolution with the main-lobe of the window spectrum with the desired response $H_D(e^{j\omega})$. With the idea that a reduction will occur in the final lowpass or bandpass filter magnitude response, the desired ideal magnitude response $|H_D(e^{j\omega})|$ should have its cutoff frequencies extended into the transition regions of the specification, as shown in Fig. 9.22. The size of this extension depends upon the main-lobe width of the final window spectrum, which is related to the duration of the final unit-sample response and window type employed. Since this duration is not known at the beginning of the design process, a guess must be made. A good starting value for this extension is equal to 30% of the transition region.

The magnitude response $|H_D(e^{j\omega})|$ generated in this fashion is then sampled at N_d points over $0 \le \omega < 2\pi$ to generate the magnitude response sequence $\{|H_D(k)|\}$. The value N_d is much larger then the anticipated duration of the final filter, with $N_d = 128$ usually being sufficient for most practical filters.

The procedure to define the magnitude response sequence that is least prone to error is to sample $|H_D(e^{j\omega})|$ at $(N_d/2) + 1$ points over $0 \le \omega \le \pi$ and apply the even symmetry property of the magnitude sequence to obtain the entire N_d-point sequence. Doing this, we have

$$
\begin{aligned}
|H_D(k)| &= |H_D(e^{j\omega})|_{\omega = 2\pi k/N_d} & \text{for } 0 \le k \le N_d/2 \\
|H_D(k)| &= |H_D(N_d - k)| & \text{for } (N_d/2) + 1 \le k \le N_d - 1.
\end{aligned}
\tag{9.33}
$$

SPECIFYING THE APPROPRIATE LINEAR-PHASE RESPONSE SEQUENCE. From $\{|H_D(k)|\}$ and a specified phase sequence $\{\text{Arg}[H_D(k)]\}$,

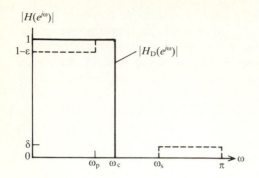

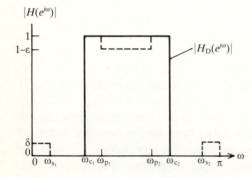

FIGURE 9-22
The desired magnitude response for lowpass and bandpass filters is obtained by extending the passband into the transition regions. The specification is shown as dashed lines.

the unit-sample response $\{h_d(n)\}$ can be obtained by performing an inverse discrete Fourier transform. The causal unit-sample response, having N_d points, of a linear-phase filter should show a point of symmetry (or antisymmetry) about the half-way point, or at $N_s = (N_d - 1)/2$. As shown in Chapter 3, four possible cases of symmetry can occur: the sequence can be symmetric or antisymmetric, and, for each of these, the point of symmetry can either fall on a sample point or between sample points. When a conventional N_d-point FFT algorithm is employed to compute the inverse DFT, N_d is an even integer. Hence, the point of exact symmetry cannot fall on a sampling point in general. In the ideal lowpass filter case considered in the previous section, when the phase response sequence was zero-valued, the N_d-point unit-sample response was almost symmetric about $n = 0$, except that a point without a partner occurred at $n = N_d/2$. When the N-point window is applied, for $N < N_d$, this extra point is deleted and the unit-sample response $\{h_N(n)\}$ can be made to be symmetric about a sampling point.

The appropriate linear-phase sequence that is assigned to match the magnitude response is determined by the symmetry properties of the unit-sample response. The phase responses corresponding to the different symmetries in $\{h_d(n)\}$ are shown in Table 9.2. The entire phase value, rather than its principal value, is given. In computing the real and imaginary parts from the magnitude and phase, this entire phase value is sufficient.

TABLE 9.2
Linear phase responses for the various types of symmetry of the N_d-point unit-sample response $\{h_d(n)\}$ (N_d is even integer)

Even symmetric with the point of symmetry on sample point $N_S = N_d/2$

$$\mathrm{Arg}[H_D(e^{j\omega})] = -\omega N_d/2 \quad \text{for } 0 \leq \omega \leq \pi$$

Even symmetric with the point of symmetry between sample points, $N_S = (N_d - 1)/2$

$$\mathrm{Arg}[H_D(e^{j\omega})] = -\omega(N_d - 1)/2 \quad \text{for } 0 \leq \omega \leq \pi$$

Antisymmetric with the point of symmetry on sample point $N_S = N_d/2$

$$\mathrm{Arg}[H_D(e^{j\omega})] = -\omega N_d/2 + \pi/2 \quad \text{for } 0 \leq \omega \leq \pi$$

Antisymmetric with the point of symmetry between sample points, $N_S = (N_d - 1)/2$

$$\mathrm{Arg}[H_D(e^{j\omega})] = -\omega(N_d - 1)/2 + \pi/2 \quad \text{for } 0 \leq \omega \leq \pi$$

The phase at frequency points for which the magnitude response is zero is undefined and can be set to an arbitrary value. To simplify the specification of the phase, we set it equal to the value of the linear trend. Then, at $\omega = 0$, the phase value of antisymmetric sequences is defined to be equal to $\pi/2$. As shown in Chapter 5, even symmetric sequences that have an even number of elements have a z-transform that always produces a zero at $z = -1$, or a magnitude response that is zero-valued at $\omega = \pi$. This condition occurs when the point of symmetry falls between sample points at $N_S = (N_d - 1)/2$. Then, at $\omega = \pi$, the phase value of even symmetric sequences is defined to be equal to $-(N_d - 1)\pi/2$.

The procedure to define the phase sequence that is least prone to error is to sample $\mathrm{Arg}[H_D(e^{j\omega})]$ at $(N_d/2) + 1$ points over $0 \leq \omega \leq \pi$ and apply the odd symmetry property of the phase sequence to obtain the entire N_d-point sequence. Doing this, we have

$$\mathrm{Arg}[H_D(k)] = \mathrm{Arg}[H_D(e^{j\omega})]\big|_{\omega = 2\pi k/N_d} \quad \text{for } 0 \leq k \leq N_d/2$$
$$\mathrm{Arg}[H_D(k)] = -\mathrm{Arg}[H_D(N_d - k)] \quad \text{for } (N_d/2) + 1 \leq k \leq N_d - 1.$$

$$(9.34)$$

As an example, let us consider the ideal lowpass filter magnitude response with $\omega_c = \pi/4$, which is sampled over $0 \leq \omega < 2\pi$ at N_d points for $N_d = 32$. Since the transfer functions of antisymmetric unit-sample responses have a zero at $\omega = 0$, they are not appropriate for the lowpass filter. The two types of symmetric unit-sample responses from a 32-point inverse DFT are shown in Fig. 9.23.

APPLYING A WINDOW. The above procedure provides an N_d-point approximation $\{h_d(n)\}$ to the desired unit-sample response $\{h_D(n)\}$. We do not want to use $\{h_d(n)\}$ directly, because it does not take advantage of the allowed

$\{H_D(k)\}$

(a)

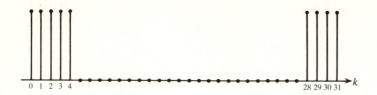

(b) Point of symmetry at $N_s = N_d/2$

$\{h_d(n)\}$

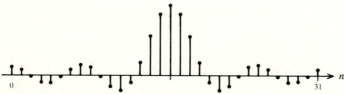

(c) Point of symmetry at $N_s = (N_d-1)/2$

$\{h_d(n)\}$

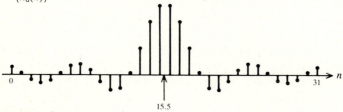

FIGURE 9-23
Effect of phase response slope on unit-sample response. (*a*) 32-point magnitude response used in the inverse DFT computations; (*b*) unit-sample response sequence result when phase slope is -16; (*c*) unit-sample response sequence result when phase slope is -15.5.

tolerances in the specification. However, by reducing the duration of the N_d-point sequence with a window sequence to N points, an FIR filter can be implemented that meets the desired specification.

For the final FIR filter to have a linear-phase response, the windowed unit-sample response $\{h_N(n)\}$ must be symmetric. To maintain the symmetry that is present in $\{h_d(n)\}$, an appropriate N-point window must be chosen accordingly. If the point of symmetry of $\{h_d(n)\}$ was chosen to fall on a sample point, the window duration N must be an odd integer. If the point of symmetry falls between sample points, N must be an even integer. Examples are shown in Fig. 9.24.

POINT OF SYMMETRY N_s
ON SAMPLING POINT

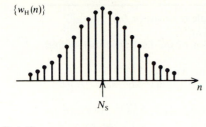

$\{w_H(n)\}$

N_s

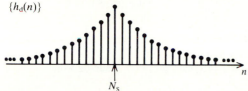

$\{h_d(n)\}$

N_s

POINT OF SYMMETRY N_s
BETWEEN SAMPLING POINTS

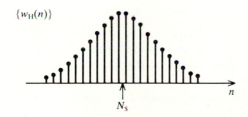

$\{w_H(n)\}$

N_s

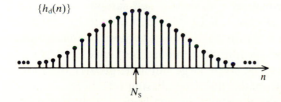

$\{h_d(n)\}$

N_s

FIGURE 9-24
Positioning a Hamming window to maintain the symmetry that is present in the original sequence.

Having described these separate parts, we can now integrate them into a procedure to design an FIR filter.

STEPS IN DESIGNING A FILTER USING THE DFT/WINDOWING METHOD. The steps to design a linear-phase FIR filter with the DFT/windowing method are summarized in Table 9.3. To illustrate this

TABLE 9.3
Steps in performing DFT/windowing FIR filter design

Step 1. Determine the approximate N_d-point unit-sample response $\{h_d(n)\}$
    (a) Sketch the desired magnitude response $|H_D(e^{j\omega})|$.
    (b) Choose N_d to be greater than the duration of the final filter unit-sample response. $N_d = 128$ is a good choice.
    (c) Assign the appropriate linear phase response $\text{Arg}[H_D(e^{j\omega})]$.
    (d) Sample $|H_D(e^{j\omega})|$ and $\text{Arg}[H_D(e^{j\omega})]$ to obtain N_d-point sequences $\{|H_D(k)|\}$ and $\{\text{Arg}[H_D(k)]\}$.
    (e) Compute the N_d-point real and imaginary sequences $\{H_R(k)\}$ and $\{H_I(k)\}$.
    (f) Evaluate the N_d-point inverse DFT. The sequence $\{h_d(n)\}$ is the real-valued result.

Step 2. Determine the N-point unit-sample response $\{h_N(n)\}$.
    (a) Choose N (good initial guess: $N = N_d/2$).
    (b) Multiply $\{h_d(n)\}$ with N-point window to obtain $\{h_N(n)\}$.
    (c) Compute the N_d-point DFT of $\{h_N(n)\}$ to obtain $\{H_{N_d}(k)\}$ (padding with zeros is necessary).
    (d) Compare $\{|H_{N_d}(k)|\}$ with desired specification.
    (e) If specification is satisfied, try to reduce N further, if specification is not satisfied increase N. Repeat (b) through (e) until minimum value for N is found.

Step 3. Linear-phase FIR filter implementation
    (a) Find the maximum value in $\{|H_{N_d}(k)|\}$, denoted by HMAX.
    (b) Scale $\{h_N(n)\}$ by 1/HMAX to find final filter coefficients.
    (c) Shift coefficient sequence to start at $n = 0$ for causal filter. For non-causal filter, shift $\{h_N(n)\}$ to be symmetric about $n = 0$.
    (d) Implement the filter using special linear-phase FIR structure.

---

procedure, we consider the N_d-point sequence $\{h_d(n)\}$ to be symmetric about the point $n = N_d/2$. In this case, the set of variable-duration windows are also all symmetric about this point. This feature simplifies the specification and application of the window, as shown below. The case of $\{h_d(n)\}$ whose point of symmetry lies between sample points can be handled by making small modifications to the procedures given below.

Step 1. Determine the approximate desired unit-sample response $\{h_d(n)\}$. The desired magnitude response, $|H_D(e^{j\omega})|$ is determined by extending the cutoff frequency ω_C into the transition region. The desired magnitude response is sampled to produce the magnitude response sequence $\{|H_D(k)|\}$. The value of N_d should be chosen to be some large number, greater than the duration of the final filter design, to make the time-aliasing effects in $\{h_d(n)\}$ negligible. Most of the FIR filter specifications in this book will be satisfied with durations of less that sixty samples. Hence, $N_d = 128$ is sufficient. The linear-phase response sequence $\{\text{Arg}[H_D(k)]\}$ is specified according to Table 9.2. For ease of specifying and applying windows, the point of symmetry of $\{h_d(n)\}$ should be specified to be $n = N_d/2$.

In evaluating the inverse DFT, the real and imaginary parts of the DFT sequence, $\{H_R(k)\}$ and $\{H_I(k)\}$, must be computed. These are given by

$$H_R(k) = |H_D(k)| \cos(\text{Arg}[H_D(k)])$$
$$\text{and} \qquad H_I(k) = |H_D(k)| \sin(\text{Arg}[H_D(k)]) \qquad \text{for } 0 \le k \le N_d - 1 \quad (9.35)$$

Since both the cosine and sine functions are periodic, it is not necessary to express the phase in the principal value range of $[-\pi, \pi]$ in the above procedure for determining the phase response.

Step 2. Determine the values of the N-point unit-sample response $\{h_N(n)\}$. By choosing $\{h_d(n)\}$ to be symmetric about $n = N_d/2$, the variable-duration windows can be specified and applied in a direct fashion, as shown in Fig. 9.25. The window is then defined to be symmetric about $n = N_d/2$ and has an odd number of points to maintain the even symmetry in the N-point windowed sequence $\{h_N(n)\}$. For example, the N-point Hamming window that is symmetric about $N_d/2$ is given by

$$w_H(n) = \begin{cases} 0.54 + 0.46 \cos[2\pi(n - N_d/2)/(N-1)] \\ \qquad \text{for } N_d/2 - (N-1)/2 \le n \le N_d/2 + (N-1)/2 \quad (9.36) \\ 0 \qquad \text{otherwise.} \end{cases}$$

The product of $\{w_H(n)\}$ and $\{h_d(n)\}$ is computed to determine the shorter N-point windowed sequence $\{h_N(n)\}$, where $N < N_d$, which is a candidate for the FIR filter unit-sample response. A reasonable first choice is $N = N_d/2$. The goal is to find the minimum value of N such that $\{h_N(n)\}$ is the unit-sample response of the FIR filter that satisfies the magnitude response specification.

The magnitude response of the resulting FIR filter can be verified by computing the N_d-point DFT of the N-point windowed unit-sample sequence $\{h_N(n)\}$, denoted $\{H_{N_d}(k)\}$. Let $\{h_r(n)\}$ and $\{h_i(n)\}$ be the real and imaginary N_d-point sequences to be applied to the DFT. To compute the DFT, $\{h_r(n)\}$ contains $\{h_N(n)\}$ and the zero values that were produced by the windowing procedure, or $h_r(n) = h_d(n) w_H(n)$ for $0 \le n \le N_d - 1$. The imaginary sequence $\{h_i(n)\}$ is set to zero.

The magnitude response sequence $\{|H_{N_d}(k)|\}$ is then compared with the specification to verify that both the passband and stopband specifications are met. Because the N_d-point sequence $\{|H_{N_d}(k)|\}$ is computed from a sequence having only N points, where $N < N_d$, it appears to define a smooth curve. In this case, we can be fairly confident that the digital filter magnitude response $|H(e^{j\omega})|$ is accurately represented by the magnitude response sequence $\{|H_{N_d}(k)|\}$.

If $\{|H_{N_d}(k)|\}$ satisfies the specification, an attempt to reduce the value of N should be made. This will insure that a shorter duration unit-sample response does not also satisfy the specification and that the minimum duration sequence has been found. If $\{|H_{N_d}(k)|\}$ does not satisfy the specification the

$\{h_{\mathrm{d}}(n)\}$

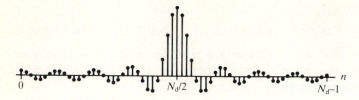

49-point window

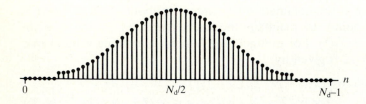

33-point window

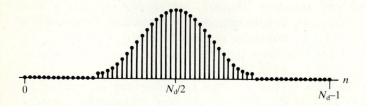

17-point window

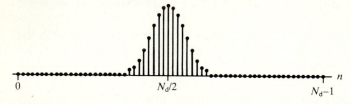

FIGURE 9-25
By locating point of symmetry of $\{h_{\mathrm{d}}(n)\}$ at $n = N_{\mathrm{d}}/2$, variable length windows can be applied directly. When $N_{\mathrm{d}}/2$ is an integer, windows having an odd-integer duration must be used.

value of N should be increased. If N becomes greater than $N_d/2$, half of the initial number of samples, this initial value was too small. Then, N_d should be increased by two and the entire procedure repeated.

Step 3. Linear-phase FIR filter implementation. Because of the windowing procedure, the scale of the magnitude response may have been reduced. It must then be adjusted to meet the specification. For example, let $|H_D(e^{j\omega})|$ have a maximum value equal to 1. Because the duration and values of $\{h_d(n)\}$ are reduced in the windowing operation, the maximum value of the filter magnitude response $|H(e^{j\omega})|$, as approximated by the final $\{|H_{N_d}(k)|\}$, may no longer be equal to 1. To make it equal to 1, the maximum value of $\{|H_{N_d}(k)|\}$, denoted by H_{MAX}, should be determined and the values of the unit-sample response should be multiplied by $1/H_{MAX}$.

To implement a causal filter, the windowed unit-sample response that is symmetric about $n = N_d/2$ must be shifted, so that the first nonzero value occurs at $n = 0$. This eliminates unnecessary delays in the filter structure. For non-causal filters, $\{h_N(n)\}$ is shifted so that its point of symmetry occurs at $n = 0$. Because of the time symmetry of $\{h_N(n)\}$, the special linear-phase FIR filter structure can be used to reduce the number of multiplications required to accomplish the filtering operation.

To illustrate this procedure, let us consider the design of a filter by combining the DFT and windowing methods.

Example 9.1. Design of linear-phase FIR filter with the DFT/windowing procedure. Let us consider the bandpass specification shown in Fig. 9.26(a), given by

passband: $-1 \le |H(e^{j\omega})|_{dB} \le 0$ for $0.42\pi \le \omega \le 0.61\pi$
stopband: $|H(e^{j\omega})|_{dB} < -60$ for $0 \le \omega \le 0.16\pi$ and $0.87\pi \le \omega \le \pi$.

Step 1. Determine $\{h_d(n)\}$.

Since the passband response will be reduced when a window is applied, the cutoff frequencies of the desired magnitude response are extended into the transition region by 30%, so that $\omega_{C1} = 0.34\pi$ and $\omega_{C2} = 0.69\pi$, as shown in Fig. 9.26(b).

The value $N_d = 64$ was chosen to make the figures easier to interpret. A larger value of N_d could have been chosen, but the time sequences are more easily drawn for $N_d = 64$.

To show the four possible different 64-point $\{h_d(n)\}$, the corresponding four possible phase characteristics given in Table 9.2 were all invoked. The point of symmetry was chosen to be either $N_s = 32$ or $N_s = 31.5$. The magnitude and phase functions were defined over $0 \le \omega \le \pi$. The N_d-point real and imaginary DFT sequences are constructed from these in a way to insure their symmetry properties, or

$$H_R(k) = |H(k)| \cos(\text{Arg}[H(k)]) \quad \text{for } 0 \le k \le N_d/2$$
$$H_R(k) = H_R(N_d - k) \quad \text{for } (N_d/2) + 1 \le k \le N_d - 1$$

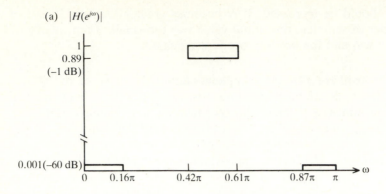

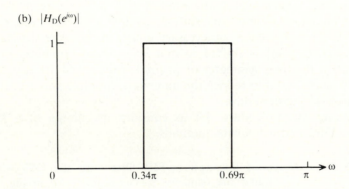

FIGURE 9-26
Initial frequency curves for Example 9.1. (*a*) Specification; (*b*) desired magnitude response function.

and

$$H_I(k) = |H(k)| \sin(\text{Arg}[H(k)]) \qquad \text{for } 0 \leq k \leq N_d/2$$
$$H_I(k) = -H_I(N_d - k) \qquad \text{for } (N_d/2) + 1 \leq k \leq N_d - 1$$

The inverse DFT was evaluated with the four possible phase responses to produce the four unit-sample responses shown in Fig. 9.27. To take advantage of the zero-valued coefficient occurring at $n = 32$, thus eliminating a multiplication in the digital filter, we can use the antisymmetric sequence that has its point of symmetry on a sampling point as $\{h_d(n)\}$. The magnitude response $|H(e^{j\omega})|$ of this sequence is shown in Fig. 9.28. This magnitude response was obtained by computing the 512-point DFT of $\{h_d(n)\}$. The overshoots and leakage effects due to the 64-point rectangular window trunction are evident.

Step 2. Determine $\{h_N(n)\}$.

A 33-point Hamming window was applied to $\{h_d(n)\}$ to obtain $\{h_{33}(n)\}$. The corresponding magnitude response is shown in Fig. 9.29(*a*). We note that it is much smoother than that of $\{h_d(n)\}$, being convolved with the spectrum of the Hamming window. The passband requirement is easily met, but the minimum

$\{h_d(n)\}$

(a)

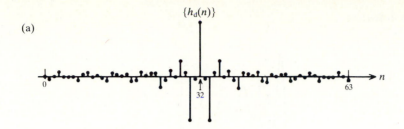

(b)

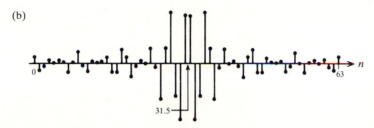

(c)

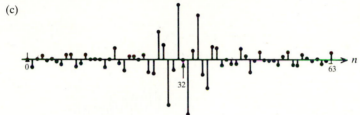

(d)

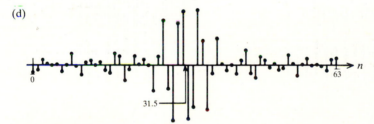

FIGURE 9-27
Four unit-sample response sequences resulting from the four possible phase response types.
(*a*) Even symmetry with point of symmetry falling on sample point; (*b*) even symmetry with
point of symmetry falling between sample points. (*c*) Odd symmetry with point of symmetry
falling on sample point; (*d*) odd symmetry with point of symmetry falling between sample
points.

stopband attenuation is approximately -60 dB. Shorter Hamming windows
exceeded the stopband attenuation specification.

Since the stopband attenuation is governed by the sidelobe magnitude of
the window spectrum, the 33-point Blackman window was considered because of
its sidelobe characteristics. The magnitude response of the 33-point $\{h_N(n)\}$
obtained with the Blackman window is shown in Fig. 9.29(*b*). Both the passband
and stopband specifications are met. In an attempt to find the minimum duration
window, shorter Blackman windows were tried. A 31-point window was found to
be the minimum duration that satisfies the specification, with results similar to

PASSBAND:

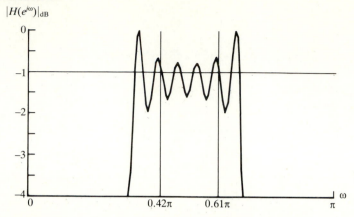

STOPBAND:

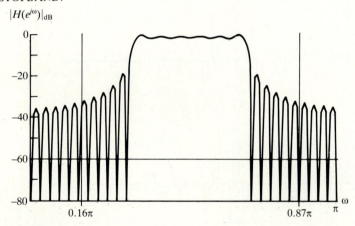

FIGURE 9-28
Magnitude response corresponding to $\{h_d(n)\}$ shown in Fig. 9-27(c).

Fig. 9.29(b). Reducing the window further caused the low-frequency stopband to be exceeded. Hence, $N = 31$ is the minimum duration, to give $\{h_{31}(n)\}$.

Step 3. FIR filter implementation.

The maximum value of $|H(e^{j\omega})|$ for $\{h_{31}(n)\}$ was found to be equal to HMAX $= 0.9998$. This value was close enough to one so that scaling the coefficients by 1/HMAX was not necessary. The causal FIR filter structure is shown in Fig. 9.30. Note that the output of the fifteenth delay is not connected to a multipler because $h_{31}(15) = 0$. However, this savings of one multiplier is bought at the cost of multiplying the outputs of 15 delays by -1 to achieve the antisymmetric unit-sample response. If an even-symmetric $\{h_d(n)\}$ was used, these negations would not be necessary. This filter would be efficient, however, if the digital hardware allowed a subtraction to be performed as efficiently as an addition.

In the next section, we consider another FIR filter design procedure that results in a structure which allows the processing to occur in parallel.

9.5 DESIGNING FIR FILTERS WITH THE FREQUENCY-SAMPLING METHOD

The frequency-sampling method is similar to the DFT method in that the desired magnitude response is sampled and a linear-phase response is

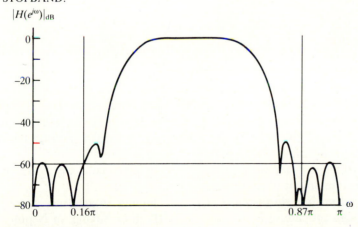

(a)   33-point Hamming window results

PASSBAND:

33-point Hamming window results

STOPBAND:

FIGURE 9-29
Magnitude response corresponding to windowed version of $\{h_d(n)\}$ shown in Fig. 9-27(c). (a) 33-point Hamming window results; (b) 33-point Blackman window results.

(b)   33-point Blackman window results

PASSBAND:

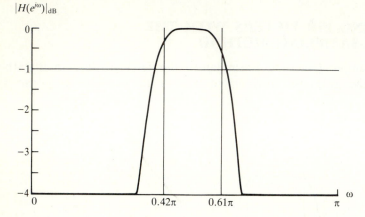

33-point Blackman window results

STOPBAND:

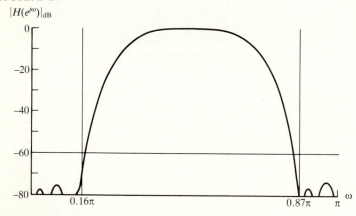

FIGURE 9-29b

specified. The difference is that, rather than computing the inverse DFT to determine the unit-sample response of the FIR filter, a special parallel filter structure is used to implement each frequency sample directly. For a given specification, the frequency-sampling procedure may produce a filter that is faster and easier to implement in hardware.

Given a desired magnitude response $|H_D(e^{j\omega})|$ having an arbitrary shape, the goal of this design procedure is to design an FIR filter having an N-point unit-sample response and having a magnitude response $|H(e^{j\omega})|$, such that

$$|H(e^{j\omega_k})| = |H_D(e^{j\omega_k})|$$

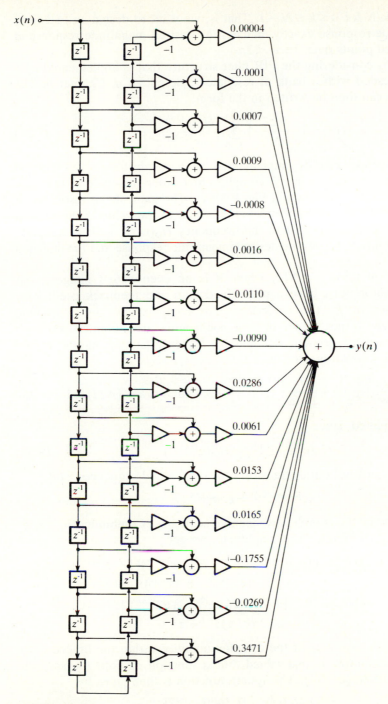

FIGURE 9-30
Linear-phase FIR filter structure produced by 31-point Blackman window.

where $\omega_k = 2\pi k/N$ for $0 \le k \le N-1$. That is, we want to design an FIR filter whose magnitude response exactly matches the desired magnitude response at N equally spaced points over $0 \le \omega < 2\pi$.

We start by considering the FIR filter structure that is implemented by a comb filter cascaded with a bank of resonators, described in Chapter 6. The system function can then be written in the form:

$$H(z) = \left(\frac{1 - z^{-N}}{N}\right) \sum_{k=0}^{N-1} \frac{H_D(e^{j2\pi k/N})}{1 - e^{j2\pi k/N}z^{-1}} \tag{9.37}$$

This system function can be implemented as a comb filter, the term in parentheses, in cascade with N complex first-order resonators that are connected in parallel, as shown in Fig. 9.31(a). The gain of the resonator at frequency $\omega_k = 2\pi k/N$, for $0 \le k \le N-1$, is equal to the value of the desired complex-valued transfer function at that frequency $H_D(e^{j2\pi k/N})$. This complex-valued gain is computed from the desired magnitude response and the assigned phase response.

For $k = 0$ (and for $k = N/2$, when N is an even-valued integer), the first-order resonator must have a real-valued gain and feedback coefficient. For other values of k, the coefficients are complex, but by combining complex-conjugate terms, a set of real coefficients are obtained for each *second-order resonator* section, denoted by $H_{R,k}(z)$. Combining the k and $N-k$ first-order resonators, we have

$$H_{R,k}(z) = \frac{H_D(e^{j2\pi k/N})}{1 - e^{j2\pi k/N}z^{-1}} + \frac{H_D(e^{j2\pi(N-k)/N})}{1 - e^{j2\pi(N-k)/N}z^{-1}} \qquad \text{for } 1 \le k \le (N/2) - 1 \tag{9.38}$$

This can be simplified, since

$$H_D(e^{j2\pi(N-k)/N}) = H_D^*(e^{j2\pi k/N}) \tag{9.39}$$

by the symmetry of the Fourier transform for real-valued sequences, and

$$e^{j2\pi(N-k)/N} = e^{-j2\pi k/N}$$

by the periodicity of the complex exponential sequence. Combining Eq. (9.38) over a common denominator and simplifying, we get

$$H_{R,k}(z) = \frac{H_D(e^{j2\pi k/N})(1 - e^{-j2\pi k/N}z^{-1}) + H_D^*(e^{j2\pi k/N})(1 - e^{j2\pi k/N}z^{-1})}{1 - (e^{j2\pi k/N} + e^{-j2\pi k/N})z^{-1} + z^{-2}}$$

$$= \frac{2\,\text{Re}[H_D(e^{j2\pi k/N})] - 2\,\text{Re}[H_D(e^{j2\pi k/N})\,e^{-j2\pi k/N}]z^{-1}}{1 - 2\cos(2\pi k/N)z^{-1} + z^{-2}} \tag{9.40}$$

where $\text{Re}[\,\cdot\,]$ is the real part of the complex quantity within the brackets. As expected, the numerator is real-valued, but a function of both the real and imaginary parts of $H_D(e^{j2\pi k/N})$. The system function is then given by

$$H(z) = \frac{1 - z^{-N}}{N}\left[\frac{H_D(e^{j0})}{1 - z^{-1}} + \frac{H_D(e^{j\pi})}{1 + z^{-1}} + \sum_{k=1}^{(N/2)-1} H_{R,k}(z)\right]. \tag{9.41}$$

(a)

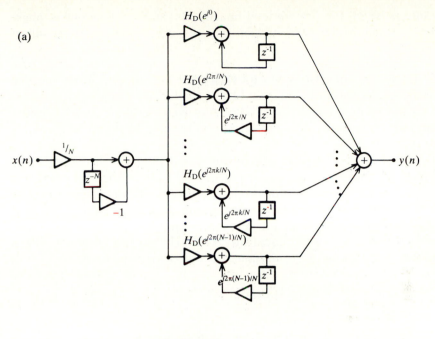

(b)

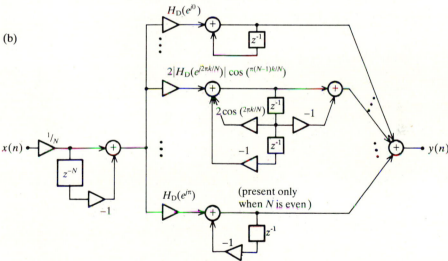

FIGURE 9-31
Digital filter structures used when applying the frequency-sampling method. (*a*) General filter; (*b*) simplified structure for linear-phase response.

This result is valid for any desired magnitude and corresponding phase response, but can be further simplified for the class of linear-phase filters below.

CHOICE OF EVEN OR ODD VALUES FOR THE UNIT-SAMPLE RESPONSE DURATION N. As shown in Chapter 6, the comb filter generates zeros that are equally spaced on the unit circle. When N is even, zeros are located at both $z = 1$ and $z = -1$, as shown in Fig. 9.32. For odd-valued N, there is a zero only on the positive real axis at $z = 1$.

Not all of the zeros generated by the comb filter can be canceled by resonators, since the pole/zero pattern is governed by the symmetry properties of the unit-sample response. For example, it was shown previously that antisymmetric sequences, since they have a zero-valued mean, always produce a zero at $\omega = 0$ ($z = 1$). Hence, the resonator gain of the desired magnitude response at $\omega = 0$ is always equal to zero. The system function for a linear-phase filter having an antisymmetric unit-sample response $H_{AS}(z)$ is given by

$$H_{AS}(z) = \frac{1 - z^{-N}}{N} \left[\frac{H_D(e^{j\pi})}{1 + z^{-1}} + \sum_{k=1}^{(N/2)-1} H_{R,k}(z) \right] \tag{9.42}$$

Hence, if a zero is *not* desired at $\omega = 0$ in the final filter transfer function, as in a lowpass filter for example, an antisymmetric unit-sample should not be chosen.

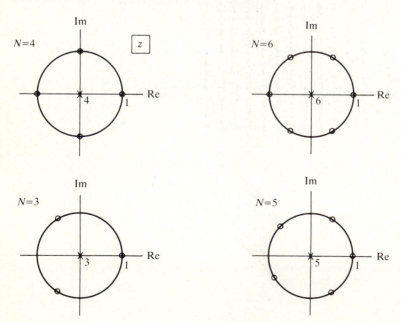

FIGURE 9-32
Pole/zero patterns in the z plane of comb filter for different values of N.

By a similar argument, symmetric sequences produce a zero at $\omega = \pi$ ($z = -1$) when N is even. Then the resonator gain of the desired magnitude response at $\omega = \pi$ is equal to zero. The system function for a linear-phase filter having a symmetric unit-sample response $H_S(z)$ is given by

$$H_S(z) = \frac{1 - z^{-N}}{N} \left[\frac{H_D(e^{j0})}{1 - z^{-1}} + \sum_{k=1}^{(N/2)-1} H_{R,k}(z) \right] \qquad (9.43)$$

Hence, if a zero is *not* desired at $\omega = \pi$ in the final filter transfer function, as in a highpass filter for example, a symmetric unit-sample with an even duration should not be chosen.

A SIMPLIFICATION FOR LINEAR-PHASE FILTERS. A linear-phase filter can have a unit-sample response that is either symmetric or antisymmetric about some point in time. The associated phase responses were given in Table 9.2.

If the unit-sample response is symmetric about $N_S = (N-1)/2$, the complex-valued samples of the resulting linear-phase transfer function can be written as

$$H_D(k) = |H_D(e^{j\omega})| \, e^{-j(N-1)\omega/2} \, \big|_{\omega = 2\pi k/N} \qquad \text{for } 0 \le k \le N/2 - 1 \qquad (9.44)$$

where the linear-phase slope $(N-1)/2$ corresponds to the point of symmetry of the unit-sample response.

As shown in Chapter 5, an even symmetric sequence with an even duration has a z-transform with a zero at $z = -1$. The corresponding magnitude response then has a zero at $\omega = \pi$. Hence, if N is an even integer, $H_D(N/2) = 0$ and there is no resonator for $k = N/2$ or $\omega = \pi$. If N is an odd integer, $N/2$ is not an integer and $\omega = \pi$ is not a sample point, as shown in Fig. 9.32. So again, there is no resonator corresponding to $\omega = \pi$. To implement a highpass filter, a zero at $\omega = \pi$ is undesirable, so either N must be chosen to be odd or an antisymmetric unit-sample response must be chosen. The simplification for the antisymmetric case is considered in the Problems.

Inserting Eq. (9.44) into the system function of the kth second-order resonator Eq. (9.40), we get

$$H_{R,k}(z) = \frac{2\,|H_D(e^{j2\pi k/N})|\,[\cos(\pi(N-1)k/N) - \cos(\pi(N-1)k/N + 2\pi k/N)\,z^{-1}]}{1 - 2\cos(2\pi k/N)\,z^{-1} + z^{-2}}$$

$$(9.45)$$

The numerator can be simplified by adding $2k\pi$ to the argument of the first cosine term, which does not change the value of the cosine function when k is an integer. This gives

$$\cos(\pi(N-1)k/N) = \cos(\pi(N-1)k/N + 2k\pi) = \cos(\pi(N+1)k/N) \qquad (9.46)$$

The second cosine term simplifies to the same value since

$$\cos(\pi(N-1)k/N + 2\pi k/N) = \cos(\pi(N+1)k/N)$$

The system function for the second-order resonators is then given by

$$H_{R,k}(z) = B_k \frac{1 - z^{-1}}{1 - 2\cos(2\pi k/N)z^{-1} + z^{-2}} \quad \text{for } 1 \le k \le (N/2) - 1 \quad (9.47)$$

where

$$B_k = 2\,|H_D(e^{j2\pi k/N})|\,\cos(\pi(N-1)k/N) \tag{9.48}$$

is the real-valued gain of the kth second-order resonator. Note that the cosine factor makes some of these gains negative. Including this result in the total system function Eq. (9.37), we get the filter shown in Fig. 9.31b and

$$H(z) = \frac{1 - z^{-N}}{N}\left[\frac{H_D(e^{j0})}{1 - z^{-1}} + \sum_{k=1}^{(N/2)-1} H_{R,k}(z)\right] \tag{9.49}$$

Before describing the frequency-sampling design procedure and presenting examples, we describe the procedure for choosing the gains of the resonators in the transition region.

SPECIFYING THE MAGNITUDE RESPONSE IN THE TRANSITION REGION. With the frequency-sampling procedure, the filter magnitude response $|H(e^{j\omega})|$ is equal to the values of the desired response $|H_D(e^{j\omega})|$ at the sampling frequency points. The values of $|H(e^{j\omega})|$ at other frequencies are interpolated from these sample point values. Because the comb-filter/resonator combination generates an N-point duration of a sinusoid, that could be modeled as being obtained with an N-point rectangular window applied to an infinite-duration sinusoid, the interpolation is achieved with $\sin(N\omega)/\sin(\omega/2)$ functions, as shown previously in Fig. 9.21(b). As such, sharp discontinuities in $|H_D(e^{j\omega})|$ produce overshoots and leakage effects in the filter magnitude response $|H(e^{j\omega})|$. To reduce these effects, the samples in the transition region are used to taper the desired response gradually between the passband and stopband. Since each nonzero sample value, $|H_D(e^{j2\pi k/N})|$, means the addition of a resonator, the number of such samples in the transition region should be minimized. Unfortunately, determining these transition sample values is a complicated procedure involving linear programming optimization procedures (see Rabiner and Gold). The values of the transition points also depend on the number of samples N and the cutoff frequency. It is claimed that, in lowpass filter designs, a maximum stopband attenuation between 44 and 54 dB can be achieved when one transition point is allowed. With two transition points, from 65 to 75 dB can be obtained, and with three transition points between 85 to 95 dB is possible. Rather than do the complicated optimization, or to rely on tabulated values, we present an alternate procedure that is more intuitive. The cost involved with using this procedure is that it may require more resonators that the optimum procedure.

To avoid the problems associated with discontinuities and to take advantage of the tolerances in the specification, a smooth transition from the passband to the stopband is introduced to define $|H_D(e^{j\omega})|$. From our experience with windows, the raised-cosine function is a good choice, as shown in Fig. 9.33. To take advantage of the passband tolerance and to allow some margin, the raised cosine curve passes through the tolerance interval at the passband frequency. To obtain some margin from leakage effects, the desired response is set to zero before the edge of the stopband. The resonator gains are then determined from the $(N/2) + 1$ samples of $|H_D(e^{j\omega})|$ over $0 \le \omega \le \pi$.

In practice, determining N, the number of sample values of $|H_D(e^{j\omega})|$, is not very difficult. A rule of thumb using the above procedure indicates that one transition point is needed in the transition region for each 10 dB of attenuation. If N_T transition points are needed, the value for N, the number of samples of $|H_D(e^{j\omega})|$ over $0 \le \omega < 2\pi$, can be determined. The frequency sampling period, denoted by $\delta\omega$, is then equal to

$$\delta\omega = |\omega_S - \omega_P|/N_T \tag{9.50}$$

and

$$N = 2\pi/\delta\omega. \tag{9.51}$$

If there is more than one transition region, the smallest one determines the value for N. Once N has been determined, the sample values of $\{|H_D(e^{j2\pi k/N})|\}$ falling in the stopband are set to zero. Initially, all the sample values of

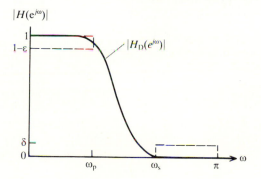

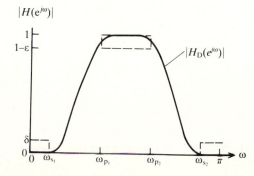

FIGURE 9-33
The desired magnitude response for lowpass and bandpass filters is obtained by connecting the passband and stopband with a raised-cosine function.

$\{|H_D(e^{j2\pi k/N})|\}$ falling in the passband are set to one. Then, the value at the inside edge of the passband is reduced to fall on the raised-cosine taper, but still within the limits. This allows us to take advantage of the passband tolerance. For passband to stopband transitions, let $k_p + 1$ be the number of the first sample inside the edge of the passband and $k_s + 1$ be that of the first sample inside the stopband. Then the cosine taper should be equal to 1 for $k = k_p$ and zero for $k = k_s$. If we let $N_{ps} = k_s - k_p$, then the cosine sequence traverses π radians, including the end points, in N_{ps} samples. The downward transition sample values are then defined by

$$|H_D(e^{j2\pi k/N})| = 0.5\left(1 + \cos\left(\frac{\pi(k - k_p)}{N_{ps}}\right)\right) \qquad \text{for } k_p \leq k \leq k_s \qquad (9.52)$$

For stopband to passband transitions, let $k_p - 1$ be the number of the first sample inside the edge of the passband and $k_s - 1$ be that of the first sample inside the stopband. Then the cosine taper should be equal to 1 for $k = k_p$ and zero for $k = k_s$. If $N_{sp} = k_p - k_s$, then the cosine sequence traverses π radians, including the end points, in N_{sp} samples. The upward transition sample values are then defined by

$$|H_D(e^{j2\pi k/N})| = 0.5\left(1 - \cos\left(\frac{\pi(k - k_s)}{N_{sp}}\right)\right) \qquad \text{for } k_s \leq k \leq k_p. \qquad (9.53)$$

Example 9.2. Determining transition region values for a bandpass filter. Let us consider the bandpass specification shown in Fig. 9.34, given by

passband: $-1 \leq H(e^{j\omega})|_{dB} \leq 0$ for $0.42\pi \leq \omega \leq 0.61\pi$

stopband: $|H(e^{j\omega})|_{dB} < -60$ for $0 \leq \omega \leq 0.16\pi$ and $0.87\pi \leq \omega \leq \pi$.

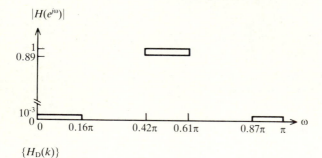

FIGURE 9-34
Frequence curves for Example 9.2. (a) Specification; (b) desired magnitude response sequence.

The two transition region sizes are

$$0.42\pi - 0.16\pi = 0.26\pi \quad \text{and} \quad 0.87\pi - 0.61\pi = 0.26\pi.$$

Since 60 dB attenuation is required, the number of transition points is $N_T = 6$. Then the frequency sampling period is $\delta\omega = 0.26\pi/6 = 0.043\pi$, and $N = 2\pi/0.043\pi = 46.2$. We will take the value rounded to the nearest integer, or $N = 46$.

We can now determine the sample numbers in the stopbands and passband. The sampling period with $N = 46$ is equal to $\delta\omega = 2\pi/46 = 0.043\pi$.

The passband samples extend from

$$k = 0.42\pi/0.043\pi = 9.77, \text{ which is rounded to } k = 10$$

to

$$k = 0.61\pi/0.043\pi = 14.19, \text{ which is rounded to } k = 14.$$

The stopband samples extend from $k = 0$ to $k = 0.16\pi/0.043\pi = 3.72$, or $k = 4$ for the low frequency stopband, and from $k = 0.87\pi/0.043\pi = 20.2$, or $k = 20$ to $k = N/2 = 23$.

The values of $\{|H_D(k)|\}$ can now be evaluated.

Low-frequency stopband:

$$|H_D(k)| = 0 \quad \text{for } 0 \le k \le 4.$$

Stopband to passband transition:

For the upward transition, $k_s - 1 = 4$, or $k_s = 5$, and $k_p - 1 = 10$, or $k_p = 11$. Then $N_{sp} = 6$ and

$$|H_D(k)| = 0.5\left(1 - \cos\left(\frac{\pi(k-5)}{6}\right)\right) \quad \text{for } 5 \le k \le 11.$$

As a check, $|H_D(10)|$ should fall within the passband limits, or

$$|H_D(10)| = 0.5(1 - \cos(5\pi/6)) = 0.933 \quad (= -0.60 \text{ dB}).$$

Passband:

$$|H_D(12)| = 1$$

Passband to stopband transition:

For the downward transition, $k_p + 1 = 14$, or $k_p = 13$, and $k_s + 1 = 20$, or $k_s = 19$. Then $N_{ps} = 6$ and

$$|H_D(k)| = 0.5\left(1 + \cos\left(\frac{\pi(k-13)}{6}\right)\right) \quad \text{for } 13 \le k \le 19.$$

As a check, $|H_D(14)|$ should fall within the passband limits, or

$$|H_D(14)| = 0.5(1 + \cos(\pi/6)) = 0.933 \quad (= -0.60 \text{ dB}).$$

High-frequency stopband:

$$|H_D(k)| = 0 \quad \text{for } 20 \le k \le 23.$$

TABLE 9.4
Steps for performing frequency-sampling FIR filter design

Step 1. Choose the value of N, the number of samples of the desired magnitude response $|H_D(e^{j\omega})|$ over $0 \le \omega < 2\pi$

Step 2. Determine the resonator gains

Step 3. Implement the second-order resonators and compute the unit-sample response

Step 4. Verify the filter design

Step 5. Compare the number of computations with the linear-phase FIR structure

FREQUENCY-SAMPLING FILTER DESIGN PROCEDURE. We now give the step-by-step procedure for designing a linear-phase FIR filter having a unit-sample response duration of N samples. These steps are summarized in Table 9.4.

Step 1. Choose the value of N. The desired magnitude response $|H_D(e^{j\omega})|$ is plotted over $0 \le \omega \le \pi$. For responses having discontinuities and transition regions, raised-cosine curves are drawn in the transition regions to connect the passband and stopband. The value of N is determined by the width of the transition region and the desired attenuation in the stopband. A rule of thumb is that one transition point is needed for each 10 dB of attenuation. For a continuous magnitude response having an arbitrary shape, the number of samples should be large enough that a good approximation to the desired response is obtained by connecting the sample points with a *smooth* curve. Since the number of resonators in the digital filter increases with N, the smallest possible value should be chosen. If N is an even integer, zeros in the transfer function are permitted at both $\omega = 0$ and $\omega = \pi$. If N is odd, a zero cannot appear at $\omega = \pi$, since it is not a sample point.

Step 2. Determine the resonator gains. Since the simplifying equations have been worked out for the linear-phase response corresponding to an N-point unit-sample response that is even-symmetric about $n = (N-1)/2$, we will use this phase response. For this case, the resonator gains are given by Eq. (9.48).

Step 3. Implementation of the second-order resonators. Each nonzero value of $|H_D(k)|$, for $0 \le k \le N/2 - 1$, results in a resonator. If the desired magnitude response is zero for a particular frequency sample, no resonator is necessary. The filter is implemented by cascading the comb filter with the parallel set of resonators. The filter unit-sample response is determined by applying a unit-sample sequence to the input. The linear-phase response of the filter can be verified in the time domain by noting that the unit-sample response $\{h_N(n)\}$, for $0 \le n \le N-1$, exhibits a point of symmetry at $n = (N-1)/2$. If a problem occurs, each resonator and comb filter combination should be examined separately to determine that its unit-sample response is symmetric about $n = (N-1)/2$.

Step 4.   Verify the filter design. The values of the filter magnitude response at the frequencies other than the sampling points should be determined. This is done by computing the magnitude of the N_d-point DFT of the N-point unit-sample response $\{h_N(n)\}$, where $N_d > N$. The value of N_d should be chosen large enough that the DFT samples, denoted by $\{H_{N_d}(k)\}$, are a good representation of the continuous-frequency transfer function $H(e^{j\omega})$. For example, if $N = 32$ then $N_d = 128$ is a good choice. The N_d-point DFT is computed by padding the N-point unit-sample response with $N_d - N$ zeros.

Upon verification, the magnitude response specification may not be met because the allowed limits are exceeded at the frequencies between the sampling points. If this happens, the specification must be sampled more densely by increasing the value of N and the design procedure repeated. If the specification is exceeded only slightly, the value of N may need to be increased by only a small number.

Step 5.   Comparison of computation counts. Count the number of multiplications that are required to determine if the frequency-sampling filter structure is more efficient than using the linear-phase nonrecursive implementation, determined from $\{h_N(n)\}$. However, the parallel structure of the frequency-sampling implementation allows the computations to be performed in parallel, which may mean that it is a faster implementation.

> **Example   9.3.   Design   of   FIR   bandpass   filter   with   frequency-sampling method.** Let us consider the bandpass specification given by
>
> passband:
> $$-1 \le |H(e^{j\omega})|_{dB} \le 0 \qquad \text{for } 0.42\pi \le \omega \le 0.61\pi$$
>
> stopband:
> $$|H(e^{j\omega})|_{dB} < -60 \qquad \text{for } 0 \le \omega \le 0.16\pi \text{ and } 0.87\pi \le \omega \le \pi.$$
>
> From Example 9.2, we found $N = 46$. The resonator gains are then given by
> $$B_k = 2|H_D(k)| \cos(45\pi k/46)$$
>
> from Eq. (9.48). Using the fifteen nonzero values from Example 9.2, we get
> $$B_k = \left(1 - \cos\left(\frac{\pi(k-5)}{6}\right)\right) \cos(45\pi k/46) \qquad \text{for } 6 \le k \le 11.$$
>
> $$B_k = 2\cos(11.74\pi) = 1.37 \qquad \text{for } k = 12$$
>
> $$B_k = \left(1 + \cos\left(\frac{\pi(k-13)}{6}\right)\right) \cos(45\pi k/46) \qquad \text{for } 13 \le k \le 19.$$
>
> The digital filter was implemented and the unit-sample response $\{h_{46}(n)\}$ was computed and is shown in Fig. 9.35(a). As expected, it has even symmetry about $n = 22.5$.
>
> As a verification, the 512-point DFT of $\{h_{46}(n)\}$ was computed by padding with zeros and the magnitude response was determined. The passband and stopband results are shown to satisfy the specification in Fig. 9.35(b).

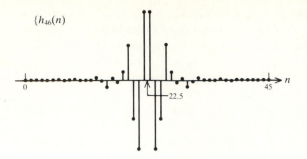

FIGURE 9-35
Unit-sample response sequence of filter satisfying specification in Example 9.3.

To implement the filter 15 second-order sections are needed, one for each nonzero $|H_D(k)|$. If multiplications by -1 are ignored, two multiplications are required for each resonator. An additional multiplication is required by the comb-filter for each output point. Hence, the frequency-sampling structure requires 31 multiplications per output point. Since $\{h_{46}(n)\}$ is even symmetric, the linear-phase FIR structure can be implemented to provide the same result with 23 multiplications per output point. In this case, the frequency-sampling structure requires more multiplications.

In comparing the results of the previous example with those from Example 9.1, we find that a 31-point unit-sample response is sufficient to satisfy the specification. Since the frequency-sampling method produces a 46 element unit-sample response, it is not as efficient. In part, this inefficiency is due to the procedure for specifying the values in the transition region. For lowpass, highpass, bandpass and bandreject filters, the DFT/windowing method is the method of choice. The frequency-sampling method really shines, however, when a magnitude response having a specific shape is to be implemented, as described in the next example.

Example 9.4. Designing a linear-with-frequency magnitude response with the frequency-sampling method. Let us consider the magnitude response given by

$$|H_D(e^{j\omega})| = \omega \qquad \text{for } 0 \leq \omega \leq \pi.$$

The N-point unit-sample response will be specified to have even symmetry, so that the resonator gain equation (9.48) can be employed. To apply the frequency-sampling procedure, we need only sample this magnitude response at $(N/2) + 1$ points, when N is an even integer, or at $(N + 1)/2$ points when N is odd. Doing this, we get

$$|H_D(k)| = 2k\pi/N, \text{ for } 0 \leq k \leq N/2 \text{ and } N \text{ even}$$

or

$$|H_D(k)| = 2k\pi/(N + 1), \text{ for } 0 \leq k \leq (N - 1)/2 \text{ and } N \text{ odd}.$$

In this example, we explore the differences produced by even and odd N, and by varying the value of N.

$N = 64$: Recall that if $\{h_N(n)\}$ has even symmetry and N is an even

PASSBAND:

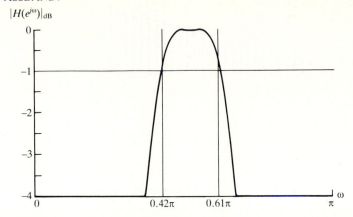

STOPBAND:

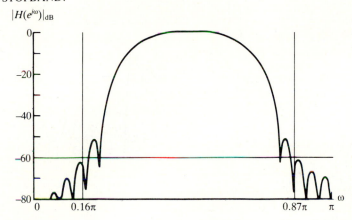

FIGURE 9-36
Magnitude response of filter satisfying specification in Example 9.3.

number, then its magnitude response has a zero at $\omega = \pi$. Proceeding blindly without this knowledge, we sample $|H_D(e^{j\omega})|$ at 33 points over $0 \le \omega \le \pi$. The sample values are shown in Fig. 9.37(a). The resonator gains are then given by

$$B_k = 4k\pi \cos(63\pi k/64)/N \qquad \text{for } 0 \le k \le 32.$$

This filter has been implemented and the unit-sample response is shown in Fig. 9.37(b). The even symmetry about $n = 31.5$ is apparent. However, when the magnitude response of this filter is determined, it is linear over most of the frequency range, but becomes zero at $\omega = \pi$. This is due to the zero dictated by the symmetry of the unit-sample response. Upon closer examination, we also find that

$$B_{32} = 2\pi \cos(63\pi/2) = 2\pi \cos(3\pi/2) = 0.$$

Since an even symmetric sequence having an odd number of points does not have a zero at $\omega = \pi$, we should try $N = 63$.

(a)

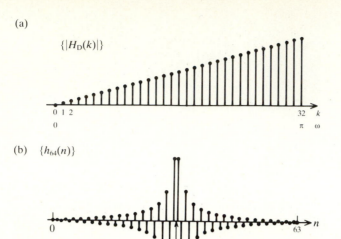

(b) $\{h_{64}(n)\}$

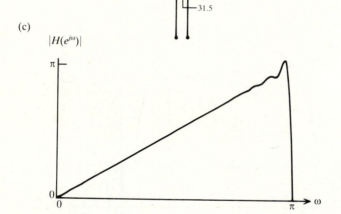

(c)

$|H(e^{j\omega})|$

FIGURE 9-37
Results for Example 9.4 when $N = 64$. (a) Input samples of magnitude response sequence; (b) unit-sample response of filter; (c) filter magnitude response.

$N = 63$: $|H_D(e^{j\omega})|$ was sampled at 32 points over $0 \le \omega < \pi$. These sample values are shown in Fig. 9.38(a). The resonator gains are then given by

$$B_k = 4k\pi \cos(62\pi k/63)/(N + 1) \qquad \text{for } 0 \le k \le 31.$$

This filter has been implemented and the unit-sample response is shown in Fig. 9.38(b). The even symmetry about $n = 31$ is apparent. The magnitude response of this filter, shown in Fig. 9.38(c), is more linear over the entire frequency range.

Having been convinced that an odd N is appropriate, we next observe the results for $N = 33$ and 15, shown in Fig. 9.39. As N is reduced, the deviation of $|H(e^{j\omega})|$ from the desired linear function is more apparent, especially near $\omega = 0$ and $\omega = \pi$, where $|H_D(e^{j\omega})|$ has sharp corners. The minimum acceptable value for N is determined by the specified tolerances, which may be in terms of maximum deviation of $|H(e^{j\omega})|$ from a linear function, for example. This case is examined in the projects.

(a) $\{|H_D(k)|\}$

(b) $\{h_{63}(n)\}$

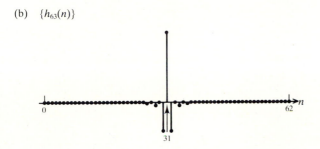

(c) $|H(e^{j\omega})|$

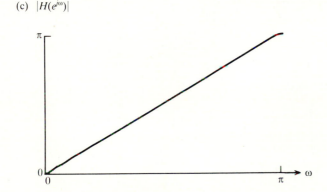

FIGURE 9-38
Results for Example 9.4 when $N = 63$. (a) Input samples of magnitude response sequence; (b) unit-sample response of filter; (c) filter magnitude response.

OBSERVATIONS ABOUT THE FREQUENCY-SAMPLING FILTER DESIGN METHOD. The system function of each second-order resonator in the frequency-sampling structure contains a $(1 - z^{-1})$ term in the numerator. This can be factored out of the total system function to simplify the filter. This cannot be done in general, because the first-order resonator, when it exists, does not contain this term. If the filter is implemented with a hardware second-order section, this term comes automatically, and since it does not require a multiplication, is implemented with little loss in time. If the filter is implemented as a program, factoring out this common term may save an instruction per resonator, and speed up the execution a little.

(a) $N=33$ results

$\{h_{33}(n)\}$

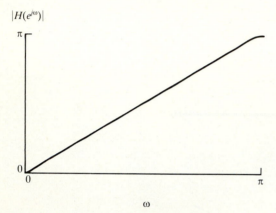

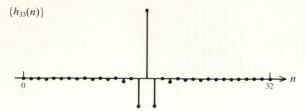

(b)

$N=15$ results

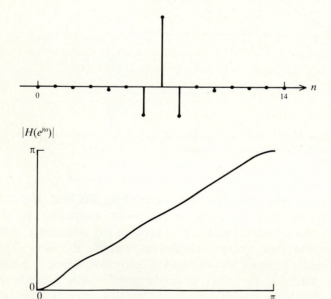

FIGURE 9-39
Results for Example 9.4 for different values of N. (a) $N = 33$; (b) $N = 15$.

One advantage of the frequency-sampling filter structure is that the computations can be performed in parallel. If it is important to minimize the processing time, each of the resonators can be implemented by a separate hardware circuit and all such circuits can process the incoming data simultaneously.

One disadvantage of the frequency-sampling structure is that the cancellation performed by the comb filter must be exact. It is this exact cancellation at $n = N$ that causes $\{h_N(n)\}$ to have a finite duration. Otherwise, the residual values circulate in the memory of the lossless resonator forever. With time, these residuals can accumulate and introduce significant errors in the filter output. To prevent this from happening in practice, the zeros of the comb filter are placed slightly inside the unit circle, at radius $r = 1 - \alpha$, for some very small $\alpha > 0$. The system function of the comb filter is then equal to

$$H_C(z) = 1 - r^N z^{-N} \tag{9.54}$$

The poles of the resonator are then also located inside the unit circle to cancel these zeros. With their poles inside the unit circle, these resonators become lossy. This loss allows any residuals that occur to decay to zero eventually.

No discussion of FIR filtering techniques is complete without a description of how to perform the filtering operation with the FFT algorithm. This topic is described in the next section.

9.6   USE OF THE FFT ALGORITHM TO PERFORM FIR FILTERING

In the preceding sections, we described the design of FIR filters and found that, to satisfy a demanding specification, the number of points in the FIR unit-sample response may be quite large. Since this large number is directly related to the number of multiplications required per output point, FIR filters typically require large amounts of computation time. In this section, rather than implement a digital filter structure, we perform the filtering operations in the frequency domain with the DFT. To be preferable to digital filter structures, this DFT approach must be faster. The fast Fourier transform (FFT) algorithm, described in Chapter 4 as an efficient computational procedure for evaluating the DFT, is the main reason that much current signal processing is performed in the frequency domain.

To make the comparison between an FIR filter and the FFT algorithm, let us consider the FIR filter to have the finite-duration unit-sample response, denoted by $h(n)$, for $0 \le n \le N_H - 1$. To implement the filter in a direct form structure, the output value at time n is determined by the convolution

$$y(n) = \sum_{k=0}^{N_H - 1} h(k) x(n - k) \tag{9.55}$$

For each output point, N_H multiplications and additions are required. If the input sequence $\{x(n)\}$ contains N_X elements, then the output sequence $\{y(n)\}$ contains $N_Y = N_H + N_X - 1$ points. Since a multiplication is typically the single operation that requires the most time to perform in a general-purpose computer, we will use the *number of real-valued multiplications* required to perform a filtering operation as a measure of the efficiency of a filtering method. To perform the direct convolution above, N_H multiplications are required for each output value, or $N_Y N_H$ multiplications for the entire output sequence.

The convolution above can be converted into a multiplicative operation by taking the N_Y-point discrete Fourier transform of the data and unit-sample response and computing

$$Y(k) = H(k)\,X(k) \qquad \text{for } 0 \le k \le N_Y - 1 \tag{9.56}$$

The output sequence values $y(n)$, for $0 \le n \le N_Y - 1$, are then determined from the inverse DFT of $\{Y(k)\}$. The following steps must be performed to accomplish the filtering operation in the frequency domain.

Step 1. Computation of the N_Y-point DFTs of the unit-sample response and the input sequence:

$$H(k) = \sum_{n=0}^{N_H - 1} h(n)\, e^{-j2\pi nk/N_Y} \qquad \text{for } 0 \le k \le N_Y - 1 \tag{9.57}$$

and

$$X(k) = \sum_{n=0}^{N_X - 1} x(n)\, e^{-j2\pi nk/N_Y} \qquad \text{for } 0 \le k \le N_Y - 1 \tag{9.58}$$

Step 2. Calculation of the product of the two N_Y-point DFTs:

$$Y(k) = H(k)\,X(k) \qquad \text{for } 0 \le k \le N_Y - 1 \tag{9.59}$$

Step 3. Computation of the N_Y-point inverse DFT:

$$y(n) = \frac{1}{N_y} \sum_{k=0}^{N_Y - 1} Y(k)\, e^{j2\pi nk/N_Y} \qquad \text{for } 0 \le n \le N_Y - 1 \tag{9.60}$$

Under special circumstances, the above three-step procedure can be modified to reduce the multiplication count. For example, if the same filter is to be applied to many different input sequences, the transformation needed to find $\{H(k)\}$ is performed only once. However, when processing very long, possibly infinite-duration, input sequences, the proper value for N_Y is not evident. For this case, the input sequence can be broken up into blocks of data having a constant size. The input data can then be processed block by block. We now examine this procedure and determine the size of the block that minimizes the number of computations per output point.

FILTERING WITH THE FFT BY USING THE OVERLAP-AND-ADD METHOD. When performing a filtering operation in the frequency domain, the size of the DFT to be computed was determined by the duration of the output sequence $N_Y = N_H + N_X - 1$. The DFTs of both the input and unit-sample response had to be computed to this size. For long input data sequences, such as the sampled audio signals from CD systems, N_X is typically much greater than N_H. To compute the N_Y-point DFTs, $\{h(n)\}$ would then have to be padded with $N_X - 1$ zeros, a large number that requires a long computation time. As an alternate procedure, we can divide the input sequence into smaller blocks of data and determine the block size that minimizes the number of multiplications required per output point. When processing data with digital filter structures, this problem does not arise, since the number of computations per output point is a constant, being determined by the structure of the filter.

Let the sequence $\{x(n)\}$ have an arbitrarily long duration, such that $N_X \gg N_H$, and break it up into blocks of N_Q samples, as shown in Fig. 9.40. The mth block is then denoted by $\{x_m(n)\}$ and defined by

$$
\begin{aligned}
x_m(n) &= x(n - mN_Q) \qquad \text{for } 0 \leq n \leq N_Q - 1 \text{ and } m = 0, 1, 2, \ldots \\
&= 0, \quad \text{otherwise}
\end{aligned}
\tag{9.61}
$$

When processed by the FIR filter having a unit-sample response containing N_H samples, the output for the mth block, denoted by $\{y_m(n)\}$ is given by

$$
y_m(n) = \sum_{k=0}^{N_Q - 1} h(n - k)\, x_m(k) \qquad \text{for } 0 \leq n \leq N_H + N_Q - 1 \tag{9.62}
$$

Since the processing is linear, the superposition principle can be applied in an interesting way. The entire input sequence can be considered to be the superposition of the individual data blocks, or

$$
x(n) = \sum_m x_m(n - mN_Q) \tag{9.63}
$$

If $\{y(n)\}$ is the linear convolution of $\{h(n)\}$ and the entire sequence $\{x(n)\}$, then it is equal to the sum of the outputs for each block, or

$$
y(n) = \sum_m y_m(n - mN_Q) \tag{9.64}
$$

Since $N_H + N_Q - 1$ is greater than N_Q, a time overlap occurs in the successive $\{y_m(n)\}$ sequences, as shown in Fig. 9.40. The superposition principle, by Eq. (9.64), states that the overlapping sections should be added to produce the desired output. Hence, this method is called the *overlap-and-add* method.

To compute the number of multiplications required per output point using this method, we let $M_Y = N_H + N_Q - 1$ and assume that the M_Y-point sequence $\{H(k)\}$ has been already computed. This is reasonable for applications in which the same filter will be used repeatedly on input data. Then the steps listed in Table 9.5 must be performed. The corresponding number of real multiplications for each step is also listed.

Each application of these three steps generates M_Y samples. But, because

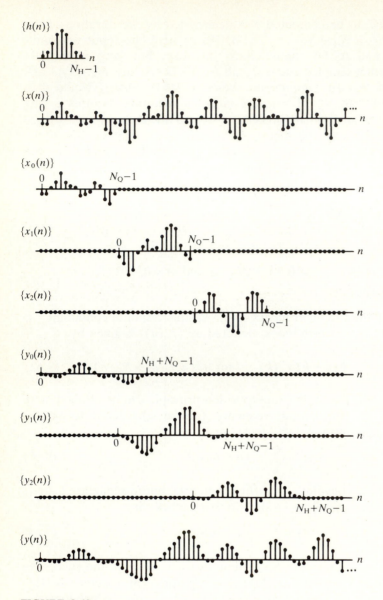

FIGURE 9-40
Relationships of sequences used in the overlap-and-add method.

of the overlap, the *valid* output sequence, or those samples that require no further manipulation, increases by N_Q points. If we let R denote the number of multiplications per valid output point, we have

$$R = [4M_Y \log_2 M_Y + 4M_Y + 4M_Y \log_2 M_Y]/N_Q$$
$$= [4(N_H + N_Q - 1)(1 + 2 \log_2 (N_H + N_Q - 1))]/N_Q \qquad (9.65)$$

TABLE 9.5

Steps and multiplication count for performing a filtering operation with the FFT

Step	Number of real multiplications
1. Compute the M_Y-point DFT of $\{x_m(n)\}$	$4M_Y \log_2 M_Y$
2. Compute the product $Y_m(k) = H(k) X_m(k)$	$4M_Y$
3. Compute the M_Y-point inverse DFT of $\{Y_m(k)\}$	$4M_Y \log_2 M_Y$

Examining this last equation, we find that for a given value of FIR filter duration N_H, there is a block size N_Q that minimizes the value of R. The curves of the R values for different values of N_Q are shown in Fig. 9.41 for different values of N_H.

To determine whether the FIR filter or the FFT approach is more efficient, the value of R should be compared to N_H, the number of computations per output point required by the general Direct Form FIR filter structure. For linear-phase filters, the value of R should be compared with $N_H/2$ to account for the linear-phase structure.

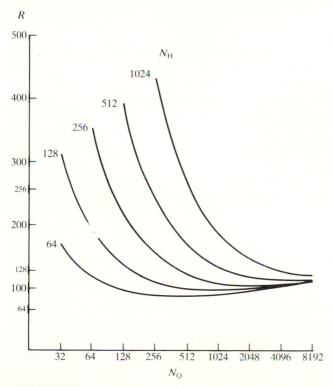

FIGURE 9-41

curves showing R, number of multiplications per output point, as a function of the block size N_Q, with size of filter duration N_H as a parameter.

Figure 9.41 should be interpreted in the following manner. For a given N_H, the value of R is observed to have a minimum for some value N_Q. For applying this overlap-and-add procedure with the *FFT*, N_Q should be chosen such that

$$N_Q + N_H - 1 = M_Y = 2^K \qquad \text{for some integer } K. \tag{9.66}$$

The input data sequences $\{x_m(n)\}$ must then be padded with $M_Y - N_H + 1$ zeros to compute the M_Y-point DFT sequence $\{X_m(k)\}$. For small values of N_H (<64), the computational overhead required to calculate the DFT sequences makes this procedure less efficient than using the FIR filter structure directly. For large values of N_H, the efficiency of the FFT algorithm is put to full advantage, but the choice for the value N_Q becomes important. If N_Q is much smaller than N_H, then an excessive number of zeros must be padded to $\{x_m(n)\}$ and the computation of the M_Y-point DFT yields only a few points at the output. As N_Q approaches infinity, Eq. (9.65) indicates that R also becomes infinite, so that a minimum R exists for some finite value of N_Q. The curves indicate that $N_Q = 4N_H$ is a reasonable choice that will provide savings in the computations when $N_H > 256$.

9.7 SUMMARY

In this chapter, three methods for synthesizing FIR filters were presented. The first, called the *windowing method,* synthesizes the filter from the unit-sample response sequence. This approach is most useful for designing an FIR filter from a *known* unit-sample response sequence. One application of this case occurs when replacing an existing analog filter with a digital FIR filter. The impulse response of the analog filter can be sampled to produce the desired unit-sample sequence. The second approach, call the *DFT/windowing method,* the unit-sample response is determined from the inverse discrete Fourier transform of the samples of the magnitude response and an assigned phase response. A window sequence can then be applied to shorten the duration and take advantage of the tolerances in the specification. When a linear-phase response is specified, the special FIR structure can be implemented to reduce the multiplication count by almost one-half. The third approach, the *frequency-sampling method,* synthesizes the filter directly from samples of the magnitude response. This method results in a filter structure that consists of a comb filter followed by a bank of parallel resonators. This parallel structure can perform the computations on the input data simultaneously, thereby reducing the computation time. This approach is most useful for designing a filter to have a magnitude response with a specific shape.

We also considered performing the filtering operations in the frequency domain with the FFT. The overlap-and-add method for performing a filter operation on long data sequences divides the input data sequence into blocks. The block size that minimizes the number of multiplications per output point was determined. The number of multiplications can then be compared to that required for digital filter structures to perform the same function. It was observed that the FFT method is not always the most efficient approach to processing data.

FURTHER READING

Antoniou, Andreas: *Digital Filters: Analysis and Design,* McGraw-Hill, New York, 1979.

Bergland, G. D.: "A guided tour of the fast Fourier transform," *IEEE Spectrum,* **6:** 41–52 (1969).

Burrus, C. S., and T. W. Parks: *DFT/FFT and Convolution Algorithms,* Wiley-Interscience, New York, 1985. [This book provides a description of the various algorithms to implement a discrete Fourier transform. Program listings are given in FORTRAN and assembly language for the Texas Instruments TMS32010 signal processing chip.]

Cooley, J. W., and J. W. Tukey: "An algorithm for the machine computation of complex Fourier series," *Math. Comp.,* **19:** 297–301 (1965).

Gold, Bernard, and Charles M. Rader: *Digital Processing of Signals,* McGraw-Hill, New York, 1969.

Hamming, R. W.: *Digital Filters,* 2d ed., Prentice-Hall, Englewood Cliffs, NJ, 1983.

McClellan, J. H., T. W., Parks, and L. R. Rabiner: "A computer program for designing optimum FIR digital filters," *IEEE Trans. Audio Electroacoust.,* **AU-21:** 506–526 (1973).

Oppenheim, Alan V., and Ronald W. Schafer: *Digital Signal Processing,* Prentice-Hall, Englewood Cliffs, NJ, 1975.

Peled, A., and B. Liu: *Digital Signal Processing: Theory, Design and Implementation.* Wiley, New York, 1976.

Rabiner, Lawrence R., and Bernard Gold: *Theory and Application of Digital Signal Processing,* Prentice-Hall, Englewood Cliffs, NJ, 1975.

Taylor, F. J.: *Digital Filter Design Handbook,* Marcel Dekker, New York, 1983. [Advanced topics, also covering ladder and wave filters. Includes hardware and software considerations]

Williams, Charles S.: *Designing Digital Filters,* Prentice-Hall, Englewood Cliffs, NJ, 1986. [Beginning level text for juniors and seniors, relying very little on previous mathematical knowledge.]

PROBLEMS

9.1. Consider the infinite-duration unit-sample responses given below. We have determined that either 5-point or 6-point rectangular windows are sufficient to implement linear-phase FIR filters that satisfy the specifications. For each response, indicate the appropriate window duration and draw the causal linear-phase FIR filter structure that performs multiplications most efficiently.

(a) $h(n) = 0.9^n$ for $n \geq 0$
    $h(n) = h(-n)$ for $n < 0$

(b) $h(n) = 0.9^n$ for $n > 0$
    $h(0) = 0$
    $h(n) = -h(-n)$ for $n < 0$

(c) $h(n) = 0.9^n$ for $n \geq 0$
    $h(n) = h(-n + 1)$ for $n < 0$

9.2. The triangular window can be analyzed as the convolution of two smaller-duration rectangular windows. Compute the spectrum of the 5-point triangular window by performing the multiplication of the spectra of the two 3-point rectangular windows given by

$$w_R(n) = \begin{cases} 1 & \text{for } -1 \leq n \leq 1 \\ 0 & \text{otherwise} \end{cases}$$

Sketch and compare the spectra of the triangular and rectangular windows.

9.3. Consider the ideal lowpass filter having the cutoff frequency $\omega_c = \pi/2$.
   (a) Compute the desired unit-sample response $\{h_D(n)\}$.
   (b) Use a 3-point rectangular window and compute the spectrum of the windowed sequence. Compare with the ideal response.
   (c) The same sequence is obtained when a 5-point rectangular window is applied (why?). Indicate by using sketches of the spectra why the same windowed magnitude response is obtained.

9.4. Give simple examples that illustrate the four linear-phase responses given in Table 9.2 by sketching representative unit-sample response sequences and their corresponding phase responses.

9.5. Indicate the possible symmetries that the unit-sample response can have to approximate a lowpass filter, a highpass, a bandpass filter and a bandreject filter. Consider the locations of desirable and undesirable zeros of the transfer function.

9.6. Write the analytic expression for the set of variable-duration Hamming windows, having duration N (even integer) and having the point of symmetry at $N_s = (N_d - 1)/2$, where N_s is not an integer ($N < N_d$).

9.7. Consider the desired magnitude response given by $|H_D(e^{j\omega})| = |\sin \omega|$. Using the DFT method, determine the following 8-point unit-sample responses.
   (a) Even-symmetric about $n = 3.5$
   (b) Anti-symmetric about $n = 3.5$
   (c) Even about $n = 0$ (except for unmatched point at $n = 4$)
   (d) Odd about $n = 0$ (except for unmatched point at $n = 4$)

9.8. Design the four possible 8-point linear-phase FIR filters having the magnitude response sequence given in Prob. 9.7 by using the frequency-sampling method.

9.9. Derive the transfer function of the second-order resonators used in the frequency-sampling method for the case of an antisymmetric unit-sample response.

COMPUTER PROJECTS

9.1. The Gaussian window

Object. Investigate the spectrum of the Gaussian window.

Analytic results. Consider the Gaussian window function defined by

$$w(t) = \exp[-t^2/(2\sigma^2)] \qquad \text{for all } t$$

Generate a 31-point window sequence $\{w_G(n)\}$ that is symmetric about $n = 0$. Find the value of σ such that these 31 points are samples of the Gaussian curve for $-3\sigma \leq t \leq 3\sigma$.

Computer results

(a) Compute the log-magnitude spectrum of this window and compare the main-lobe width and sidelobe level with the windows given in Table 9.1.

(b) Try to invent your own window sequence that has better characteristics than those above. Do this by modifying the values of one of the window sequences given in this chapter. An analytic form for the window sequence values is not necessary.

9.2. Variable-duration windows

Object. Write a subroutine to generate a variable-duration window sequence.

This subroutine will be used in the following project. Write the subroutine

```
HAM(WH, NW)
```

where

WH (real output array) of size 128 that contains a Hamming window sequence centered at $N = 64$ and has duration equal to NH. The rest of the WH array is equal to zero.

NW (integer input) is the duration of the window. NW must be an odd integer.

This WH array is appropriate for windowing unit-sample sequences that are symmetric about a point. The windowing operation is accomplished by multiplying HD(N), the 128-point approximation to the desired unit-sample response by the 128-point WH(N) array.

Repeat for the Blackman window by writing

```
BLACK(WB, NW)
```

Computer results Verify the operation of HAM and BLACK by computing and plotting the window sequences for NW = 31, 63 and 127.

9.3. Linear-phase FIR filter design using the DFT/windowing method

Object. Design a causal linear-phase FIR filter that satisfies one of the magnitude response specifications given in Appendix B using the DFT/windowing procedure.

Analytic results

(a) Sketch the (linear, not log) magnitude response that is obtained by extending the passbands.

(b) Choose the type of symmetry (even or odd) and the point of symmetry (on a sample point or between sample points) of the *causal* unit-sample response. Justify your choice.

(c) Sketch the appropriate linear-with-frequency phase response curve for the filter. The principal value of the phase is not needed and should not be sketched.

Computer results

(a) Compute and plot the 128-point approximate desired unit-sample response $\{h_d(n)\}$ that is symmetric about $n = 63$. $H_D(N)$ is symmetric about $N = 64$.

(b) Truncate $\{h_d(n)\}$ by using an N_w-point Hamming or Blackman window. Justify your choice or compare the two.

Starting with $N_w = 64$, truncate $H_D(N)$. Compute and plot the log-magnitude response of the windowed unit-sample response. Reduce N_w until the specification is no longer met. Provide log-magnitude response plots for all the values of N_w used, in the order that they were tried, to indicate your strategy.

(c) For the filter design that meets the specification and has the smallest unit-sample response duration, do the following:
    (i) determine HMAX and scale the coefficients so that the maximum of the magnitude response is equal to one.
    (ii) plot the unit-sample response and indicate the symmetry on the plot.
    (iii) plot the phase response and sketch the linear trend on the plot.
    (iv) sketch the block diagram of the FIR filter structure.

9.4. Linear-phase FIR filter design using frequency-sampling method

Object. Design a causal linear-phase FIR filter that satisfies one of the magnitude response specifications given in Appendix B using the frequency-sampling proce- dure.

Analytic results
(a) Sketch the (linear, not log) magnitude response that is obtained by assigning the response in the transition region to be a raised-cosine function that connects the passband and the stopband.
(b) Determine the value for N, the duration of the filter unit-sample response.

Computer results
(a) Implement the comb filter and resonators. Compute and plot the unit-sample response. Indicate the symmetry on the plot.
(b) Compute and plot 65 elements of the 128-point log-magnitude response of the filter. *Circle* the points on the response curve that were specified in the frequency sampling procedure.
(c) Sketch the filter block diagram. Compare the number of multiplications of this frequency-sampling structure with that required for the nonrecursive FIR linear-phase structure having the same unit-sample response.

9.5. Designing a filter with specific magnitude response shape with the frequency-sampling method

Object. Design a filter whose magnitude response approximates a given analytic form using the frequency-sampling method.
    Implement the filter that approximates the desired magnitude response given by one of the following:
(1) Linearly decreasing response

$$|H_D(e^{j\omega})| = \alpha(1 - \omega/\pi) \qquad \text{for } 0 \le \omega \le \pi$$

(2) Exponential response

$$|H_D(e^{j\omega})|_{dB} = e^{-\alpha\omega} \qquad \text{for } 0 \le \omega \le \pi$$

(3) Gaussian response centered at $\omega = \alpha\pi$

$$|H_D(e^{j\omega})| = \exp[-(\omega - \alpha\pi)^2] \qquad \text{for } 0 \le \omega \le \pi$$

    Let α be the number determined by placing a decimal point before the first nonzero digit in your student ID number. For example if ID = 0123, then $\alpha = 0.123$.

Error criterion. The filter magnitude response is allowed to deviate from this specification by $\pm 2\%$ of the maximum of the magnitude response.

Analytic results

Sketch the desired magnitude response and indicate the limits that must contain the final filter magnitude response.

Computer results

(a) Compute and plot

$$|H_D(k)| = |H_D(e^{j2\pi k/128})| \qquad \text{for } 0 \le k \le 64.$$

(b) Starting with $N = 32$, implement the comb filter and resonators. Compute and plot the unit-sample response.

(c) Reduce the value of N. For each reduction, compute the 128-point magnitude response and compare with the specification by calculating the error sequence given by

$$\varepsilon(k) = |H_D(k)| - |H(k)| \qquad \text{for } 0 \le k \le 64.$$

The error element having the largest magnitude, denoted by $|\varepsilon|_{max}$, should be found. The filter design is successful when the minimum N is found, for which $|\varepsilon|_{max} < 0.02\alpha$.

(d) For the filter design that meets the specification and has the smallest number of resonators, sketch the filter block diagram. Compute and plot the 128-point magnitude response of the filter. *Circle* the points on the response curve that were specified in the frequency sampling procedure. Compute and plot the unit-sample response. Compare the number of multiplications with that required for the nonrecursive FIR linear-phase structure.

CHAPTER
10

FINITE-PRECISION EFFECTS

10.1 INTRODUCTION

Finite-precision effects are caused by the errors due to representing numbers to a finite accuracy. These effects were previously noted in Chapter 8, when the computed pole locations of IIR filters were different from those desired, the difference being due to round-off effects in the calculations. These effects are also important when implementing a filter structure on a computer or digital hardware, which represents numbers as a finite set of binary values. In this case, the input sequence and coefficients values must be represented in such a *finite-precision* number system, as well as the results of the arithmetic operations of addition and multiplication.

We start by defining the two most common forms of representing numbers in the computer, fixed-point and floating-point notation. Then, three effects that occur in finite-precision number systems are considered. The first is the conversion of continuous-valued, or infinite-precision, signal values into finite-precision values. This occurs in the analog-to-digital conversion operation. This effect is shown to be similar to adding noise to the input signal values. The second effect occurs when the filter coefficient values, computed to infinite precision in theory, are constrained to equal one of the allowable numbers represented in the computer for the filter implementation. As a result, the singularities in the z plane are limited to lie on a grid of allowable locations. The deviations that are produced in the resulting transfer function are considered. The third effect is the truncation that results when the sum or product of two finite-precision numbers is itself represented as a finite-precision number. If the filter being implemented is a nonrecursive structure, this effect is similar to adding noise to the output of each multiplier. However, if the filter has a recursive structure, the truncation of the product may cause the output to exhibit undesired offsets or oscillations, called limit cycles.

Since these finite-accuracy effects are nonlinear and random, their analysis is very complicated. To provide some insight, special cases are analyzed to indicate the type of behavior that can be expected. Procedures for simulating and observing these effects on a computer are described. The intent of this chapter is to present some points to be aware of when designing filters in a finite-precision number system, but not to provide a comprehensive analysis of these effects. Such an analysis is most difficult and beyond the scope of this text. In the cases when the analysis can be performed, it requires results from nonlinear system theory and probability theory. The analysis can be found in the references at the end of the chapter.

10.2 FINITE-PRECISION NUMBER REPRESENTATION WITHIN THE COMPUTER

In a computer, or any other digital hardware structure, numbers are represented as combinations of a finite number of *binary digits,* or *bits* that take on the values of 0 and 1. These bits are usually organized into *bytes* containing 8 bits, or *words* containing 16 bits, but with 32-bit words becoming increasingly common. We consider two common forms that are used to represent numbers in a digital computer: *fixed-point* and *floating-point notation.*

FIXED-POINT NOTATION. In fixed-point notation, the partition between whole numbers and fractions, the *binary-point,* is constrained to lie at particular, or fixed, position in the bit pattern, as shown in Fig. 10.1. The first bit, called the *sign bit,* is reserved for indicating the sign of the number, usually 0 for positive and 1 for negative. The magnitude of the number is then expressed in powers of 2, with the binary point separating the positive

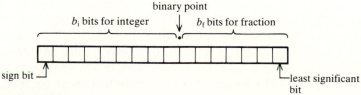

FIGURE 10-1
Allocation of bits in a word for fixed-point representation (16-bit word is shown).

exponents, including zero, and the negative exponents. For example, the decimal value of the binary number 01.101_2 is determined as

$$01.101_2 = (+)(1 \times 2^0) + (1 \times 2^{-1}) + (0 \times 2^{-2}) + (1 \times 2^{-3}) = 1.625_{10}.$$

The *precision* of the number system is defined as the increment between two consecutive numbers and is determined by the value of the right-most bit, called the *least significant bit*. In the number system given in the previous example, a number can be expressed to the nearest 2^{-3} or 0.125. The *range* of the number system is defined as the interval between the most negative and most positive numbers that can be expressed. The binary point determines a tradeoff between precision and range. When the binary point is defined to lie at the right of the bit pattern, there is no fraction and only integer values are allowed. This corresponds to the INTEGER variable in FORTRAN. For digital filters, we will find that it is most convenient to have the binary point lie one or two bit positions to the right of the sign bit, thus allowing fractions to occur.

Conversion of decimal numbers into binary numbers. In converting a decimal number into a binary number, the integer part and fraction must be converted separately. Only positive numbers will be considered, with the procedure to represent negative numbers having a particular magnitude being described later. Converting the integer part involves repeated division by two, with the remainder at each division step forming the binary number. The binary number is formed by starting at the binary point and proceeding left-ward, with the final nonzero remainder being the most significant bit. Every finite decimal integer can be represented in a finite number of bits.

The conversion of the decimal fractional part involves repeated multiplications by two, with the integer part of the product at each step forming the binary number and the fractional part being used in the continued multiplications. The binary number is formed by starting at the binary point and proceeding right-ward, with the final entry being the least significant bit in the binary representation in the number system. Not all decimal fractions can be represented in a finite number of bits. This process may need to be terminated when the number of bits available to express the fraction is exceeded.

In converting a decimal fraction into a finite-precision binary number, the least significant bit can be generated by *truncation* or by *round-off*. In performing truncation, the magnitude of the infinite-precision number is converted into finite-precision by using the available number of bits. The process terminates after the least significant bit has been determined. In round-off, the conversion proceeds for an additional bit beyond the least significant bit. If this additional bit is a one, the least significant bit is incremented. If the additional bit is zero, the process terminates and produces the same result as truncation. The round-off representation is more accurate, but requires more operations.

Example 10.1. Decimal to binary conversion. Convert 6.86_{10} into binary notation having 8-bits of which four bits are used to represent integers, including sign, and four bits represent fractions.

Integer part:

		remainder
$\frac{6}{2} = 3$		0
$\frac{3}{2} = 1$		1
$\frac{1}{2} = 0$		1

Hence, $6_{10} = 0110_2$

Fraction. To compare the results obtained with truncation and round-off, we will convert the number to 5 bits:

	integer part	
$0.86 \times 2 = 1.72$	1	
$0.72 \times 2 = 1.44$	1	
$0.44 \times 2 = 0.88$	0	
$0.88 \times 2 = 1.76$	1	(least significant bit)
$0.76 \times 2 = 1.52$	1	

Hence,

$$0.86_{10} = 0.1101_2 (= 0.8125_{10}) \text{ by truncation}$$

$$= 0.1110_2 (= 0.875_{10}) \text{ by rounding}$$

Combining these two results, we have

$$6.86_{10} = 0110.1101_2 \text{ by truncation}$$

$$= 0110.1110_2 \text{ by rounding}$$

2's complement notation. One common fixed-point number system currently used is *2's complement notation,* because of its efficiency for performing computations. The bit pattern of a negative number is obtained by starting with the magnitude expressed as a positive number. Then all the bits are complemented and a binary one is added to the least significant bit.

Example 10.2. Decimal to binary conversion of negative numbers using 2's complement notation. Convert -6.86_{10} into binary notation having 8-bits of which four bits are used to represent integers, including sign, and four bits represent fractions.

First, the magnitude of the number is converted into binary notation. Using the round-off results from Example 10.1, we have

$$6.86_{10} = 0110.1110_2$$

Complementing this result bit-by-bit, we get 1001.0001_2. Adding 0.0001_2 to this result, we have

$$-6.86_{10} = 1001.0010_2 (= -6.875_{10})$$

Table 10.1 shows the entire 2's complement repertoire of patterns and their decimal equivalents for 2, 3 and 4-bit number systems.

TABLE 10.1
Numbers expressed in 2's-complement fixed-point notation

Binary number	Integer values (binary point at right)	Fractions (binary point after sign bit)
2-bit number system		
01	1	0.5
00	0	0
11	−1	−0.5
10	−2	−1
3-bit number system		
011	3	0.75
010	2	0.5
001	1	0.25
000	0	0
111	−1	−0.25
110	−2	−0.5
101	−3	−0.75
100	−4	−1
4-bit number system		
0111	7	0.875
0110	6	0.75
0101	5	0.625
0100	4	0.5
0011	3	0.375
0010	2	0.25
0001	1	0.125
0000	0	0
1111	−1	−0.125
1110	−2	−0.25
1101	−3	−0.375
1100	−4	−0.5
1011	−5	−0.625
1010	−6	−0.75
1001	−7	−0.875
1000	−8	−1

FLOATING-POINT NOTATION. The second form of expressing numbers is *floating-point* notation, also called *real* notation. In this case, a number x is expressed in the form

$$x = m2^e \tag{10.1}$$

where m is the *mantissa* and e is the *exponent*. The notation is conventionally defined to have $1/2 \le |m| < 1$. For example, the number 1.5 is expressed as $(0.75)2^1$. Both m and e are expressed as fixed-point numbers, using a constant

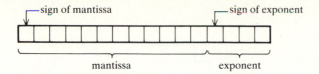

FIGURE 10-2
Allocation of bits in a word for floating-point representation (16-bit word is shown).

pattern within the word, as shown in Fig. 10.2. If the word contains b bits, let b_m be assigned to the mantissa and b_e to the exponent. Since $b = b_e + b_m$, a hardware designer must decide on the bit allocation between the mantissa and exponent. The number of bits assigned to b_e determines the range of the number system, while the number of bits assigned to b_m determines the precision. A common compromise is to set $b_m = 3b/4$. In many minicomputers, two 16-bit words are combined to express a floating-point number as a 32-bit word, of which $b_m = 24$ and $b_e = 8$.

In the next section, we consider the effect of representing the infinite-precision values of a discrete-time sequence in such a finite-precision system.

10.3 QUANTIZATION ERROR IN ANALOG-TO-DIGITAL CONVERSION

As described in Chapter 3, a sampling process converts a continuous-time signal $\{x(t)\}$ into a discrete-time sequence $\{x(n)\}$. In practice, this sampling is performed with an analog-to-digital converter (ADC) that represents the sampled value with a finite number of bits, usually in fixed-point notation. Hence, the ADC also simultaneously converts the continuous-valued sample $x(n)$ into one that has been quantized, denoted by $x_q(n)$. The value of $x_q(n)$ is assigned to one of the numbers represented by the finite number of bits in the ADC. The process of converting the $x(n)$ into $x_q(n)$ can be modeled by the quantizer characteristic shown in Fig. 10.3. The quantizer range is expressed in terms of 2's-complement numbers that span the amplitude range from $-M$ to $M - 1$ in K steps, each of size $\Delta = 2M/K$. If the quantizer contains b bits, $K = 2^b$ and

$$\Delta = \frac{2M}{2^b} \tag{10.2}$$

For a given quantizer range, increasing the number of bits reduces the step size, making the staircase pattern approach a straight line.

There are two types of errors that can be introduced by such a quantizer. The first is *granular error*, which occurs when $|x(n)| < M$, and is denoted by $\varepsilon_g(n)$, where

$$\varepsilon_g(n) = x(n) - x_q(n) \tag{10.3}$$

Note that the magnitude of this error is always less than the step size Δ. The

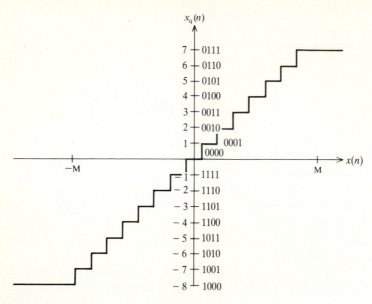

FIGURE 10-3
Quantizer characteristic for a 4-bit quantizer.

second error occurs when $|x(n)| > M$, when the amplitude exceeds the quantizer limits, and is called *clipping error*, denoted by $\varepsilon_c(n)$, where

$$\varepsilon_c(n) = \begin{cases} x(n) - M - 1 & \text{for } x(n) > M - 1 \\ x(n) + M & \text{for } x(n) < -M \end{cases} \tag{10.4}$$

Very large, theoretically infinite, clipping amplitude errors are possible, making the analysis of these errors rather complicated. In the analysis below, we set the quantizer limits large enough, or equivalently scale the input sequence to have small enough values, so that this clipping error does not occur. In this case, only the granular error is important.

To make optimum use of the quantizer, it is usually desired to have the signal values distributed throughout the entire range of numbers represented by the bits in the ADC. This is typically done by adjusting the maximum value of the input signal magnitude so that it is equal to the maximum magnitude of the number system. Signals encountered in practice, such as speech and sensor outputs, are typically random in that their exact analytic form is not known. Some random signals tend to have a large dynamic range, such that the signal power is large at some times and very small at others. For such signals, a different quantization strategy is usually employed, in which the quantizer range is matched to the average power level of the signal. The relevant parameter in this case, is the *root-mean-square* (*rms*) value of the signal, denoted by X_{rms}. The value of X_{rms}^2 can be calculated from the power in the

N-point sequence $\{x(n)\}$ by

$$X_{\text{rms}}^2 = \frac{1}{N} \sum_{n=0}^{N-1} x^2(n) \tag{10.5}$$

The rms values for many common signals, such as telephone speech and recorded music levels on tape are usually specified or determinable by simple measurements.

For speech processing applications, it is not uncommon to have the quantizer limits set equal to $\pm 4X_{\text{rms}}$. This setting represents a compromise in which an accurate representation of the small-amplitude signals, which occur more commonly, is achieved at the expense of large errors for more rarely occurring large-magnitude signals. Nonlinear quantizer characteristics are also common, in which the step size varies with the magnitude of the input signal. A further discussion of nonuniform and also time-varying quantization schemes can be found in the reference by Jayant and Noll.

To analyze the quantization operation, we employ the additive error model shown in Fig. 10.4. The quantized value $x_q(n)$ is considered to be equal to the true value $x(n)$ plus a random additive noise term $\varepsilon_g(n)$, or

$$x_q(n) = x(n) + \varepsilon_g(n) \tag{10.6}$$

When matched to a random signal having a known average power level, the step size Δ is equal to

$$\Delta = \frac{8X_{\text{rms}}}{2^b} \tag{10.7}$$

The quantization error $\varepsilon_g(n)$ is commonly assumed to be random and uniformly distributed between $\pm \Delta/2$. The noise power, or variance, can then be shown to be equal to $\Delta^2/12$. The ratio of the signal power to that of the noise, called the *signal-to-noise ratio*, denoted *SNR*, is then given by

$$\text{SNR} = \frac{X_{\text{rms}}^2}{\Delta^2/12} = \frac{12}{64} 2^{2b} \tag{10.8}$$

The SNR is observed to improve exponentially with the number of bits in the quantizer. Expressing the SNR value in decibels, we have

$$\text{SNR}_{\text{dB}} = 10 \log_{10} \left(\frac{12}{64} 2^{2b} \right) = 6b - 7.3 \tag{10.9}$$

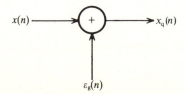

FIGURE 10-4
Additive noise model for quantizer.

This last result states that the signal-to-noise ratio improves by 6 dB for every bit that is added to the ADC. The effect of the quantization error on the SNR for a digital audio system is illustrated in the example below.

> **Example 10.3. Quantization error in a digital audio system.** Let the input be $\{x(n)\}$ with an rms value equal to X_{rms} and let the quantizer limits span $\pm 4X_{rms}$. If a 16-bit quantizer is employed, a SNR of greater than 88 dB can be achieved. The current claim for commercial digital audio systems is approximately 84 dB. The difference is due to the thermal noise in the analog amplifiers and the digital-to-analog conversion, which were not considered in the analysis.

EFFECT OF QUANTIZATION NOISE AT THE FILTER OUTPUT. The quantization error in the ADC process was modeled as the addition of a noise sequence to the infinite-precision sample values. To examine the contribution of this noise sequence to the output sequence produced by a linear digital filter, we apply the quantized signal to the filter and can express the output sequence as

$$\{y_q(n)\} = \{x_q(n)\} * \{h(n)\}$$
$$= \{x(n) + \varepsilon_g(n)\} * \{h(n)\} \tag{10.10}$$

Because the filter is linear, we can apply the superposition principle to get

$$\{y_q(n)\} = \{x(n)\} * \{h(n)\} + \{\varepsilon_g(n)\} * \{h(n)\}$$
$$= \{y(n)\} + \{v_\varepsilon(n)\} \tag{10.11}$$

The sequence $\{y(n)\}$ is the output that would have been observed for infinite-precision samples and $\{v_\varepsilon(n)\}$ is the output produced by the sequence of noise values. These output noise values can be written explicitly as the sum of the random errors scaled by the terms of the unit-sample response as

$$v_\varepsilon(n) = \sum_{k=-\infty}^{\infty} h(k)\varepsilon_g(n-k) \tag{10.12}$$

Adding a sequence of random numbers produces a random result, about which little can be said in general. However, if the random error sequence consists of independent values having the same variance σ_ε^2, then the variance of the output σ_v^2 is equal to

$$\sigma_v^2 = \sigma_\varepsilon^2 \sum_{k=-\infty}^{\infty} h^2(k) \tag{10.13}$$

This simple relationship allows us to compute the output noise power in terms of the input noise power and the filter unit-sample response.

> **Example 10.4. Noise variance at the output of the 3-sample averager.** The difference equation of 3-sample average is given by
>
> $$y(n) = \tfrac{1}{3}[x(n+1) + x(n) + x(n-1)]$$

If the input sequence has been quantized to the set of integers, then the step size is 1. If the range of the quantizer is large (>100) and the input amplitude has been matched to the quantizer, the quantization errors can be considered to be uniformly distributed. The error variance is then $\sigma_\varepsilon^2 = \frac{1}{12}$. The errors can be considered independent if consecutive values of $\{x(n)\}$ tend to occur in different levels of the quantizer in a random fashion. The noise variance at the output is then

$$\sigma_v^2 = \sigma_\varepsilon^2 \sum_{k=-1}^{1} \tfrac{1}{9} = \tfrac{1}{36}$$

Example 10.5. Noise variance at the output of the first-order recursive filter. The difference equation of the first-order recursive filter is given by

$$y(n) = ay(n-1) + x(n)$$

If the input sequence has been quantized to the set of integers, then the step size is 1. Under the same conditions as in Example 10.4, the noise variance at the output is then

$$\sigma_v^2 = \sigma_\varepsilon^2 \sum_{k=0}^{\infty} a^2 = 1/[(12)(1-a^2)]$$

Having considered the quantization of the input sequence, we now turn our attention to the quantization of the coefficient values in the digital filter.

10.4 EXPRESSING COEFFICIENTS IN FINITE PRECISION

When a digital filter is implemented in hardware, the coefficient values must be quantized to the accuracy of the digital number system employed in the hardware. In this section, we show that the effect of this quantization is to limit the locations of the poles and zeros in the z plane to a grid of allowable positions. If the desired singularity location does not lie on a grid point, then the shift that must be made in these singularity locations causes the transfer function $H(e^{j\omega})$ of the hardware filter structure to be different from that anticipated when the design was performed in infinite precision. For FIR filters, this deviation occurs in the locations of the zeros. For IIR filters there is a similar deviation in the pole locations. However, for filters having poles that lie close to the unit circle, these deviations may even cause the IIR filter to be unstable, as shown in the following example.

Example 10.6. Quantization of IIR filter coefficients. Let

$$H(z) = \frac{1}{(1 - 0.901z^{-1})(1 - 0.943z^{-1})} \quad \text{(cascade form)}$$

$$= \frac{1}{1 - 1.844z^{-1} + 0.849643z^{-2}} \quad \text{(direct form)}$$

If the coefficients are quantized by rounding to the nearest 0.05, then the cascade form becomes

$$H(z) = \frac{1}{(1 - 0.90z^{-1})(1 - 0.95z^{-1})}$$

which has poles at $z = 0.9$ and 0.95. The resulting filter is stable. However, the Direct form becomes

$$H(z) = \frac{1}{1 - 1.85z^{-1} + 0.85z^{-2}}$$

which factors into poles located at $z = 0.85$ and 1.0. The pole lying on the unit circle makes the filter unstable. This example illustrates that the sensitivity of a filter to coefficient quantization increases with the order of the filter.

FIR FILTER. Let us consider the second-order FIR filter system function factored into the form to show the zero locations in the z plane explicitly, or

$$\begin{aligned} H(z) &= A(1 - r\,e^{j\theta}z^{-1})(1 - r\,e^{-j\theta}z^{-1}) \\ &= A(1 - 2r\cos(\theta)\,z^{-1} + r^2z^{-2}) \end{aligned} \tag{10.14}$$

where A is a gain constant. When represented in a finite-precision number system, the coefficients $2r\cos(\theta)$ and r^2 can take on only certain values, determined by the precision of the number system, similar to the quantization of sample values in the ADC described in the previous section. To illustrate the effects of this quantization, let us consider the zero locations to be in the right half of the z plane, and quantized to 4 bits. The positive values are listed in Table 10.2.

The effect of quantizing the coefficients on the system function is most evident in the z plane. Recall that the coefficient r^2 indicates that the singularity lies on a circle defined by $|z| = r$. Quantizing r^2 to have only a set of values defines a set of circles in the z plane on which the singularity must lie, as shown in Fig. 10.5(a). The coefficient $2r\cos(\theta)$ defines a vertical line in the

TABLE 10.2
Allowable coefficient values for second-order filter with 4-bit quantization

Bit pattern	Coefficient values $(2r\cos(\theta)$ or $r^2)$	Radius of circle in z plane r
0000	0	0
0001	0.125	0.354
0010	0.25	0.5
0011	0.375	0.611
0100	0.5	0.707
0101	0.625	0.791
0110	0.75	0.866
0111	0.875	0.935
(multiplication by 1)	1.0	1.0

(a)

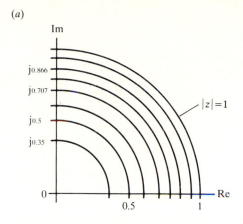

(b)

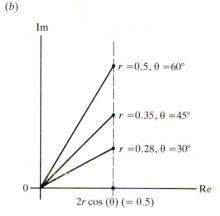

FIGURE 10-5
Coefficient quantization effects in the z plane.
a) Quantizing r^2 produces circles of radius r.
b) Quantizing $2r \cos(\theta)$ produces vertical lines.

z plane passing though $z = 2r \cos(\theta)$, as shown in Fig. 10.5(b). This vertical line indicates the pairs of allowable r and θ values that can occur. Quantizing this coefficient produces a set of corresponding vertical lines. The locations of the allowable singularities for the second-order section can then fall only at the points of intersection between these circles and vertical lines. The points of intersection for the 4-bit quantization are shown in Fig. 10.6. Note that quantizing the coefficients in this manner produces a non-uniform distribution of the locations that tend to cluster around $z = \pm j$. Alternate filter structures have been proposed that produce a uniform distribution of the allowable locations in the z plane. These are described in the references.

Because of the coefficient quantization, the position of the zero must be shifted from that desired by theory to one of the four adjacent allowable positions. The filter designer must choose one of these positions or increase the number of bits in the finite-precision representation. This latter choice comes with an increase in hardware cost and possibly slower operation. The following example illustrates the implication of these results for FIR filter design.

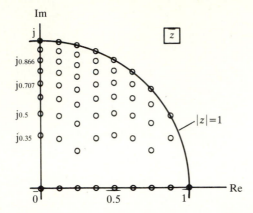

FIGURE 10-6
Allowable locations in the right-hand z plane for singularities when 4-bit quantization is employed. Complex-conjugate singularity locations are not drawn.

Example 10.7. Effects of FIR filter coefficient quantization. Let us assume that we want to implement a second-order FIR filter having 4-bit precision. Let the desired filter transfer function be given by

$$H_{\text{FIR}}(z) = (1 - 0.9e^{j70°}z^{-1})(1 - 0.9e^{-j70°}z^{-1})$$

The desired zero location at $z = 0.9e^{j70°}$ is shown in Fig. 10.7(a). Because of coefficient quantization, the zero position must be moved to one of the four adjacent grid locations, labeled 1 through 4 in the figure, that are located at $z = 0.866e^{j64°}$, $z = 0.866e^{j73°}$, $z = 0.935e^{j66°}$ and $z = 0.935e^{j74°}$. The magnitude responses of these four zeros are shown in Fig. 10.7(b). All the zeros produce a valley in the magnitude response, but the location of the valley is related to the angle of the zero, while the depth is related to the radius of the zero and the interaction with the complex-conjugate zero.

IIR FILTERS. For IIR filters, a similar effect occurs to that described above for the FIR filters. However, since IIR filters can become unstable, an additional consideration is required. As discussed in Chapter 5, an IIR filter can be configured either as a cascade of second-order sections or in a direct form implementation. In any case, the poles must be inside the unit circle for a stable filter. For the cascade configuration of second-order sections, the locations of the poles of each section can be easily verified or be made stable by placing the poles at the set of allowable locations determined by the quantization that lie strictly inside the unit circle. The desired pole must then be transferred to the closest allowable pole location in the hardware implementation. Because of this shift in the singularity location, a change in either the resonant frequency or bandwidth occurs. As described in Chapter 6, as the poles approach the unit circle, the changes in the magnitude response become increasingly drastic.

Example 10.8. Effects of IIR filter coefficient quantization. Let us assume that we want to implement a second-order IIR filter having 4-bit precision. Let the

(a)

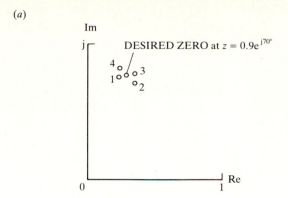

(b)

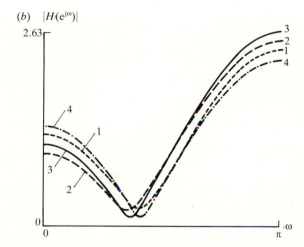

FIGURE 10-7
Effect of quantization on zero placement. *a*) Desired zero location must be shifted to one of four closest neighbors when filter is implemented. *b*) Four magnitude response curves corresponding to the four allowable pole locations.

desired transfer function be given by

$$H_{\text{IIR}}(z) = \frac{1}{(1 - 0.9e^{j70°}z^{-1})(1 - 0.9e^{-j70°}z^{-1})}$$

The desired pole location at $z = 0.9e^{j70°}$ is shown in Fig. 10.8(*a*). Because of coefficient quantization, the pole position must be moved to one of the four adjacent grid locations, labeled 1 through 4 in the figure, that are located at $z = 0.866e^{j64°}$, $z = 0.866e^{j73°}$, $z = 0.935e^{j66°}$ and $z = 0.935e^{j74°}$. The magnitude responses of these four poles are shown in Fig. 10.8(*b*). All the poles produce a peak in the magnitude response, but the location of the peak is related to the angle of the pole, while the height is related to the radius of the pole and the interaction with the complex-conjugate pole.

For both FIR and IIR filters, as the order of the filter increases, the placement of the singularities is complicated by their interactions. In practice, an optimization procedure is usually performed to find the *combination* of singularity positions that produces the best approximation to the desired

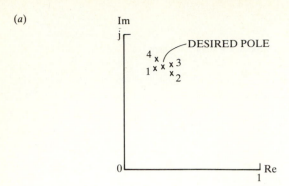

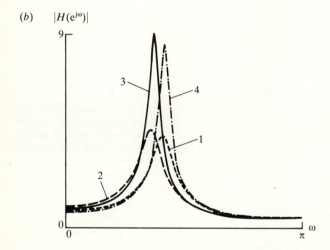

FIGURE 10-8
Effect of quantization on pole placement. *a*) Desired pole location must be shifted to one of four closest neighbors when filter is implemented. *b*) Four magnitude response curves corresponding to the four allowable pole locations.

specification. Such an optimization procedure is beyond the scope of this book, but can be found in the references.

10.5 PERFORMING ARITHMETIC IN FINITE-PRECISION NUMBER SYSTEMS

Having considered the effects of quantization of the discrete-time sequence and coefficient values, we now consider the effects of performing the arithmetic operations of addition and multiplication needed to implement a digital filter. We describe both fixed-point and floating-point representations and discuss the advantages and disadvantages of performing these operations next.

FIXED-POINT ARITHMETIC. When expressing the input sequence in fixed-point notation, each value in the sequence must be scaled so that the result does not fall outside, or overflow, the number range when additions and multiplications are performed. We consider each operation separately.

Addition in fixed-point arithmetic. The addition of two fixed-point numbers expressed in 2's-complement notation is simple: The two numbers are simply added bit by bit starting from the right, with the carry bit being added to the next bit location. The final carry bit is discarded. A subtraction is performed by first negating the number to be subtracted and then adding.

> **Example 10.9. Addition of two fixed-point numbers.** Using the 4-bit number representation xxx.x, perform the following additions:
>
> $1.5_{10} + 1_{10}$:
>
> $$\begin{array}{r} 001.1 \\ 001.0 \\ \hline 010.1 \end{array} \qquad (= 2.5_{10})$$
>
> $2.5_{10} + (-2_{10})$:
>
> $$\begin{array}{r} 010.1 \\ 110.0 \\ \hline 1\ \ 000.1 \end{array} \qquad (= 0.5_{10})$$
>
> Note that the final carry bit is ignored.
>
> $-2.5_{10} + 2_{10}$:
>
> $$\begin{array}{r} 101.1 \\ 010.0 \\ \hline 111.1 \end{array} \qquad (= -0.5_{10})$$
>
> $3 + 3.5$:
>
> $$\begin{array}{r} 011.0 \\ 011.1 \\ \hline 110.1 \end{array} \qquad (= -1.5_{10})$$
>
> Overflow error occurs in the last result because 6.5_{10} cannot be represented in this 4-bit number system.

When two numbers of b bits each are added and the sum cannot be represented by b bits, an *overflow* is said to occur. To describe a procedure that prevents it from occurring, let us consider an FIR filter structure that contains a total of M additions, as shown in Fig. 10.9. Let the number system have its binary point immediately to the right of the sign bit, so that only fractions are represented. The values of the input sequence $\{x(n)\}$ can then be normalized to have magnitudes less than or equal to one. The multiplier coefficient values can also be scaled to be less than 1. Then the sum of these M terms is always less than M. To prevent overflow when the sum is computed, the input sequence must be scaled so that all the values have magnitudes that are less than $1/M$. This scaling then ensures that the maximum possible total is less than or equal to 1.

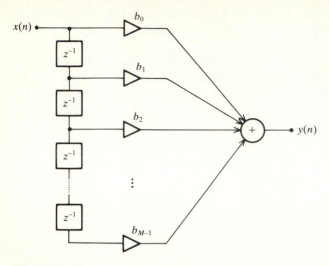

FIGURE 10-9
Filter structure having M inputs to adder.

The above procedure to prevent overflow indicates the limitations of implementing a high-order filter, or one having a large M. If the filter is implemented in one of the direct forms, a large M necessitates having a large number of bits to ensure that the quantization error in the data sequence is sufficiently small after the sequence has been scaled to have a maximum of $1/M$. The filter can also be implemented as a cascade of second-order sections in which a maximum of three additions is performed at any one time. This reduces the severity of the scaling operation.

Multiplication in fixed-point arithmetic. As in any number system, multiplication is a three step procedure. First the magnitude and sign components are separated, then the magnitudes are multiplied and, finally, the sign of the product is determined and applied to the result. These steps are especially apparent in 2's-complement notation, since the magnitudes of positive and negative numbers have different bit patterns. In finite-precision number systems, the number of bits and position of the binary point must remain constant. When a b-bit number multiplies another b-bit number, the product may contain $2b$-bits. For example, $11_2 \times 11_2 = 1001_2$. If the b-bits are organized into $b = b_i + b_f$, where b_i represents the integer part and b_f represents the fraction, then the product may contain $2b_i + 2b_f$ bits. The result is represented as a b-bit number by retaining the b_f bits to the right of the binary point, through truncation or rounding, and the b_i bits to the left.

Example 10.10. Multiplication of two fixed-point numbers. Using the 4-bit number representation xxx.x, perform the multiplication given below. In this case, $b = 4$, $b_f = 1$ and $b_i = 3$.

$$2.5_{10} \times 1.5_{10}: \quad 010.1 \times 001.1$$

$$
\begin{array}{r}
0101 \\
\times\, 0011 \\
\hline
0101 \\
0101 \\
0000 \\
0000 \\
\hline
0001111 \\
\end{array}
$$

$0001111 = 00011.11$ when the binary point is restored

Restoring the 4-bit representation, the product becomes

$0011.1_2 (= 3.5_{10})$ by truncation, and

$0100.0_2 (= 4.0_{10})$ by rounding

The truncation that occurs when the $2b$-bit product is represented as a b-bit result is one source of error that will be analyzed below.

Another alternative is to implement the filter in floating-point representation. We consider the problems associated with floating-point representations next.

FLOATING-POINT ARITHMETIC. Floating-point representation of numbers is a partial solution to the overflow problem, because a larger range of numbers can be represented. This representation reduces the need for scaling the input and the worry about overflow errors. Let us consider performing additions and multiplications with the floating-point representation to determine the disadvantages.

Addition of two floating-point numbers. The addition of two floating-point numbers is more complicated than that of two fixed-point numbers. To perform an addition, the exponents of the two numbers must be set equal. To avoid exceeding the range of the mantissa, the exponent of the smaller number must be increased to equal that of the larger number. In doing this, the mantissa of the smaller number must fall outside the range $\frac{1}{2} \le |m| < 1$. The two mantissas are then added using the fixed-point procedure. The sum is re-normalized to be in the magnitude range $[\frac{1}{2}, 1)$ by adjusting the exponent.

Example 10.11. Addition of two floating-point numbers. Let us consider an 8-bit number representation with 5 bits to represent the mantissa and 3 bits to represent the exponent. We want to perform the addition $5.5_{10} + (-1.25)_{10}$.

Expressing 5.5_{10} in floating-point notation, we get 0.6875×2^3 or $01011\ 011_2$.

Expressing -1.25_{10} in floating-point notation, we get -0.625×2^1. To set the exponents equal, we must express -1.25_{10} as -0.15625×2^3 or $111011\ 011_2$.

Performing the addition of the 6-bit expanded mantissas, we get

$$\begin{array}{r} 010110 \\ 111011 \\ \hline 1 \quad 010001 \end{array} \quad (= 0.53125_{10})$$

The result obtained by truncation is $01000\,011_2 = 4_{10}$, and the result obtained by rounding is $01001\,011_2 = 4.5_{10}$.

The floating-point representation is costly in terms of the time required to compute a sum, since first the exponents must be made equal before the sum is computed. This increase in computation time slows the filter operation and may prevent it from operating in real time.

Multiplication of two floating-point numbers. In performing a multiplication of two floating-point numbers, the two mantissas are multiplied using the fixed-point multiplication procedure and the two exponents are added. The exponent value is then adjusted, so that the magnitude of the product of the mantissas falls within $[\frac{1}{2}, 1)$.

> **Example 10.12. Multiplication of two floating-point numbers.** Let us consider an 8-bit number representation with 5 bits to represent the mantissa and 3 bits to represent the exponent. We want to perform the multiplication
>
> $$2.735_{10} \times (-1.25)_{10}$$
>
> Expressing 2.735_{10} in floating-point notation, we get 0.68375×2^2 or $01011\,010_2$.
> Since the second multiplier is negative, the product will be negative. Expressing the magnitude 1.25_{10} in floating-point notation, we get 0.625×2^1 or $01010\,001_2$.
> Performing the multiplication of the mantissas, we get
>
> $$\begin{array}{r} 01011 \\ 01010 \\ \hline 00000 \\ 01011 \\ 00000 \\ 01011 \\ 00000 \\ \hline 001101110 \end{array}$$

which is equal to $0.01101110_2 (= 0.4296875_{10})$ when the binary point is restored. By truncation, we get $0.0110_2 (= 0.375_{10})$, and by rounding we get $0.0111_2 (= 0.4375_{10})$.

Performing the addition of the exponents, we get $010_2 + 001_2 = 011_2 (= 3_{10})$. After converting the mantissa into a negative number, the final result is

$$11010\,011_2 (= -3_{10}) \text{ by truncation or}$$
$$11001\,011_2 (= -3.5_{10}) \text{ by rounding}$$

As the examples above indicate, truncations must be performed after both addition and multiplication operations with floating-point arithmetic. Moreover, these truncations have a variable magnitude, depending on the value of the exponent. This variability makes the analysis of these effects very difficult. To give an indication of the effects that can occur with truncation, we will consider only fixed-point filter implementations.

10.6 EFFECTS OF FINITE-PRECISION ARITHMETIC ON DIGITAL FILTERS

In this section, we describe the effects of finite-precision arithmetic operations on the performance of FIR and IIR filters. To simplify the analysis, only fixed-point notation is considered. The input sequence and coefficient values can be scaled so that overflow cannot occur in the addition operation. Hence, the only truncation error associated with computing the product in a finite-precision number system need be considered. This truncation can be modeled as having an additive error source at the output of each multiplier, as shown in Fig. 10.10. We first consider the simpler FIR filter.

FINITE-PRECISION ARITHMETIC EFFECTS IN FIR FILTERS. Let us consider the FIR filter implemented in the direct form nonrecursive structure shown in Fig. 10.11. The truncation errors appear as an additive noise source at the output of each multiplier. If rounding is used, the noise value at each multiplier can be assumed to be uniformly distributed between $\pm E/2$, where $E = 2^{-b}f$ is the precision of the number system of the computer. The variance, or power, of this noise component is then

$$\sigma_n^2 = E^2/12 = 2^{-2b}f/12 \tag{10.15}$$

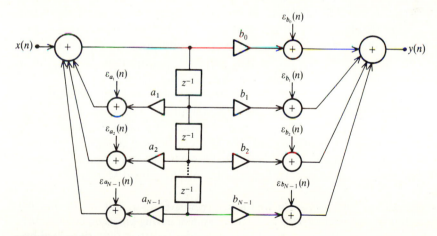

FIGURE 10-10
Multiplication truncation effect is modeled as an additive noise source at the output of each multiplier for the general Direct Form II filter structure.

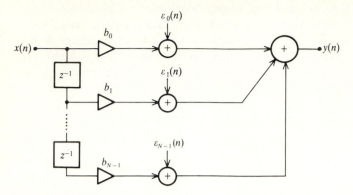

FIGURE 10-11
Multiplication truncation effect is modeled as an additive noise source at the output of each multiplier for the Direct Form II FIR filter structure.

If the noise values for the different multipliers can be considered to be independent of each other, then it can be shown that the noise powers add. For N multipliers, the output noise power is equal to

$$\sigma^2_{\text{out}} = N\frac{E^2}{12} = N2^{-2b_f}/12 \tag{10.16}$$

Since the noise power increases with N, and N is also equal to the duration of the FIR filter unit-sample response $\{h(n)\}$, this is another reason why it is desirable to make the duration of $\{h(n)\}$ as short as possible.

FINITE-PRECISION ARITHMETIC EFFECTS IN IIR FILTERS. The analysis of truncation effects for IIR filters is more complicated than for FIR filters because of the feedback structure. Because of this feedback, the errors associated with finite-precision multiplications and additions circulate back into the filter and affect future values of the output. As described below, these errors may produce unwanted oscillations, called *limit cycles*.

In general, the analysis of these limit cycles is very difficult. To demonstrate their existence, we consider the simple special case of the limit cycles that occurs in response to a unit-sample sequence. Since the input to the filter is zero for values of n other than $n = 0$, these limit cycles are commonly called *zero-input limit cycles*. These cycles exist when the unit-sample response is prevented from decaying to zero because of multiplication truncation.

> **Example 10.13. Limit cycle in first-order recursive filter.** Let us assume the digital filter is implemented with the number system having precision equal to 0.1. Consider the unit-sample response of the first-order recursive filter described by
>
> $$y(n) = -0.9\,y(n-1) + x(n)$$
>
> The unit-sample response of this hardware implementation is shown in Fig. 10.12.

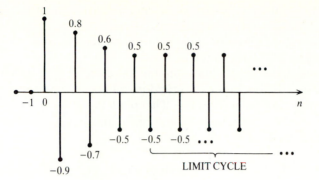

FIGURE 10-12
Example of a zero-input limit cycle in the unit-sample response of a first-order recursive filter implemented with finite-precision arithmetic.

The unit-sample response begins to follow an exponential decay, as expected, until the magnitude of 0.5 is reached. After that point in time, the product $0.9 \times 0.5 = 0.45$ is rounded back to 0.5 at the output of the multiplier. The minus sign in the feedback loop causes the output to alternate between $+0.5$ and -0.5. This is one example of a zero-input limit cycle.

These limit cycles also appear as unwanted noise at the output. But they are different from those previously considered in that, rather than being random (producing a "hiss"), they are correlated (producing a "hum" or "buzz"). Even for such simple cases, the complete analysis of these limit cycles, which determines under what conditions they exist and what their amplitude and frequency characteristics are, is beyond the scope of this book, but is included in the references.

Parallel versus cascade implementation of filters. Parallel combinations are easier to analyze because they can be considered separately. For example, for parallel implementations, all sections can exhibit zero-input limit cycles independently. Cascade implementations are more difficult to analyze because the output of the first section is then the input to the second. Hence only the first section in a cascade connection can exhibit a zero-input limit cycle. The second section then has this limit cycle as an input and, by definition, cannot have a zero-input limit cycle. Since the nature of the limit cycle depends on the first or second-order filter characteristics, and only the first section can exhibit a zero-input limit cycle, the sequence observed at the output of the cascade combination depends on the ordering of the sections in the filter implementation. This effect is considered in the projects.

Since the analysis of these quantization effects is very difficult, we now turn to programming techniques to simulate these effects and provide insights into their behavior.

10.7 PROGRAMS TO SIMULATE QUANTIZATION EFFECTS

We now describe two subroutines that can be added to the filter subroutines to study the behavior of digital filters under finite-precision conditions. The first subroutine performs the quantization of an infinite-precision number to a finite-precision value. The second subroutine scales a discrete-time sequence to a particular range to match an ADC quantizer characteristic and also to prevent overflow errors in the computations.

To generate more interesting input sequences than the unit-sample sequence, we also describe the operation of a random number generator that is usually available on most computer systems. The random input sequence is used along with the unit-sample sequence and the sinusoidal sequence to observe the operation of the filters in the projects.

QUANTIZER SUBROUTINE. To simulate finite-precision systems, it is useful to have a subroutine that transforms an infinite-precision value into the closest finite-precision value represented in the number system. This subroutine can be applied to values of the discrete-time sequence to simulate a quantizer. It can also be applied to the results of multiplication and addition to simulate rounding and overflow effects.

Let X be an infinite-precision number that is represented as a floating-point number on the computer. The value of X is to be converted into a second floating-point number XQ, that is equal to the finite-precision value, closest to X, that can be represented by a 2's-complement fixed-point number having NBIT bits. Of these NBIT bits, NI are used to represent the integer part, including the sign, and NF ($=$ NBIT $-$ NI) bits are used to represent the fractional part. To perform this operation we want to write the subroutine

```
QUANT(X,NI,NF,XQ)
```

where X (real input), NI and NF (integer inputs) and XQ (real output) are defined above. The precision is equal to 2^{-NF}, and the range of integer values is then from -2^{NI-1} to $2^{NI-1} - 2^{-NF}$.

The simplest and fastest way of implementing this conversion is to use the truncation function IFIX in FORTRAN. If Y is a floating-point number, then $N = \text{IFIX}(Y)$ is the integer part. For example, if $Y = 2.8$, then $N = \text{IFIX}(2.8) = 2$. If $Y > 0$, by adding 0.5 to Y, we get a rounding to the nearest integer, or $N = \text{IFIX}(Y + 0.5)$. For example, $N = \text{IFIX}(2.8 + 0.5) = 3$. If $Y < 0$, we must subtract 0.5. For example, if $Y = -3.9$, then $N = \text{IFIX}(-3.9 - 0.5) = -4$. Using this IFIX function, the quantizer can be implemented by first multiplying the value of X by a constant C, performing $\text{IFIX}(CX + 0.5)$, and then dividing the result by C. This sequence of steps produces the desired value for XQ when $C = 2^{NF}$. The value of NI is used to check whether the number X is within the allowable range. If X exceeds the range, XQ should be set to the maximum or minimum value in the number system, depending on whether X exceeds the positive or negative limit. The following example illustrates the conversion process.

Example 10.14. Illustration of the QUANT conversion process. Consider an 8-bit digital hardware system using 2's-complement notation with NI $= 5$ and NF $= 3$. Quantize the values of X given below.

Since NF $= 3$, the precision is $2^{-3} = 0.125$. Then $C = 2^3 = 8$. Hence, the quantized value can be expressed as

$$XQ = \frac{\text{IFIX}(8X + 0.5)}{8} \qquad \text{for } X \geq 0,$$

and

$$XQ = \frac{\text{IFIX}(8X - 0.5)}{8} \qquad \text{for } X < 0$$

Since NI $= 5$, the range is from -16 (1000 000) to 15.875 (01111 111).

For $X = 3.1415926$, $XQ = [\text{IFIX}(25.1327408 + 0.5)]/8 = 25/8 = 3.125$.

For $X = -2.87$, $XQ = [\text{IFIX}(-22.96 - 0.5)]/8 = -23/8 = -2.875$.

For $X = 37$, $XQ = [\text{IFIX}(296 + 0.5)]/8 = 296/8 = 37$

$$= 15.875 \text{ (maximum limit)}.$$

For $X = -37$, $XQ = [\text{IFIX}(-296 - 0.5)]/8 = -296/8 = -37$

$$= -16 \text{ (minimum limit)}.$$

SCALE SUBROUTINE. Before QUANT can be applied to the elements of an arbitrary discrete-time sequence, the sequence values must be scaled, or multiplied by a factor F, so that the range of the signal values matches the range of the finite-precision number system. As discussed previously, one scaling factor is appropriate for deterministic signals (sinusoids), while a different factor is appropriate for random signals (speech). For deterministic signals, the range of the signal is known. Let the values of the signal be in the range $[-R, R]$. If the maximum magnitude of the quantizer is equal to M, which occurs at the negative limit, the signal is scaled by $F = M/R$. For random signals, the rms value X_{rms}, rather than the maximum value, is the important parameter. One common scale factor is then $F = M/(4X_{\text{rms}})$.

The subroutine to perform this function is given by

    SCALE(X,NX,F)

where

X   (real input/output array) is the discrete-time sequence to be scaled

NX   (integer input) is the size of the array

F   (real input) is the scale factor

The operation that is performed is given by

$$X(N) = F * X(N) \qquad \text{for } 1 \leq N \leq NX$$

We next describe the sequences that can be used as the inputs to filters implemented with finite-precision number systems.

INPUT SEQUENCES. Previously, the unit-sample sequence and sinusoidal sequences have been employed as inputs to the various filters. These can also be used for finite-precision filters. The unit-sample sequence shows the existence of the zero-input limit cycle for recursive filters. The sinusoidal quence having the frequency corresponding to the location of the filter magnitude response maximum value can be used to ensure that the range of the number system is not exceeded. We also consider a third type of input sequence, the *random* sequence, that can be used to observe the filter output behavior of signals that are more representative of actual signals. We first consider the *white* random sequence, which on average has a spectrum that is constant with frequency. We then filter this white sequence with a lowpass filter to generate a *lowpass* correlated random sequence, and with a highpass filter to generate a *highpass* correlated random sequence. A random number generator is described next that can be used to generate the random discrete-time sequence.

In FORTRAN, a random number can be generated by a call to the pseudorandom number generator. On the VAX/VMS system, this is done with the statement

    A=RAN(NRAN1,NRAN2),

where NRAN1 and NRAN2 are *seeds* used by the generator. The values for these seeds must be specified to equal two arbitrary integers at the beginning of the program, before the random number generator is called. Each time this statement is executed, the variable A takes on a new value of a uniformly-distributed random number between 0 and 1. The average value of these random numbers is then 0.5 and the variance is equal to $\frac{1}{12}$. The nth element of the N-point *white random sequence* $\{w(n)\}$ can be set to a zero-mean random number with the same variance with the statement

$$w(n) = \text{RAN(NRAN1,NRAN2)} - 0.5, \quad \text{for } 0 \le n \le M - 1 \qquad (10.17)$$

The meaning of *white* will be discussed in terms of the spectrum of $\{w(n)\}$ below. The sequence containing 64 random values is shown in Fig. 10.13(a). The random number generators in most languages and on most computers are similar to that described above.

The *power spectrum sequence,* denoted by $\{P(k)\}$, of any N-point sequence $\{x(n)\}$ is obtained by computing the N-point DFT sequence $\{X(k)\}$ and then summing the squares of the real and imaginary parts, or

$$P(k) = \frac{1}{N}[X_R^2(k) + X_I^2(k)] \qquad \text{for } 0 \le k \le N - 1 \qquad (10.18)$$

If this is performed M times and the ith result is denoted by $\{P_i(k)\}$; the average value of the sequence, $\{P_{av}(k)\}$, is given by

$$P_{av}(k) = \frac{1}{M}\sum_{i=1}^{M} P_i(k) \qquad \text{for } 0 \le k \le N - 1 \qquad (10.19)$$

The random process $\{x(n)\}$ is said to be *white,* when

$$\lim_{M \to \infty} P_{av}(k) = \alpha \qquad \text{for all } k \qquad (10.20)$$

where α is a constant. That is, the white sequence has equal components at each frequency, on average. The power spectral sequence $\{P(k)\}$ of a single

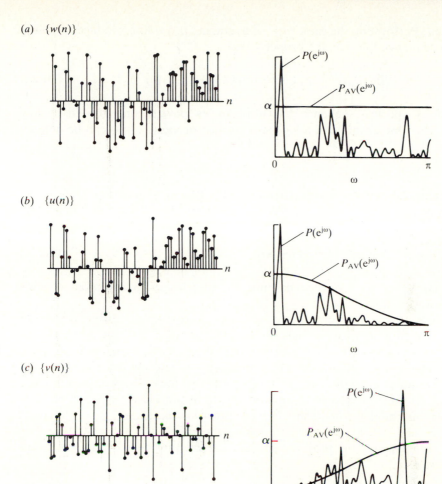

FIGURE 10-13

Examples of random sequences. The discrete-time sequences represent a typical realization of the random process. The frequency curves represent the power spectrum of the realization and the average power spectrum. a) White random process. b) Lowpass random process. c) Highpass random process.

N-point realization of $\{w(n)\}$, however, usually appears quite random itself, as shown in Fig. 10.13(a).

Correlated random sequences are obtained by filtering the white process $\{w(n)\}$. One example of a *lowpass* random process $\{u(n)\}$ can be defined as

$$u(n) = \tfrac{1}{2}[w(n) + w(n+1)] \qquad \text{for } 0 \le n \le N - 1 \qquad (10.21)$$

An example of this process is shown in Fig. 10.13(b). To determine an N-point sequence for $\{u(n)\}$, $N+1$ points of the white sequence $\{w(n)\}$ must be generated. Performing the same type of averaging procedure produces a power spectral sequence that has the property that

$$\lim_{M\to\infty} P_{av}(k) = \alpha(1 + \cos(2\pi k/N))/2 \qquad \text{for all } k \qquad (10.22)$$

This power spectral sequence has most of its power concentrated at low frequencies. One example of a *highpass* random process $\{v(n)\}$ can be defined as

$$v(n) = \tfrac{1}{2}[w(n) - w(n+1)] \qquad \text{for } 0 \le n \le N-1 \qquad (10.23)$$

An example of this process is shown in Fig. 10.13(c). Performing the same type of averaging procedure produces a power spectral sequence that has the property that

$$\lim_{M\to\infty} P_{av}(k) = \alpha(1 - \cos(2\pi k/N))/2 \qquad \text{for all } k \qquad (10.24)$$

This power spectral sequence has most of its power concentrated at high frequencies.

These white and correlated sequences represent interesting classes of signals that approximate signals that occur in practice. To observe the finite-precision effects, these sequences are first applied to a digital filter that is implemented in an infinite-precision number system, approximated by the REAL variable in a 16-bit or 32-bit computer system. The output sequence produced in this case can be compared with that produced by the same filter structure but implemented with a finite-precision hardware. The effects of finite-precision can be observed by taking the difference between these two output sequences. The projects at the end of this chapter illustrate this procedure.

10.8 SUMMARY

Quantization effects are caused by the errors that are due to representing numbers with a finite precision. Two finite-precision number systems have been described, the fixed-point and floating-point notation. Three effects due to finite-precision were considered. The first occurs when the continuous-valued signal amplitude is represented with a finite number of bits in the analog-to-digital conversion operation. This effect was shown to be similar to adding noise to the signal. The second effect occurs when the filter coefficient values are expressed in finite precision. This effect was shown to limit the locations of the singularities in the z plane to lie on a grid. The corresponding deviations of the magnitude response were illustrated for poles and zeros. The third effect is the truncation that results when the sum or product of two finite-precision numbers is itself represented as a finite-precision number. For FIR filters implemented in the direct-form nonrecursive structure, this effect was shown to introduce additive noise to the output sequence. For IIR filters, however, these errors circulate in the memory of the filter and can produce undesirable oscillations, called limit cycles. The chapter ends by describing programming techniques to simulate these finite-precision effects, which are very difficult to determine analytically.

These finite-precision effects are most important in three practical cases:

1. when specifying the number of bits in an ADC;

2. when implementing a filter in special purpose hardware to minimize the number of bits in the finite-precision number system;

3. when implementing filters whose poles lie near the unit circle.

REFERENCES TO TOPICS

Zero-input limit cycles

Blackman, R. B.: *Linear Data-smoothing and Prediction in Theory and Practice,* Addison-Wesley, Reading, MA, 1965.

Jackson, L. B.: "Roundoff-noise analysis for fixed-point digital filters realized in cascade or parallel form," *IEEE Trans. Audio Electroacoust.,* **AU-18:** 107–122, (1970).

Jackson, L. B.: "Roundoff noise analysis in cascade realizations of finite impulse response digital filters," *Bell Sys. Tech. J.,* **52:** 329–345 (1973).

Filter optimization procedures

Avenhaus, E.: "On the design of digital filters with coefficients of limited word length," *IEEE Trans. Audio Electroacoust.,* **AU-20:** 206–212 (1972).

Oppenheim, Alan V., and Ronald W. Schafer: *Digital Signal Processing,* Prentice-Hall, Englewood Cliffs, NJ, 1975.

Rabiner, Lawrence R., and Bernard Gold: *Theory and Application of Digital Signal Processing,* Prentice-Hall, Englewood Cliffs, NJ, 1975.

Taylor, F. J.: *Digital Filter Design Handbook,* Marcel Dekker, New York, 1983. Advanced topics also covering ladder and wave filters. Includes hardware and software considerations.

Analog-to-digital conversion

Jayant, N. S. and Peter Noll: *Digital Coding of Waveforms,* Prentice-Hall, Englewood Cliffs, NJ, 1984.

Mano, M. Morris: *Digital Logic and Computer Design,* Prentice-Hall, Englewood Cliffs, NJ, 1979.

PROBLEMS

10.1. **Range and precision using 2's-complement fixed-point notation.** Consider a computer with an 8-bit word. Determine the range of values and the increment between numbers that can be expressed when the binary point is placed at the following locations:
  (a) after the right-most bit
  (b) after the sign bit
  (c) in the middle of the word

10.2. Repeat Problem 10.1 for a 16-bit word and a 32-bit word.

10.3. **Range and precision using floating-point notation.** Determine the range of values and the increment between numbers that can be expressed for a b-bit word containing b_m bits for the mantissa and b_e bits for the exponent for the following values of b, b_m and b_e:

(a) $b = 8$, $b_m = 6$ and $b_e = 2$
(b) $b = 8$, $b_m = 2$ and $b_e = 6$
(c) $b = 16$, $b_m = 12$ and $b_e = 4$
(d) $b = 32$, $b_m = 24$ and $b_e = 8$

10.4. Convert the following decimal numbers to binary representation using 16-bit notation in which the fraction is represented by 8 bits. Indicate the results obtained both by rounding and truncation.

(a) 117
(b) 0.0726
(c) 43.34
(d) −18.68

10.5. Perform the following additions in binary representation using an 8-bit notation in which the fraction is represented by 4 bits.

(a) $2.5_{10} + 3.25_{10}$
(b) $2.5_{10} + -3.25_{10}$

10.6. Perform the following multiplications in binary representation using an 8-bit notation in which the fraction is represented by 3 bits. Indicate the results obtained both by rounding and truncation.

(a) $2.5_{10} \times 3.25_{10}$
(b) $2.5_{10} \times -3.25_{10}$

10.7. Perform the following calculations with binary numbers using 2's-complement fixed-point with 8-bit words in which three bits are assigned to the fraction. Perform the truncations by rounding.

(a) $1.5 - 3$
(b) $2 \times \pi$.
(c) 0.33×0.8
(d) -0.9×5.6

10.8. Perform the calculations given in Problem 10.7 by using floating-point notation with 8-bit words in which 4 bits are assigned to the exponent (including the sign bit of the exponent).

10.9. Express π as an 8-bit word using 2's complement notation, when the range of the number system is $[-8, 7.9375]$.

10.10. Derive the SNR value similar to Eq. (10.8) for a sinusoidal signal whose amplitude is equal to the maximum level of the quantizer.

10.11. Consider the all-pass system function given by

$$H(z) = \frac{1 - \rho^{-1}z^{-1}}{1 - \rho z^{-1}} \qquad \text{for } 0 < \rho < 1$$

If the filter coefficients are quantized to 8-bits in 2's-complement notation, does the resulting filter still represent an all-pass system?

10.12. Consider the digital filter defined by the difference equation

$$y(n) = 0.9y(n-1) + x(n)$$

Determine the range to which the input sequence must be scaled such that overflow is avoided for any input that is to operate in the number range $[-M, M]$.

COMPUTER PROJECTS

10.1. Spectral whiteness of random number generator

Object. Test that the random number generator on your computer produces a white process.

Computer results. Generate a 128-point sequence of the white random sequence $\{w(n)\}$.

To show the random shape of one realization of the power spectral sequence, compute and plot 65 points of the power spectral sequence corresponding to $0 \le \omega \le \pi$.

To show the limiting behavior of the averaging process, generate 10 independent $\{w(n)\}$ sequences, compute the power spectral sequence of each, compute and plot the average. To save on memory, compute a running average by accumulating the individual spectral sequences and then dividing by the number. Repeat with 100 independent sequences.

Comment on the "whiteness" of the spectral averages.

10.2. Spectral content of correlated random processes

Object. Test that the correlated random processes have lowpass and highpass spectral shapes.

Computer results

(1) Generate a 128-point sequence of the lowpass random sequence $\{u(n)\}$ defined by Eq. (10.21). To show the random shape of one realization of the power spectral sequence, compute and plot 65 points of the power spectral sequence corresponding to $0 \le \omega \le \pi$. To show the limiting behavior of the averaging process, generate 10 independent $\{u(n)\}$ sequences, compute the power spectral sequence of each, compute and plot the average. To save on memory, compute a running average by accumulating the individual spectral sequences and then dividing by the number. Repeat with 100 independent sequences. Comment on the agreement of the computer results with Eq. (10.22) as the number of sequences increases.

(2) Repeat Part (1) for the highpass random process $\{v(n)\}$.

10.3. ADC conversion errors

Object. Compute the quantization errors produced by an analog-to-digital converter and to compare the noise power to the predicted value.

Consider a 5-bit quantizer characteristic that has the range $[-8, 7.5]$. Implement the subroutines QUANT, to perform the quantization, and SCALE, to match the input sequence to the quantizer. Two 128-point input sequences are to be generated: $\{\sin(\pi n/32)\}$ and $\{w(n)\}$ (the white random sequence).

Analytic results
(1) Determine the scale factors to be applied to the two input sequences to match each to the quantizer.
(2) Compute the variance of the additive noise that is predicted if the quantization errors are uniformly distributed and independent.

Computer results. For each sequence:
(a) Compute and plot the original infinite-precision sequence. Use PLOT.
(b) Scale the sequence to match the quantizer. Denote this input sequence by $\{x(n)\}$. Compute and plot the output sequence of the quantizer $\{x_q(n)\}$ when $\{x(n)\}$ is applied.
(c) Compute and plot the quantization error sequence $\varepsilon(n)$ by

$$\varepsilon(n) = x_q(n) - x(n) \qquad \text{for } 0 \le n \le 127$$

(d) Estimate the power of the error sequence, denoted by $\hat{p}$, given by

$$\hat{p} = \frac{1}{128} \sum_{n=0}^{127} \varepsilon^2(n)$$

and compare with the predicted value.

10.4. Deviation of magnitude response due to coefficient quantization

Object. Observe qualitatively the degradation of the magnitude response due to finite-precision effects on the filter coefficients.

For this project use an IIR and/or FIR filter that you have designed in a previous project and for which you have the coefficient values.

Computer results
(1) For comparison, plot the magnitude response of the filter with the coefficients expressed in infinite-precision (approximated by the FLOATING POINT variable in the computer).
(2) Apply the coefficients to the QUANT subroutine that simulates a 16-bit word with the integer part being represented by 4 bits. Compute and plot the magnitude response. Repeat while reducing the word size by one bit for each iteration, until the magnitude response is grossly distorted or, for IIR filters, an instability is encountered (overflow error occurs).
(3) Determine the minimum number of bits needed to satisfy the magnitude specification.

10.5. Zero-input limit cycle for an IIR filter

Object. Observe the limit cycle behavior of an IIR filter.

For this project use one of the IIR filters that you have designed in a previous project. Determine the coefficients for both the direct form and cascade of second-order sections. Multiplication truncation is accomplished by using the QUANT subroutine after any multiplication of two terms is performed. This can be done most easily by implementing a subroutine SUMPQ, which is identical to SUMP in Chapter 2 except for including the QUANT subroutine.

Analytic results. Determine the proper range of the quantizer to prevent overflow errors from occurring when a unity-amplitude sinusoidal sequence is applied to the filter. Generating the true unit-sample response may be helpful.

Computer results

(1) *Limit-cycle of direct form implementation.*

    (a) Starting with 8-bits in the quantizer, matched to the range determined above, plot the unit-sample response that is produced when all the products are truncated. Demonstrate the existence of a zero-input limit cycle.

    (b) Reduce the quantizer word size by one bit, while maintaining the same range, and repeat Part (a).

    (c) Repeat part (b) until the quantizer has four bits.

(2) *Limit-cycle of cascade of second-order section implementation. Repeat Part* (1) *under the following conditions*:

    (a) by using only the first section of the CSOS structure.

    (b) by using the entire CSOS structure.

    (c) by using a different section of the CSOS structure.

    (d) by using the entire CSOS structure with this second section in the first position of the cascade structure.

Since only the first section can have a zero-input limit cycle, the two total filter limit cycles observed in Parts (*b*) and (*d*) should be different.

CHAPTER
11

INVERSE
FILTERING

11.1 INTRODUCTION

In this chapter, we discuss the interesting and important topic in signal processing called inverse filtering. This topic integrates many of the digital signal processing concepts presented in the previous chapters to solve the problem of removing distortions from observed signals that are introduced by measuring instruments or by physical transmission media. This inverse filter is also commonly called a *deconvolution* filter in the fields of geophysics and bioengineering. Inverse filters work best when applied to a system that can be modeled by a system function $H_m(z)$ that has the minimum-phase property. This property is defined in terms of the pole/zero pattern of $H_m(z)$, the phase response $\text{Arg}[H_m(e^{j\omega})]$ and the unit-sample response $\{h_m(n)\}$. Two procedures for designing the inverse filter are presented. The first performs the design directly in the z plane by placing the poles of the inverse filter at the locations of the model zeros and the filter zeros at the location of the model poles. This procedure is limited to models for which either the system function or the pole/zero pattern is known. In the second design procedure, an FIR filter approximation to the ideal inverse filter is obtained. Although not exact, these approximations have useful applications in practice. Modifications to this design procedure are discussed to include additive noise in the observed signal.

422

A common alternate procedure for designing inverse filters *in the time domain* requires concepts that are not difficult, but different from the topics covered in this text. These time-domain design procedures are not discussed here, but can be found in the references.

11.2 APPLICATIONS OF INVERSE FILTERS

In many signal processing applications, it is common to encounter three sources of distortion in the observed signals. The first involves the limitations of the signal detector. If the detector magnitude response is not flat and the phase response is not linear with frequency, then the detector acts like a filter that distorts the signal being sensed. Most transducers, such as microphones and antennas, exhibit this effect. The second source of distortion is due to the physical medium through which the signal propagates. For example, acoustic signals can pass through air (speech signals), water (SONAR signals), biological tissues (diagnostic ultrasound signals) or through the earth (geoseismic signals). Each medium has its own effect on the acoustic signals. The effect of the medium on the signal can be modeled as a filtering operation. Inverse filters serve to compensate for the filtering actions produced by the detector and/or the medium. The third source of distortion is additive noise. Since its effect on inverse filters is complicated, its description will be delayed to the end of the chapter. We now illustrate two representative cases of the first two sources of distortion.

RADAR PROBLEM. Consider the radar problem shown in Fig. 11.1. The range to an airplane is determined by transmitting an electromagnetic pulse and detecting the arrival time of the echo. If t_0 is the elapsed time from pulse transmission, the range is equal to $ct_0/2$, where c is the pulse propagation speed.

For individual airplanes, this procedure works quite well. The radar

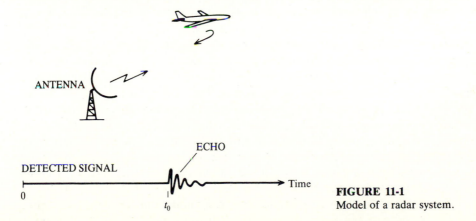

FIGURE 11-1
Model of a radar system.

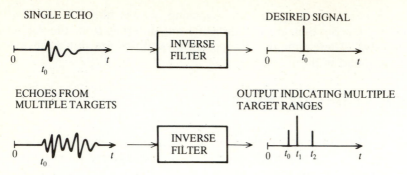

FIGURE 11-2
Desired operation of an inverse filter for the radar problem.

system is limited, however, by the finite bandwidth of the antenna system, which prevents the transmitted pulse from being impulsive. As a consequence, when two airplanes are in close proximity, the two echoes may not appear as separate waveforms, but merge into one as shown in Fig. 11.2. In this case, an inverse filter is applied to the detected signals in an attempt to perform *pulse compression,* or to compress the pulse waveform produced by the system into one having a shorter duration, to improve the resolution. Since the more complicated signal from multiple aircraft can be considered to be the superposition of many pulse waveforms, the inverse filter is applied to the observed waveform to transform it into multiple shorter-duration waveforms, thus resolving the individual airplanes. This example can easily be extended to typify many problems in bioengineering and geophysics in which an overlap in identical pulse waveforms occurs.

REVERBERANT TRANSMISSION CHANNEL. To illustrate the distortion produced by a physical medium, let us consider the data transmission channel between two modems shown in Fig. 11.3. When the receiving modem is terminated with an impedance value that does not match the characteristic impedance of the channel, a part of the received signal is reflected back into the channel, causing an echo to propagate back from the receiver to the transmitter. If an improper termination exists at the transmitter end of the transmission line, a reflection occurs there as well. If an impulsive signal is transmitted, the signal detected at the receiving end includes the transmitted signal and the set of echoes, as shown in Fig. 11.3. Since only a partial reflection occurs at each end, the amplitudes of the echoes decrease with time. When a nonimpulsive signal is transmitted, the received signal is simply the convolution of the transmitted signal and the impulsive received signal. In conventional communication systems, the entire received signal produced by one transmission must decay before a second signal is transmitted. This limits the data transmission rate across the channel.

To compensate for these echoes, we can model the channel as a filter

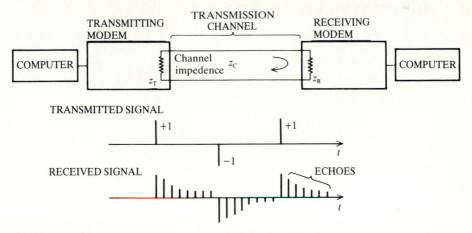

FIGURE 11-3
Model of a reverberant data transmission channel that exhibits echoes.

whose unit-sample response is identical to the detected signal when a single impulse is transmitted. In this case, the channel can be modeled as a first-order recursive filter, shown in Fig. 11.4. An inverse filter, called an *echo suppressor,* is applied in an attempt to compensate for the behavior of the channel. The input to the inverse filter is the entire set of detected echoes and the desired output is the single pulse that was transmitted, as shown in Fig. 11.5.

We will find below that an exact inverse filter can be implemented only for systems that have the *minimum-phase* property. In the next section, we define this property.

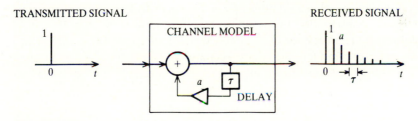

FIGURE 11-4
Digital filter model for the reverberant channel.

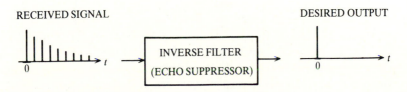

FIGURE 11-5
Desired operation of an inverse filter for the reverberant channel.

11.3 MINIMUM-PHASE SYSTEMS

In this section, we model the *causal* and *stable* detector or physical medium in terms of the z plane pole/zero pattern of its system function $H_m(z)$, its phase response $\text{Arg}[H_m(e^{j\omega})]$ and its unit-sample response $\{h_m(n)\}$. The subscript m denotes *model*. We can then define the three corresponding criteria by which we can recognize a minimum-phase model.

THE z PLANE CRITERION. In Chapter 5, the system function for a *stable* and *causal* discrete-time filter was shown to have its poles within the unit circle and its zeros anywhere in the z plane. We now define three types of model system functions $H_m(z)$, depending on the location of the zeros. The three cases are shown in Fig. 11.6. If all the zeros of $H_m(z)$ lie *inside* the unit circle, the system model is called *minimum-phase*, if all the zeros lie *outside* the unit circle, the model is called *maximum-phase*, and if zeros lie both inside and outside the unit circle, the model is called *mixed-phase*. As shown below, only for minimum-phase models is the inverse filter both stable and causal.

THE PHASE RESPONSE CRITERION. In the discussion below, we consider $\text{Arg}[H(e^{j\omega})]$ to be the *entire* phase response of the model, rather than the *principal value*. If we are given the principal value, we can construct the entire phase function by adding or subtracting 2π radians to the principal value at the points of discontinuity, as shown in Fig. 11.7. This procedure is called *phase unwrapping*.

In Chapter 5, the phase response of a filter was determined in the z plane by adding the contributions from its individual singularities. As ω went from 0 to π, a pole decreases the phase response by π radians, while a zero inside the unit circle increases the phase response by π radians. A zero outside the unit circle made no *net* contribution to the phase response as ω went from 0 to π. If all the singularities are inside the unit circle, which is the case for a causal and stable minimum-phase model, the phase response $\text{Arg}[H_m(e^{j\omega})]$ starts at zero at $\omega = 0$, is a continuous-function of ω, and returns to zero at $\omega = \pi$. Zeros on

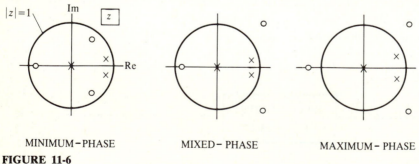

MINIMUM – PHASE   MIXED – PHASE   MAXIMUM – PHASE

FIGURE 11-6
Pole/zero pattern for minimum-phase, mixed-phase and maximum-phase systems.

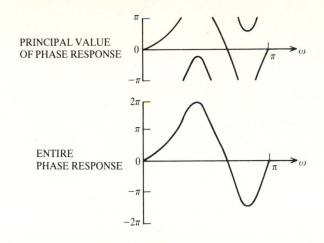

PRINCIPAL VALUE OF PHASE RESPONSE

ENTIRE PHASE RESPONSE

FIGURE 11-7
Comparison of principal value of the phase response, which is produced by the DFT, and the entire phase response.

the unit circle cause discontinuous jumps of π radians in the phase response. If M_z zeros lie outside the unit circle, the phase response is equal to $-M_z\pi$ at $\omega = \pi$.

One other point must be considered before the phase response criterion of a minimum-phase model is given. In the description of all-pass filters in Chapter 8, it was shown that when a zero of the system function, located at $z = z_0$, is moved to $z = 1/z_0$, the shape of the magnitude response of the system is unaltered, but the phase response changes. Hence, a given magnitude response can have many valid phase response functions.

From the behavior of the magnitude and phase responses described above, we can form the following definition of a minimum-phase system:

For a *given magnitude response,* the minimum-phase system is the causal system that has the smallest magnitude phase at every frequency ω.

That is, in the set of causal and stable filters having the same magnitude response, the minimum-phase response exhibits the smallest deviation from zero phase. Note that the zero-phase filters described previously in the text cannot be minimum-phase because they are noncausal and their system functions have poles outside the unit circle. By adding sufficient delay to the unit-sample response, these filters become causal and have a linear phase response. The minimum-phase filter having the same magnitude response will have a phase response that has a smaller magnitude at each frequency.

THE UNIT-SAMPLE RESPONSE CRITERION. In the time-domain, we can find the condition on a *causal* unit-sample response $\{h_m(n)\}$, by which we can recognize a minimum-phase system. Recall that for a causal system, $h_m(n) = 0$, for $n < 0$. Let us define the *energy evolution factor* $E(M)$ to be

$$E(M) = \sum_{n=0}^{M} h_m^2(n) \qquad \text{for } 0 \le M < \infty \qquad (11.1)$$

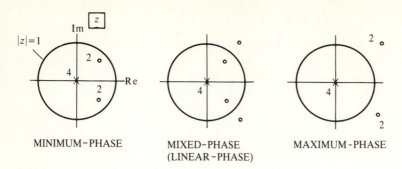

FIGURE 11-8
Pole/zero pattern for minimum-phase, mixed-phase (linear-phase) and maximum-phase systems in Example 11.1.

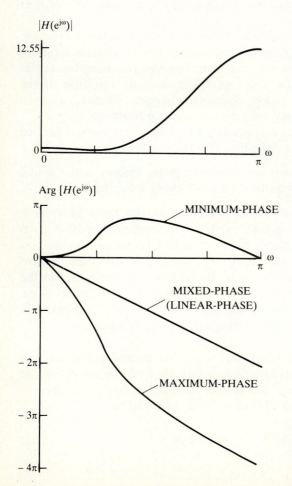

FIGURE 11-9
Magnitude and phase responses for minimum-phase, mixed-phase (linear-phase) and maximum-phase systems in Example 11.1.

For a particular value of M, $E(M)$ measures the accumulated energy in the filter unit-sample response sequence from $n = 0$ to $n = M$. For minimum-phase filters, $E(M)$ takes on the maximum possible value for each M, in the set of filters having the same magnitude response. That is, for the minimum-phase filter, the energy in the unit-sample response is concentrated toward the beginning of the sequence.

Example 11.1. Minimum-, linear- and maximum-phase systems. Since the minimum-phase property of causal and stable systems is determined by the locations of the zeros, consider the fourth-order all-zero system containing a double complex-conjugate set of zeros located at $z = 0.7e^{\pm j\pi/4}$.

The minimum-, linear- and maximum-phase system pole zero patterns having the identical magnitude response are shown in Fig. 11.8.

The magnitude response and the phase responses of the three systems are shown in Fig. 11.9. The minimum-phase system is seen to have the phase with the smallest deviation from zero at each frequency.

The unit-sample responses for the three systems is shown in Fig. 11.10. For the fourth-order FIR system, the unit-sample response has at most five nonzero values that occur for $0 \le n \le 4$. The energy evolution factors for the three different systems have the same value for $M \ge 4$. For the minimum-phase system, however, $E(M)$ takes on the maximum values for $0 \le M \le 3$.

11.4 FORMULATING THE PROBLEM FOR APPLYING AN INVERSE FILTER

To apply an inverse filter, we must first produce a discrete-time model of the distorting physical system. The inverse filter then serves to compensate this distortion. To illustrate the approach, let us consider the model of a bandlimited detector, denoted by $H_m(z)$ and shown in Fig. 11.11. To observe the limitations of the detector, we apply a unit-sample sequence $\{d(n)\}$ to the input of the model and observe the output sequence $\{y(n)\}$. In the z domain, since $D(z) = 1$, the response function is given by

$$Y(z) = H_m(z) \tag{11.2}$$

To compensate for the bandlimited detector, we apply $\{y(n)\}$ to the input of an inverse filter, having the system function $H_{if}(z)$. The *desired* output of the inverse filter is a replica of the input sequence applied to the detector, or $\{y_d(n)\} = \{d(n)\}$. In the z-domain, we have $Y_d(z) = 1$. Since the desired system response is equal to

$$Y_d(z) = H_{if}(z) H_m(z) \tag{11.3}$$

the system function of the inverse filter is equal to

$$H_{if}(z) = 1/H_m(z) \tag{11.4}$$

Hence, the system function of the inverse filter is the reciprocal, or *inverse*, of the model, from which the name *inverse filter* is derived.

Let the system function of the detector model be given by

$$H_m(z) = \frac{N(z)}{D(z)} \tag{11.5}$$

(*a*) MINIMUM-PHASE

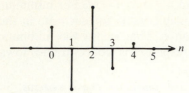

MIXED-PHASE
(LINEAR-PHASE)

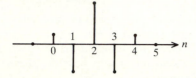

MAXIMUM-PHASE

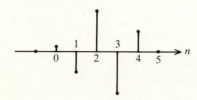

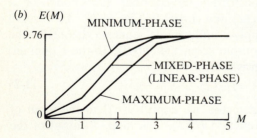

(*b*) $E(M)$

FIGURE 11-10
a) Unit-sample responses for minimum-phase, mixed-phase (linear-phase) and maximum-phase systems in Example 11.1.
b) Comparison of energy evolution factors.

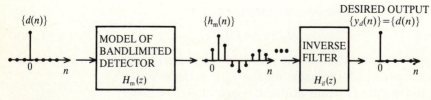

FIGURE 11-11
Model of bandlimited detector and application of inverse filter.

where $N(z)$ and $D(z)$ are polynomials in z^{-1}. The inverse filter system function is then given by

$$H_{if}(z) = \frac{D(z)}{N(z)} \tag{11.6}$$

This result indicates that the poles of the inverse filter system function correspond to the zeros of the model and the zeros of the inverse filter are the poles of the model. For a *stable* and *causal* inverse filter, the poles of the inverse filter system function must lie within the unit circle. *Hence, a causal and stable inverse filter can be implemented only if the physical system is minimum-phase.* The following examples illustrate this point.

Example 11.2. Inverse filter for the radar problem. Let us assume that the radar pulse can be accurately given by

$$h_m(n) = \frac{r^n}{\sin(\omega_0)} \sin[\omega_0(n+1)] \qquad \text{for } n \geq 0$$

Using the results of Example 5.42, the radar pulse sequence is modeled by the unit-sample response of the second-order filter defined by

$$H_m(z) = \frac{1}{1 - 2r \cos(\omega_0)z^{-1} + r^2 z^{-2}}$$

The system function of the inverse filter that compresses $\{h_m(n)\}$ into $\{d(n)\}$ is

$$H_{if}(z) = 1/H_m(z) = 1 - 2r \cos(\omega_0) z^{-1} + r^2 z^{-2}$$

Example 11.3. Inverse filter for the transmission channel problem. Let the transmission channel exhibit only one echo, having amplitude a, where $|a| < 1$, and occurring K sample periods after the transmitted signal is received. Then the unit-sample response of the channel is given by

$$h_m(0) = 1 \qquad h_m(K) = a \qquad \text{and} \qquad h_m(n) = 0, \text{ otherwise}$$

The channel model system function is then equal to

$$H_m(z) = 1 + az^{-K}$$

The system function of the inverse filter is then

$$H_{if}(z) = 1/H_m(z) = 1/(1 + az^{-K})$$

It is left as an exercise for the student to show that the inverse filter converts $\{h_m(n)\}$ into the unit-sample sequence $\{d(n)\}$.

Example 11.4. Inverse filter for nonminimum-phase system. Let us consider a physical system whose model system function is given by

$$H_m(z) = 1 + az^{-1} \qquad \text{where } |a| > 1$$

This system model is not minimum-phase because it has a zero outside the unit circle at $z = -a$. The inverse filter system function is given by

$$H_{if}(z) = 1/H_m(z) = 1/(1 + az^{-1})$$

Since the pole of $H_{if}(z)$ is outside the unit circle, the inverse filter is unstable, but it still converts $\{h_m(n)\}$ into $\{d(n)\}$ (show this). However, if this conversion is not done *exactly*, or if there is *any* noise present, the output sequence of this inverse filter will approach infinity.

The formulation presented above provides a procedure for designing inverse filters in the z plane. However, in practice, it may not be possible to determine the system function representation $H_m(z)$ of the physical system, even if the unit-sample response can be observed. We next describe how to obtain an FIR filter approximation to the inverse filter from the observed unit-sample response of the physical system.

11.5 AN FIR FILTER APPROXIMATION TO THE INVERSE FILTER

We now describe a method for determining an FIR filter approximation to the inverse filter from an N-point observation of the model unit-sample response $h_m(n)$, for $0 \le n \le N-1$, denoted by $\{h_{m,N}(n)\}$. Using $\{h_{m,N}(n)\}$, we can compute an N_d-point DFT, where $N_d \gg N$. We show below why it is desirable to make the duration of the DFT sequence much larger than the number of points in $\{h_{m,N}(n)\}$. From the DFT sequence, we can determine the N_d-point magnitude response sequence $\{|H_{m,N}(k)|\}$ and the phase response sequence $\{\text{Arg}[H_{m,N}(k)]\}$. Since the exact inverse filter is related to the model system function by $H_{if}(z) = 1/H_m(z)$, we define the magnitude and phase response sequences of the FIR approximate inverse filter by

$$|H_{ifa}(k)| = 1/|H_{m,N}(k)| \qquad \text{for } 0 \le k \le N_d - 1 \tag{11.7}$$

and

$$\text{Arg}[H_{ifa}(k)] = -\text{Arg}[H_{m,N}(k)] \qquad \text{for } 0 \le k \le N_d - 1 \tag{11.8}$$

The N_d-point unit-sample response of the FIR approximate inverse filter, denoted by $\{h_{ifa}(n)\}$ is computed by performing the inverse DFT on the real and imaginary sequences obtained from $\{|H_{ifa}(k)|\}$ and $\{\text{Arg}[H_{ifa}(k)]\}$.

This procedure generates an inverse filter that is only an approximation for the following reasons. For systems having finite-duration unit-sample responses, we can acquire the entire unit-sample response and have $\{h_{m,N}(n)\} = \{h_m(n)\}$. In this case, $H_m(z)$ has only zeros and the exact inverse filter is all-pole. Hence, the unit-sample response of the exact inverse filter $\{h_{if}(n)\}$ has an infinite duration. The unit-sample response $\{h_{ifa}(n)\}$ obtained with the above procedure is a finite-duration approximation to $\{h_{if}(n)\}$. As described in Chapter 4, since the DFT sequence $\{H_{ifa}(k)\}$ is the sampled version of the exact inverse filter transfer function $H_{if}(e^{j\omega})$, then $\{h_{ifa}(n)\}$ is the first period of the periodic extension of $\{h_{if}(n)\}$, or

$$h_{ifa}(n) = \sum_{k=-\infty}^{\infty} h_{if}(n + kN_d) \qquad \text{for } 0 \le n \le N_d - 1 \tag{11.9}$$

To minimize the associated time-aliasing effects, the value of N_d, the size of the N_d-point DFT, should be made as large as possible.

In the case that $\{h_m(n)\}$ has an infinite-duration, the observed sequence $\{h_{m,N}(n)\}$ is equal to the first N samples, or

$$h_{m,N}(n) = h_m(n) \qquad \text{for } 0 \le n \le N - 1 \qquad (11.10)$$

This is equivalent to windowing the infinite-duration sequence with an N-point rectangular window. The DFT sequence $\{H_{m,N}(k)\}$ is then equal to the samples of the convolved transfer function, described in Chapter 9, given by

$$H_{m,N}(e^{j\omega}) = \frac{1}{2\pi} \int_{-\pi}^{\pi} H_m(e^{j\theta}) W_R(e^{j(\omega-\theta)}) \, d\theta \qquad (11.11)$$

where $W_R(e^{j\omega})$ is the spectrum of the N-point rectangular window. This convolution causes the transfer function of the approximate inverse filter to deviate from that desired. In general, the agreement between $H_m(e^{j\omega})$ and $H_{m,N}(e^{j\omega})$ improves as N is increased. The required value of N depends on the application.

In practice, several problems could exist with this design procedure. First, the values of the DFT sequence $\{H_{m,N}(k)\}$ computed from the observed N-point unit-sample response may not be accurate for the spectral values that are small. In the inverse filter, these small magnitudes correspond to large gains when the reciprocal is computed. This effect is especially true for bandlimited models. A second problem occurs when only the magnitude response of the system can be observed, rather than $\{h_{m,N}(n)\}$. For example, the characteristics of many physical systems such as transmission channels, or the acoustic properties of biological tissue for diagnostic ultrasound applications, are known only in terms of their magnitude response. Another example occurs in speech processing, in which speech sounds are classified according to their power spectral properties. A third problem occurs when the observed signals are corrupted with noise. In this case, only the average power spectral sequence, which is usually modeled as the magnitude-squared response of the model $|H_m(e^{j\omega})|^2$, can be computed. An approximation to the magnitude response sequence can be found by taking the square-root of the average power spectral sequence.

In the next section, we describe the procedure to compute the phase response sequence from the magnitude response sequence for minimum-phase models. The effects of noise will be considered in the last section.

11.6 DISCRETE HILBERT TRANSFORM RELATIONSHIP FOR MINIMUM-PHASE SYSTEMS

For discrete-time *minimum-phase* systems, the log-magnitude response and phase response can be shown to form a discrete-time Hilbert transform pair.

Here we will only give the definition of this transform as it relates to designing inverse filters. The derivation and other properties of the discrete Hilbert transform are given in the references. It is sufficient to state that the phase response sequence of the minimum-phase FIR filter can be determined from the observed magnitude response sequence by employing the discrete-time Hilbert transform. The observed magnitude response and the computed phase response sequences can then be employed in the procedure previously described in Section 11.5 to implement the inverse filter approximation.

Using the results of the discrete Hilbert transform, the phase response is given by

$$\text{Arg}[H(e^{j\omega})] = \frac{1}{2\pi} P \int_{-\pi}^{\pi} \log_e |H(e^{j\theta})| \cot[(\theta - \omega)/2] \, d\theta \qquad (11.12)$$

where the symbol P indicates the Cauchy principal value of the integral. The Cauchy principal value is needed here because the cotangent function becomes infinite when its argument becomes zero, or when $\theta = \omega$. In this case, the integral must be evaluated through the limiting procedure given by

$$\text{Arg}[H(e^{j\omega})] = \lim_{\varepsilon \to 0} \frac{1}{2\pi} \left[\int_{-\pi}^{\omega - \varepsilon} \log_e |H(e^{j\theta})| \cot[(\theta - \omega)/2] \, d\theta \right. \qquad (11.13)$$

$$\left. + \int_{\omega + \varepsilon}^{\pi} \log_e |H(e^{j\theta})| \cot[(\theta - \omega)/2] \, d\theta \right]$$

The forms of the two functions that are multiplied and then integrated, for a particular value of ω, are shown in Fig. 11.12. Because of the complicated analytic forms, the evaluation of the above integral must usually be performed numerically. The log-magnitude response *sequence* is then used to determine the values of the phase response *sequence*. A simple program to evaluate this integral is given in the Projects.

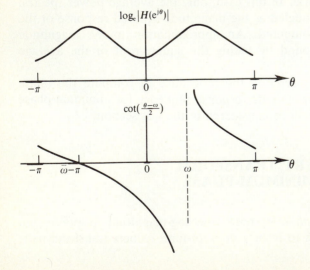

FIGURE 11-12
Relationship of the functions for evaluating the discrete Hilbert transform.

From the integral equation, one restriction is obvious: $|H(e^{j\omega})|$ cannot be equal to zero, since the logarithm would then be infinite and the integral unbounded. The Cauchy principal value argument is not applicable here because a sign change does not occur in the logarithm as its argument passes through zero. Hence, $H(z)$ cannot contain zeros on the unit circle. But for minimum-phase systems, all the singularities are inside the unit circle, and this procedure can be employed to find the minimum-phase response.

Example 11.5. Computing the minimum-phase response by using the discrete Hilbert transform. Consider the magnitude response given by

$$|H_m(e^{j\omega})| = e^{-\beta|\omega|} \qquad \text{for } -\pi \le \omega \le \pi$$

This magnitude response is a common model for the acoustic attenuation of biological soft tissue employed in diagnostic ultrasound tissue characterization described in the references.

The log-magnitude response is the linear-with-frequency function shown in Fig. 11.13 and given by

$$\log_e |H_m(e^{j\omega})| = -\beta |\omega| \qquad \text{for } 0 \le \omega \le \pi$$

(a) $\log_e |H_m(e^{j\omega})|$

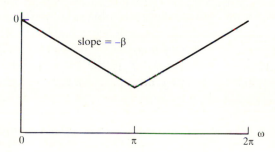

(b) $\text{Arg}\,[H_m(e^{j\omega})]$

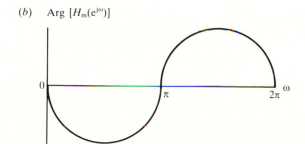

(c)

$\{h_m(n)\}$

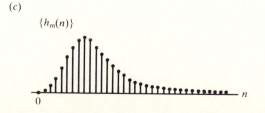

FIGURE 11-13
Results of Example 11.5. *a)* Log-magnitude response employed in Hilbert transform. *b)* Phase response produced by Hilbert transform. *c)* The unit-sample response determined by inverse DFT.

This function was sampled at 1024 points over the range $0 \le \omega \le 2\pi$ to obtain the log-magnitude response sequence. The integral to determine the phase response was performed numerically using the DHT subroutine listed in the projects. The resulting phase response $\{\text{Arg}[H_{\text{m}}(e^{j\omega})]\}$ of the minimum-phase model is shown in the figure. The phase response has the minimum-phase characteristics of being continuous-valued and returning to zero at $\omega = \pi$. The unit-sample response $\{h_{\text{m}}(n)\}$ shown in the figure was obtained by computing the inverse DFT of the real and imaginary sequences computed from these magnitude and phase response sequences.

11.7 DESIGNING INVERSE FILTERS WITH THE DISCRETE HILBERT TRANSFORM

With the aid of the discrete Hilbert transform, we can compute the minimum phase response from the magnitude response. For designing inverse filters, the magnitude response employed in Hilbert transform $|H_{\text{inv}}(e^{j\omega})|$ is equal to the inverse of the model magnitude response, $|H_{\text{inv}}(e^{j\omega})| = |H_{\text{m}}(e^{j\omega})|^{-1}$. If $\log_e |H_{\text{inv}}(e^{j\omega})|$ is employed in Eq. (11.12), then $\text{Arg}[H_{\text{inv}}(e^{j\omega})]$, the phase response of the minimum-phase inverse filter, is obtained. Performing the integration numerically produces the N-point sequence $\{\text{Arg}[H_{\text{inv}}(k)]\}$. From this sequence and $\{|H_{\text{inv}}(k)|\}$, the real and imaginary DFT sequences can be determined. The real part of the inverse DFT is the N-point unit-sample response of the inverse filter. The imaginary part should be zero. This procedure is illustrated in the following examples.

Example 11.6. Inverse filter design with the discrete Hilbert transform for all-pole system model. Let us consider the second-order all-pole system model given by

$$H_{\text{m}}(z) = \frac{1}{1 - 2r \cos(\theta)z^{-1} + r^{2z-2}}$$

with $r = 0.9$ and $\theta = \pi/8$. The magnitude response is then given by

$$|H_{\text{m}}(e^{j\omega})| = [1 + a^2 + b^2 + a(2 + b) \cos(\omega) + 2b \cos(2\omega)]^{-1/2}$$

where $a = -2r \cos(\theta) = -1.66$ and $b = r^2 = 0.81$. The inverse magnitude response function $|H_{\text{inv}}(e^{j\omega})| = |H_{\text{m}}(e^{j\omega})|^{-1}$. This inverse magnitude response is shown in Fig. 11.14(a) and $\log_e |H_{\text{inv}}(e^{j\omega})|$ is shown in Fig. 11.14(b). Applying the discrete Hilbert transform to this latter function produces the phase response shown in Fig. 11.14(c). Using this phase response and $|H_{\text{inv}}(k)|$, we find the real and imaginary DFT sequences. Performing an inverse DFT produces the unit-sample response $\{h_{\text{inv}}(n)\}$ shown in Fig. 11.14(d). The sequence values that are significantly different than zero are $h_{\text{inv}}(0) = 1$, $h_{\text{inv}}(1) = a = -1.66$ and $h_{\text{inv}}(2) = b = 0.81$, or those for the exact inverse filter. The other values in $\{h_{\text{inv}}(n)\}$ should be zero, but are not exactly zero because of errors in the finite-precision arithmetic.

(a) $|H_{inv}(e^{j\omega})|$

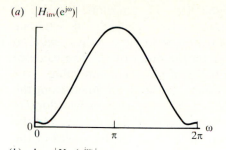

(b) $\log_e |H_{inv}(e^{j\omega})|$

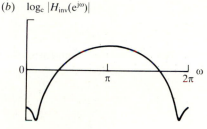

(c) $\mathrm{Arg}\,[H_{inv}(e^{j\omega})]$

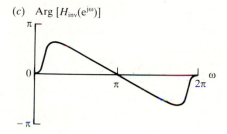

(d) $\{h_{inv}(n)\}$

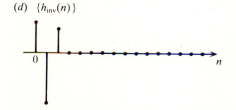

FIGURE 11-14
Results of Example 11.6. a) Magnitude response of inverse filter. b) Log-magnitude response employed in Hilbert transform. c) Phase response produced by Hilbert transform. d) The unit-sample response determined by inverse DFT. 16-point sequences were used.

The previous example illustrates than when the system model contains only poles, the inverse filter contains only zeros, and the inverse filter unit-sample response has a finite duration. The size of the sequences used in the discrete-Hilbert transform and the inverse DFT must then be equal to or larger than the duration of the unit-sample response. If, however, the system model contains a zero, then the inverse filter will contain the corresponding pole. The unit-sample response of the exact inverse filter will then have an infinite duration. To obtain a valid approximation, the size of the sequences used in the Hilbert transform and inverse DFT computations should be large enough that time-aliasing effects are not significant. To determine the size of these sequences that will make the time-aliasing effect negligible, we examine the pole in the inverse filter system function having the largest radius, equal to r_m. The unit-sample response will then decay on the order of r_m^n. The value of N should be chosen such that $r_m^N < 0.001$ or 0.1% of the maximum value.

The approximate value of the duration N can also be determined from the bandwidth of the peak in the inverse filter magnitude response, as was done in Chapter 7. As an alternate procedure, several values of N can be tried and the unit-sample response sequence can be examined for each. If the values of the unit-sample response do not change significantly as N is increased, then time-aliasing effects may be neglected.

> **Example 11.7. Inverse filter design with the discrete Hilbert transform for all-zero system model.** Let us consider the second-order all-zero system model given by
>
> $$H_m(z) = 1 - 2r\cos(\theta)z^{-1} + r^2 z^{-2}$$
>
> with $r = 0.9$ and $\theta = \pi/8$. The magnitude response is then given by
>
> $$|H_m(e^{j\omega})| = [1 + a^2 + b^2 + a(2 + b)\cos(\omega) + 2b\cos(2\omega)]^{1/2}$$
>
> where $a = -2r\cos(\theta) = -1.66$ and $b = r^2 = 0.81$. The inverse magnitude response function $|H_{inv}(e^{j\omega})| = |H_m(e^{j\omega})|^{-1}$. This inverse magnitude response is shown in Fig. 11.15(a) and $\log_e |H_{inv}(e^{j\omega})|$ is shown in Fig. 11.15(b). Applying the discrete Hilbert transform to this latter function produces the phase response shown in Fig. 11.15(c). Using this phase response and $|H_{inv}(k)|$, we find the real and imaginary DFT sequences.
>
> To determine the size of these sequences, we determine from the bandwidth of $|H_{inv}(e^{j\omega})|$ that $r_m \approx 0.9$. The value of N is then equal to
>
> $$N = \log_e[0.001]/\log_e[r_m] = 66$$
>
> Performing a 64-point inverse DFT produces $\{h_{inv}(n)\}$. The first 32 samples of $\{h_{inv}(n)\}$ are shown in Fig. 11.15(d).

INVERSE FILTERS FOR NONMINIMUM-PHASE SYSTEM MODELS. If the system model is not minimum-phase, because a zero lies outside the unit circle, the resulting filter is not a pure inverse-filter in the sense that a unit-sample sequence is produced at the output when the unit-sample response of the system model is applied. The output, however, does have a magnitude spectrum that is constant with frequency. In Chapter 5, it was shown that a zero outside the unit circle at radius $|z| = r_z > 1$, generates the same form of magnitude response, to within a constant, as a zero at radius $|z| = 1/r_z$. Since the discrete Hilbert transform employs the magnitude response to determine the minimum-phase response, the resulting inverse filter is minimum-phase and its corresponding pole appears at $|z| = 1/r_z$.

In Chapter 8, we observed that all-pass filters have the property of a flat magnitude response and a unit-sample response that is not a unit-sample

(a) $|H_{inv}(e^{j\omega})|$

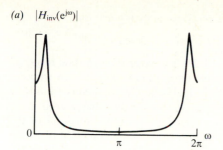

(b) $\log_e |H_{inv}(e^{j\omega})|$

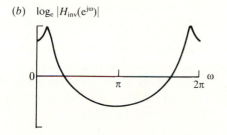

(c) Arg $[H_{inv}(e^{j\omega})]$

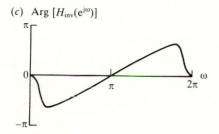

(d) $\{h_{inv}(n)\}$

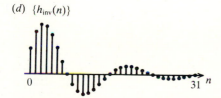

FIGURE 11-15
Results of Example 11.7. *a*) Magnitude response of inverse filter. *b*) Log-magnitude response employed in Hilbert transform. *c*) Phase response produced by Hilbert transform. *d*) The unit-sample response determined by inverse DFT. 64-point sequences were used.

sequence. The output of the inverse filter determined with the above procedure, when the unit-sample response of the nonminimum-phase system model is applied at the input, will be equal to the unit-sample response of an all-pass filter.

Example 11.8. Inverse filter design with the discrete Hilbert transform for a non-minimum-phase system model. Let us consider the second-order all-zero system model given by

$$H_m(z) = 1 - 2r \cos(\theta)z^{-1} + r^2 z^{-2}$$

with $r = 1.11$ and $\theta = \pi/8$. The magnitude response is then given by

$$|H_m(e^{j\omega})| = [1 + a^2 + b^2 + a(2 + b)\cos(\omega) + 2b\cos(2\omega)]^{1/2}$$

where $a = -2r\cos(\theta) = -2.05$ and $b = r^2 = 1.23$. The inverse magnitude response function $|H_{inv}(e^{j\omega})| = |H_m(e^{j\omega})|^{-1}$. This inverse magnitude response is to within a gain constant of that shown previously in Fig. 11.15(a). This gain constant adds a constant value to $\log_e |H_{inv}(e^{j\omega})|$. Since the cotangent function has a zero mean value, this constant additive term does not contribute to the evaluation of the integral. Applying the discrete Hilbert transform to this latter function produces the minimum-phase response sequence. Using this phase response and $|H_{inv}(e^{j\omega})|$, we find the real and imaginary DFT sequences. Performing a 64-point inverse DFT produces the same $\{h_{inv}(n)\}$ shown in Fig. 11.15(d). When the unit-sample response of the non-minimum phase system $\{h_{nm}(n)\}$ is applied to this inverse filter, it produces the $\{y(n)\}$ shown in Fig. 11.16. Note that $\{y(n)\}$ has the form of the unit-sample response of a second-order all-pass filter. As such, its magnitude spectrum is constant with frequency, but its phase spectrum is not identically zero. Hence, the inverse filter has corrected the magnitude response characteristics of the system model, but a phase distortion still exists.

In the next section, we examine the effects of additive noise in the design and performance of inverse filters.

(a) $\{h_{nm}(n)\}$

0   31

(b) $\{h_{inv}(n)\}$

0   31

(c) $\{y(n)\}$

0   31

FIGURE 11-16
Results of Example 11.8. *a*) Unit-sample response of nonminimum-phase system. *b*) The unit-sample response of inverse filter. *c*) Output of inverse filter is identical to the unit-sample response of all-pass filter.

11.8 THE EFFECTS OF NOISE ON INVERSE FILTERS

In addition to the requirement that the physical system be minimum-phase, the limitations of inverse filters for system compensation and pulse compression are usually encountered in the presence of noise. To illustrate these limitations, let us consider the noise to be *white*, implying that the power spectral sequence, defined in Chapter 10, is constant with frequency, on average. For example, consider the observed sequence $\{x(n)\}$ to be the sum of the signal sequence $\{s(n)\}$ and a white noise sequence $\{w(n)\}$, given by

$$\{x(n)\} = \{s(n)\} + \{w(n)\} \tag{11.14}$$

This case is shown in Fig. 11.17.

For random signals, we must compute *average power spectra*. Typically, an N-point data window, like the Hamming window, is applied to the observed random signal to obtain an N-point observed sequence. If $W_i(e^{j\omega})$ is the Fourier transform of the ith N-point observed sequence $\{w_i(n)\}$, and M such independent sequences are observed, then the average power spectrum, $P_W(e^{j\omega})$, is defined as

$$P_W(e^{j\omega}) = \lim_{M\to\infty} \left[\frac{1}{MN} \sum_{i=1}^{M} W_i(e^{j\omega}) \, W_i^*(e^{j\omega}) \right]$$

$$= \lim_{M\to\infty} \left[\frac{1}{MN} \sum_{i=1}^{M} |W_i(e^{j\omega})|^2 \right] \tag{11.15}$$

The division by N ensures that the Fourier transform remains finite for any value of N. Since the noise sequence $\{w(n)\}$ is random, the observed signal $\{x(n)\}$ is also random. A similar expression can be written for the power spectrum of $\{x(n)\}$, denoted $P_X(e^{j\omega})$, given by

$$P_X(e^{j\omega}) = \lim_{M\to\infty} \left[\frac{1}{MN} \sum_{i=1}^{M} X_i(e^{j\omega}) \, X_i^*(e^{j\omega}) \right]$$

$$= \lim_{M\to\infty} \left[\frac{1}{MN} \sum_{i=1}^{M} |X_i(e^{j\omega})|^2 \right] \tag{11.16}$$

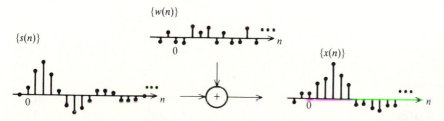

FIGURE 11-17
An example of adding random noise to a pulse waveform.

For the nonrandom sequence $\{s(n)\}$, the power spectrum $P_S(e^{j\omega})$ is equal to

$$P_S(e^{j\omega}) = S(e^{j\omega})\, S^*(e^{j\omega}) \tag{11.17}$$
$$= |S(e^{j\omega})|^2$$

In many practical cases, it can be argued that the noise is statistically independent of the signal. The power spectrum of the observed signal is then equal to

$$P_X(e^{j\omega}) = P_S(e^{j\omega}) + P_W(e^{j\omega}) \tag{11.18}$$

The inverse filter design procedures considered above can be called *noiseless*, since noise effects were not considered. The noiseless inverse filter is designed to compensate for the spectrum of the signal and neglects the noise. The magnitude response of the noiseless inverse filter is then given by

$$|H_{if}(e^{j\omega})| = 1/|S(e^{j\omega})| \tag{11.19}$$

This result shows that the noiseless inverse filter gain is a maximum at the frequency where the signal spectrum is a minimum. When we try to apply this noiseless inverse filter to the signal with noise, the noise components in the signal are amplified in the output, and the results are usually disappointing.

INCLUDING NOISE EFFECTS INTO THE DESIGN OF THE INVERSE FILTER. If the characteristics of the noise are included in the design of the inverse filter, the performance can be improved slightly. The gain of the noiseless inverse filter is a maximum at the frequencies where the signal spectrum is a minimum. If there are noise components at the frequency range where the signal spectrum is low, these will dominate the output. There is a trade off between the amount of pulse compression that can be obtained and the amount of noise that is present in the output.

To illustrate this trade off, let us denote the inverse filter magnitude response that includes the noise effects by $|H_{s+w}(e^{j\omega})|$ and let it be defined by

$$|H_{s+w}(e^{j\omega})|^2 = \frac{1}{P_S(e^{j\omega}) + \lambda P_W(e^{j\omega})} \tag{11.20}$$

where λ is a parameter that controls the amount of compression. If $\lambda = 0$, the noiseless inverse filter results, providing the maximum pulse compression, but also allowing the maximum effect of the noise to occur at the output. If $\lambda P_W(e^{j\omega}) \gg P_S(e^{j\omega})$, $|H_{s+w}(e^{j\omega})|^2 = 1/(\lambda P_W(e^{j\omega}))$ and the filter acts to suppress the noise without providing significant pulse compression. For the important practical case of white noise, the noise power spectrum is equal to a constant over frequency, or $P_W(e^{j\omega}) = W_0^2$.

The design of this inverse filter uses the same procedure as described in the previous section, the only changes being that $|H_{s+w}(e^{j\omega})|$ is used rather than $1/S(e^{j\omega})$ and that a value for λ must be specified. The following example illustrates this procedure.

Example 11.9. Designing an inverse filter for signal with noise. Let us consider the system model described by

$$H_m(z) = \frac{(1 - 0.95e^{j\pi/4}z^{-1})(1 - 0.95e^{-j\pi/4}z^{-1})}{(1 - 0.9e^{j\pi/6}z^{-1})(1 - 0.9e^{-j\pi/6}z^{-1})(1 - 0.9e^{j\pi/3}z^{-1})(1 - 0.9e^{-j\pi/3}z^{-1})}$$

Let us assume that the observed signal is equivalent to having a white noise sequence $\{w(n)\}$, that is uniformly distributed between -0.5 and 0.5, added to the output of this system. The noise spectrum is then constant with frequency and

$$P_W(e^{j\omega}) = W_0^2 = \tfrac{1}{12}$$

We will investigate the effect of choosing the value of λ by examining the performance of the filter under both noise-free and added noise conditions when λ takes on the values 0, 0.5, 2 and 8.

The first step is to determine the magnitude response of the inverse filter that is given by

$$|H_{s+w}(e^{j\omega})| = \frac{1}{[P_S(e^{j\omega}) + \lambda W_0^2]^{1/2}}$$

where the pulse spectrum is given by

$$P_S(e^{j\omega}) = |H_m(e^{j\omega})|^2$$

The forms of $|H_{s+w}(e^{j\omega})|$ for the different values of λ are shown in Fig. 11.18(a). We note that the zero in $H_m(z)$ at $z = 0.95e^{j\pi/4}$ produces a peak in these transfer functions at $\omega = \pi/4$. The gain is also large for high frequencies to compensate for the general lowpass characteristic of $H_m(e^{j\omega})$. The $\lambda = 0$ curve corresponds to the magnitude response of the noiseless inverse filter. As λ increases, the values of the maximum gains decrease to reflect the dominance of noise in those frequency ranges. As λ increases, the curves become flatter, indicating the increasing dominance of the constant spectral component λW_0^2.

These magnitude curves were sampled at 128 points to generate the sequence $\{|H_{s+w}(k)|\}$. The logarithm of this sequence was then employed in Eq. (11.12) to determine the phase response sequence $\{\text{Arg}[H_{s+w}(k)]\}$. The phase sequences for the different values of λ are shown in Fig. 11.18(b). As λ increases, the phase response tends toward zero.

Having $\{|H_{s+w}(k)|\}$ and $\{\text{Arg}[H_{s+w}(k)]\}$, the 128-point inverse DFT can be computed to determine the unit-sample response of the inverse filter $\{h_{s+w}(n)\}$. The first 32 points of the sequences obtained for the different values of λ are shown in Fig. 11.19. We note that the zero in $H_m(z)$ produces an infinite-duration component in the sequences. As λ increases, these unit-sample responses approach the unit-sample sequence $\{d(n)\}$, a result which is consistent with the flattening of the magnitude and phase spectra.

The pulse compression behavior of these filters can be examined by their response to the noiseless pulse sequence $\{h_m(n - n_0)\}$, shown in Fig. 11.20(a). The output sequences are shown in Fig. 11.20(b) for the different values of λ. For $\lambda = 0$, the unit-sample sequence occurring at $n = n_0$ is to be expected. Upon closer examination, we note a slight residual component for $n > n_0$. This residual is due to the time-aliasing effect produced with the 128-point sequences and can be reduced by increasing the size of the DFT sequences. As λ increases, the pulse

(a) $|H_{s+w}(e^{j\omega})|$

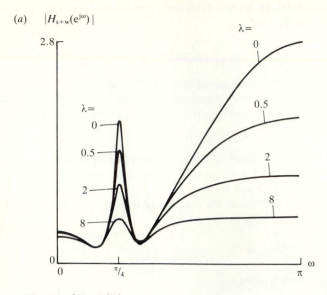

(b) $\mathrm{Arg}[H_{s+w}(e^{j\omega})]$

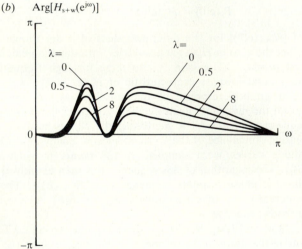

FIGURE 11-18
Response curves of inverse filter for signal in additive noise, for different values of λ. a) Magnitude responses employed in Hilbert transform. b) Phase response produced by Hilbert transform.

compression capability of the inverse filter decreases and the form of the output sequences approaches that of $\{h_m(n - n_0)\}$.

The behavior of the inverse filter in the presence of noise can now be examined. The input to the filter is the delayed pulse sequence plus noise $\{h_m(n - n_0) + w(n)\}$ shown in Fig. 11.21(a). The output sequences that are observed when this input is applied to the different filters are shown in Fig. 11.21(b). As λ decreases the effects of the noise become more apparent in the output.

To quantify this observation, the variance of the noise in the output W^2 was computed. This was done by subtracting the noiseless outputs, shown in

$\{h_{S+W}(n)\}$

$\lambda = 0$

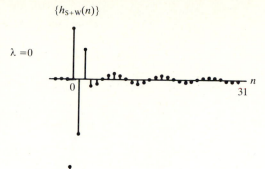

$\lambda = 0.5$

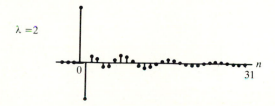

$\lambda = 2$

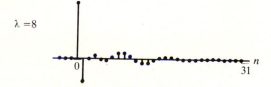

$\lambda = 8$

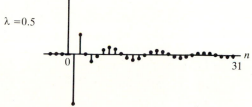

FIGURE 11-19
Unit-sample responses determined by inverse DFT for different values of λ.

Fig. 11.20(b), from the respective sequences in Fig. 11.21(b). The average of the squared residuals then provides an estimate of the noise variance W^2. The computed values of W^2 are given in the figure for the different values of λ. As λ increases, the value of W^2 approaches W_0^2, the variance of the noise in the input sequence.

Summarizing these results, we note that when noise is present there is a trade off between the ability to compress the pulse and to suppress the noise. When the maximum pulse compression, $\{h_m(n-n_0)\} \rightarrow \{d(n-n_0)\}$, is attempted, the noise component in the output has the highest power. The degree of pulse compression that should be attempted depends on the particular application and specification to be achieved.

(a) Input: $\{h_m(n-n_0)\}$

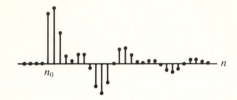

(b) Output: $\{y(n)\}$

$\lambda = 0$

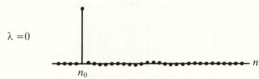

$\lambda = 0.5$

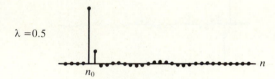

$\lambda = 2$

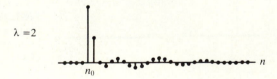

$\lambda = 8$

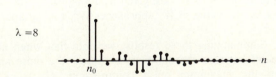

FIGURE 11-20
Noiseless input results. a) Input is a delayed unit-sample response of the minimum-phase system.
b) Output of inverse filters having unit-sample response shown in Fig. 11.19.

(*a*) Input: $\{h_m(n-n_0) + w(n)\}$

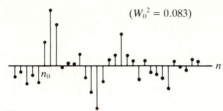

$(W_0{}^2 = 0.083)$

(*b*) Output: $\{y(n)\}$

$(W^2 = 0.24)$

$\lambda = 0$

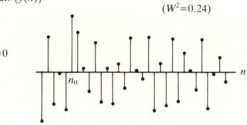

$(W^2 = 0.22)$

$\lambda = 0.5$

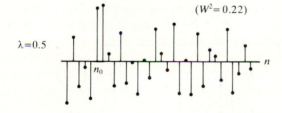

$(W^2 = 0.13)$

$\lambda = 2$

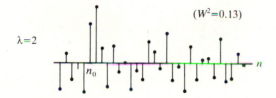

$(W^2 = 0.09)$

$\lambda = 8$

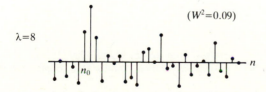

FIGURE 11-21
Results with input containing additive noise. *a*) Input is a delayed unit-sample response of minimum-phase system plus white noise. *b*) Output of inverse filters having unit-sample response shown in Fig. 11.19. The variance of the output noise is equal to W^2.

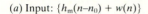

11.9   SUMMARY

This chapter began by considering two common distortions that occur to observed signals and by defining discrete-time models that generate these distortions. Inverse filters were found to work best on models whose system functions have the minimum-phase property. The minimum-phase property was defined in terms of the pole/zero pattern in the z plane of the system function of the model $H_m(z)$, the phase response $\text{Arg}\{H_m(e^{j\omega})\}$ and the unit-sample response $\{h_m(n)\}$. Two procedures for designing the inverse filter were described. The first performed the design in the z plane and the second in the frequency domain. The z plane procedure produced the system function of the inverse filter directly from the model pole/zero pattern: the poles of the inverse filter were located at the positions of the model zeros and the filter zeros at the positions of the model poles. This procedure is limited to models for which either the system function or the pole/zero pattern is known. The frequency-domain approach involved computing the DFT sequence of the observed unit-sample response of the model to implement an approximate inverse filter. The types of approximations that resulted by employing this procedure were described. In cases for which only the magnitude response of the model is known, an FIR approximation to the inverse filter can be designed by using the discrete-time Hilbert transform. For minimum-phase systems, the log-magnitude response sequence is the discrete-time Hilbert transform of the phase response sequence. This useful result was employed for designing inverse filters. The chapter ended by considering the problem of designing inverse filters for signals in the presence of noise.

A common alternate procedure for designing inverse filters *in the time domain* requires concepts that are not difficult, but sufficiently different from the topics covered in this text that they were not covered here. These time-domain design procedures can be found in the references.

REFERENCES TO TOPICS

General inverse filtering applications

Arya, V. K. and J. K. Aggarwal: *Deconvolution of Seismic Data,* Hutchinson Ross, Stroudsburg, PA, 1982.
Mendel, J. M.: *Optimal Seismic Deconvolution,* Academic Press, New York, 1983.
Anderson, B. D. O. and J. B. Moore: *Optimal Filtering,* Prentice-Hall, Englewood Cliffs, NJ, 1979.

Time-domain design of inverse filters

Makhoul, J.: "Linear prediction: A tutorial review," *Proc. IEEE,* **63,** 561–580 (1975).
Robinson, E. A., and S. Treitel: *Geophysical Signal Processing,* Prentice-Hall, Englewood Cliffs, NJ, 1980.
Schwartz, Mischa, and Leonard Shaw: *Signal Processing*: *Discrete Spectral Analysis, Detection, and Estimation,* McGraw-Hill, New York, 1975.
Treitel, S., and E. A. Robinson: "The design of high-resolution digital filters," *IEEE Trans. Geosci. Electron.,* **GE-4:** 25–38 (1966).

Discrete-Hilbert transform

Oppenheim, Alan V. and Ronald W. Schafer: *Digital Signal Processing,* Prentice-Hall, Englewood Cliffs, NJ. 1975.

Adaptive filters

Cowan, C. F. N., and P. M. Grant: *Adaptive Filters,* Prentice-Hall, Englewood Cliffs, NJ, 1985.
Goodwin, G. C., and K. S. Sin: *Adaptive Filtering, Prediction and Control,* Prentice-Hall, Englewood Cliffs, NJ, 1984.
Haykin, S.: *Adaptive Filter Theory,* Prentice-Hall, Englewood Cliffs, NJ, 1986.
Honig, M. L., and D. G. Messerschmitt: *Adaptive Filters,* Kluwer, Boston, 1984.

System modeling

Kuc, R.: "Modeling acoustic attenuation of soft tissue with a minimum-phase filter," *Ultrasonic Imaging Journal,* **6:** 24–36 (1984).

PROBLEMS

11.1. Show that a linear-phase filter cannot be minimum-phase by considering the pole/zero pattern of this system.

11.2. Determine whether the following unit-sample responses are from a minimum-phase filter. If not, indicate why not.

(a) $h(n) = 1$ for $0 \leq n \leq N - 1$
(b) $h(n) = n + 1$ for $0 \leq n \leq N - 1$
(c) $h(n) = r^n \cos(\omega_0)n$ for $n \geq 0$, $h(n) = 0$, otherwise
(d) $h(n) = r^n \sin(\omega_0)n$ for $n \geq 0$, $h(n) = 0$, otherwise

11.3. Consider a *minimum-phase* FIR filter having the unit-sample response given by $h_{min}(n)$, for $0 \leq n \leq N - 1$. Let the *maximum-phase* FIR filter, with the identical magnitude response, have the unit-sample response $h_{max}(n)$, for $0 \leq n \leq N - 1$. Show that $\{h_{max}(n)\}$ has the same values as $\{h_{min}(n)\}$, but in *time-reversed* order, or

$$h_{max}(n) = h_{min}(N - 1 - n) \quad \text{for } 0 \leq n \leq N - 1$$

11.4. Given any linear-phase FIR filter, how can this filter be transformed into a minimum-phase filter without changing the magnitude response?

11.5. Given the unit-sample response of a minimum-phase filter $\{h(n)\}$, show that a second filter having the unit-sample response $\{h_2(n)\} = \{h(n - k)\}$, where k is any nonzero integer, is not minimum-phase.

11.6. Consider the causal unit-sample response $\{h(n)\}$, for which $h(0) = 0$. Show that this filter cannot be minimum-phase.

11.7. Consider a physical system whose model system functionis given by

$$H_m(z) = 1 + az^{-1} \quad \text{where } |a| > 1$$

and whose unit-sample response is given by $\{h_m(n)\}$. This system model is not minimum-phase because it has a zero outside the unit circle at $z = -a$. The inverse filter system function given by

$$H_{if}(z) = 1/H_m(z) = 1/(1 + az^{-1})$$

is then unstable. Consider using instead the filter defined by

$$H(z) = (1/a)/[1 + (1/a)z^{-1}]$$

(a) What is the output sequence of $H(z)$ when $\{h_m(n)\}$ is the input sequence?
(b) What is the magnitude spectrum of this output sequence?

11.8. Consider the fourth-order minimum-phase FIR system $H_{\min}(z)$ having all four zeros at $z = 0.9$.
(a) Compute the magnitude response of this system.
(b) For each of the possible phase responses having this magnitude response form, calculate and sketch the phase response and the unit-sample response.

11.9. Inverse filter of an ARMA model. Let the system function of an ARMA model be given by

$$H_m(z) = \frac{1 - 0.5z^{-1}}{1 + 0.5z^{-1}}$$

Determine and sketch the following:
(a) the z plane pole/zero pattern of the ARMA model
(b) the unit-sample response of this model
(c) the pole/zero pattern and block diagram of the inverse filter

11.10. Repeat Problem 11.9 for the system function of an ARMA model given by

$$H_m(z) = \frac{1 - 0.81z^{-2}}{1 - 1.27z^{-1} + 0.9z^{-2}}$$

COMPUTER PROJECTS

11.1. Inverse filter for the radar system

Object. Simulate a radar system and to perform inverse filter design in the z plane.

 Generation of the ideal radar return. Let $r(n)$, for $0 \le n \le 63$, be the realization of an ideal radar return having the following values:

$r(1) = 1$ (isolated return)

$r(31) = 1$ $r(32) = -1$ $r(33) = 1$ (multiple return)

$r(n) = 0$ otherwise

The *observed* radar return, denoted by $\{s(n)\}$, is the first 64 samples of the convolution of $\{r(n)\}$ with a pulse waveform $\{p(n)\}$. Let $\{p(n)\}$ be the unit-sample response of the AR system described by the difference equation given by

$$y(n) = 1.8 \cos(\pi/8)y(n-1) - 0.81y(n-2) + x(n)$$

$$- 1.9 \cos(\pi/4)x(n-1) + 0.9025x(n-2)$$

The goal is to retrieve $\{r(n)\}$ from $\{s(n)\}$.

Analytic results
(a) Determine the z plane pole/zero pattern of the ARMA system.
(b) Determine and sketch the block diagram for the inverse filter.

Computer results
(a) For the pulse waveform $\{p(n)\}$, use only the first 30 samples of the unit-sample response of the ARMA system.
(b) Verify the operation of the inverse filter by computing and plotting the output when $\{p(n)\}$ is the input. Why is the output for $n = 30$ not equal to zero?
(c) Generate and plot the observed radar sequence $\{s(n)\}$.
(d) Compute and plot the output of the inverse filter when $\{s(n)\}$ is the input.

11.2. Inverse filter of the reverberant communication channel

Object. Simulate a reverberant communication channel and construct an inverse filter.

Communication channel model. Let the system function of a reverberant channel be given by

$$H_m(z) = 1 - 0.81z^{-2}$$

Analytic results. Determine and sketch the following for this channel model:
(a) the z plane pole/zero pattern
(b) the unit-sample response
(c) the pole/zero pattern and block diagram of the inverse filter

Computer results. Implement the channel model and inverse filter. Perform the following.
(a) Plot the unit-sample response of the channel model.
(b) Plot the output of the inverse filter when sequence in Part (a) is the inut.
(c) Plot the output of the channel filter when the FSK sequence, generated in Chapter 2, is the input.
(d) Plot the output of the inverse filter when the sequence in Part (c) is the input.

11.3. Approximate FIR inverse filter for an FIR system model

Object. Determine the approximations of this inverse filter design procedure.
Consider the FIR system model to be given by

$$H_m(z) = 1 - 0.9z^{-1}$$

We want to find the inverse filter using the frequency-domain procedure.

Analytic results. Determine the true inverse filter and sketch its unit-sample response $\{h_{if}(n)\}$.

Computer results
(a) Plot the model unit-sample response $\{h_m(n)\}$. Compute and plot the N_d-point magnitude and phase sequences $\{|H_m(k)|\}$ and $\{\mathrm{Arg}[H_m(k)]\}$ for $N_d = 16$.
(b) Compute and plot $\{h_{ifa}(n)\}$, the unit-sample response of the FIR approximate inverse filter determined from $\{|H_m(k)|^{-1}\}$ and $\{-\mathrm{Arg}[H_m(k)]\}$. Compare with $\{h_{if}(n)\}$. Verify that $\{h_{ifa}(n)\}$ corresponds to the first period of the periodic extension of $\{h_{if}(n)\}$.
(c) Repeat for $N_d = 64$.

11.4. Approximate FIR inverse filter for an IIR system model

Object. Determine the approximations of this inverse filter design procedure. Consider the IIR system model to be given by

$$H_m(z) = 1/(1 - 0.9z^{-1})$$

We want to find the inverse filter using the frequency-domain procedure.

Analytic results. Determine the true inverse filter and sketch its unit-sample response $\{h_{if}(n)\}$.

Computer results

(a) Compute and plot the unit-sample response $\{h_{m,N}(n)\}$, where

$$h_{m,N}(n) = h_m(n) \qquad \text{for } 0 \le n \le N - 1$$

Let $N = N_d = 16$. Compute and plot the N-point magnitude and phase sequences $\{|H_{m,N}(k)|\}$ and $\{\text{Arg}[H_{m,N}(k)]\}$.

(b) Compute and plot $\{h_{ifa}(n)\}$, the unit-sample response of the FIR approximate inverse filter determined from $\{|H_{m,N}(k)|^{-1}\}$ and $\{-\text{Arg}[H_{m,N}(k)]\}$. To observe the pulse compression performance, compute and plot 55 points of the output of this approximate inverse filter when $\{h_m(n)\}$ is the input.

(c) Repeat for $N = N_d = 32$.

11.5. Inverse filter design using the Hilbert transform for an FIR system

Object. Determine performance of the inverse filter design procedure for FIR systems. Consider the FIR system model to be given by

$$H_m(z) = 1 - 0.9z^{-1}$$

Let us assume that we only know the magnitude response of this system, from which we want to design the inverse filter.

Analytic results

(a) Determine the true inverse filter and sketch its unit-sample response $\{h_{if}(n)\}$.

(b) Calculate and sketch the magnitude response $|H_m(e^{j\omega})|$ of this system.

Computer results

(a) Compute and plot the N-point log-magnitude response sequence $\{|H_m(k)|_{dB}\}$. Let $N = 16$.

(b) Use the discrete-Hilbert transform subroutine DHT, given below, to compute the phase response. Plot the result and compare with that of the true filter.

(c) Compute and plot the unit-sample response of the FIR approximate inverse filter. Compare with $\{h_{if}(n)\}$.

(d) Repeat for $N = 64$.

```
        SUBROUTINE DHT(XLMAG, XPHS, N)
C
C       XLMAG (REAL INPUT ARRAY) CONTAINS THE LOG-MAGNITUDE
C             SEQUENCE IN DB
C       XPHS (REAL OUTPUT ARRAY) IS THE COMPUTED PHASE SEQUENCE
C       N (INTEGER INPUT) IS SIZE OF ARRAY (POWER OF 2
C             CORRESPONDING TO NUMBER OF POINT OVER 0 ≤ ω < 2π.)
```

```
C
        DIMENSION XLMAG(N),XPHS(N)
        N2=N/2
        N2P1=N2+1
        NP2=N+2
        PI=4.*ATAN(1.)
        DBTONP=1./8.69              ! CONVERSION FROM DB TO NAPERIAN LOG
        DOMGA=2.*PI/FLOAT(N)        ! FREQ INCREMENT
        XPHS(1)=0.                  ! PHASE AT ω=0 AND ω=π IS ZERO
        XPHS(N2P1)=0.
C
        DO 1 I=2,N2
        II=I
        TYPE*,II,'/',N2                  ! INDICATES PROGRAM PROGRESS
        SUM=0.
        OM=FLOAT(I-1)*DOMGA
C
C  INTEGRATION LOOP
C
          DO 2 J=1,N
          IF(J.EQ.I) GOTO 2              ! JUMP OVER COTAN(0) POINT
          THETA=FLOAT(J-1)*DOMGA
          ALPHA=(THETA-OM)/2.
          COTAN=COS(ALPHA)/SIN(ALPHA)
          SUM=SUM+DBTONP*XLMAG(J)*COTAN
2         CONTINUE
        XPHS(I)=SUM/FLOAT(N)
1       XPHS(NP2-I)=-XPHS(I)            ! PHASE IS ODD SEQUENCE
        RETURN
        END
```

11.6. Inverse filter design using the Hilbert transform for an IIR system

Object. Determine performance of the inverse filter design procedure for FIR systems.

Consider the IIR system model to be given by

$$H_m(z) = 1/(1 - 0.9z^{-1})$$

Let us assume that we only know the magnitude response of this system, from which we want to design the inverse filter.

Analytic results
(a) Determine the true inverse filter and sketch its unit-sample response $\{h_{if}(n)\}$.
(b) Calculate and sketch the magnitude response $|H_m(e^{j\omega})|$.

Computer results
(a) For $N = 32$, compute and plot the log-magnitude response sequence $\{|H_m(k)|_{dB}\}$.
(b) Use the discrete-Hilbert transform subroutine DHT to compute the phase response. Plot the result and compare with that of the true filter.
(c) Compute and plot the unit-sample response of the FIR approximate inverse filter. Compare with $\{h_{if}(n)\}$.

APPENDIX A

PROGRAM LISTINGS

The computer programs listed below evaluate the DFT using the fast Fourier transform and provide the graphics routines for displaying arrays on the terminal and transferring them to a file for printer hard copy. The FFT routines have been adapted from the Cooley–Tukey fast Fourier transform algorithm (ref: Cooley, Lewis and Welch, *IEEE Trans. Education,* **12,** 27–34, 1969). Programs to test these routines are listed and sample print-outs are presented. The routines include:

> PLOT – to display an array of arbitrary size with a vertical abscissa.
> FFT – to compute the DFT with the fast Fourier transform algorithm.
> PLOTMR – to display the 65 unique values of a 128-point magnitude response.
> PLOTPR – to display the 65 unique values of a 128-point phase response.
> PLOTLM – to display the 65 unique values of a 128-point log-magnitude response.

The FORTRAN and C-language programs have been written on a

VAX*-780 computer system using the VMS operating system. The Pascal programs have been written on a DEC Rainbow* PC with MS-DOS and using Turbo Pascal, available from Borland International, 4585 Scotts Valley Drive, Scotts Valley, CA 95066.

Inputs to programs:

X – real-valued array.

XR and XI – arrays containing the real and imaginary components of the DFT.

NX – size of array.

NPRNT (FORTRAN and Pascal) – causes graph to be printed when NPRNT = 1.

DSK (Pascal) – disk file name for data output.

LOG2N – integer equal to $\log_2(N)$ for N-point DFT.

NTYPE – integer, when = 1, DFT is performed; when = -1, IDFT is performed.

DBSCAL – (<0) defines lower limit of log-magnitude graph.

SPEC – (<0) defines specification level to be marked on log-magnitude graph.

IF1 and IF2 – two integers indicating DFT bin numbers to be marked on graph.

---

* A trademark of Digital Equipment Corporation.

FORTRAN programs.

```
      SUBROUTINE PLOT(X,NX,NPRNT)
      DIMENSION X(NX)
      BYTE ARR(71),MARK
C FIND MINIMUM AND MAXIMUM VALUES IN ARRAY
      XMAX=0.
      XMIN=0.
      DO 1 I=1,NX
      IF(X(I).GT.XMAX) XMAX=X(I)
1     IF(X(I).LT.XMIN) XMIN=X(I)
      IF(XMAX.LT.-XMIN) XMAX=-XMIN
      IF(XMAX.EQ.0.) XMAX=1.
      TYPE 200,XMAX
      IF(NPRNT.EQ.1) PRINT 200,XMAX
200   FORMAT(' MAX VAL= ',F6.2)
C IPOS IS DOTTED LINE POSITION
      IPOS=36
      IF(XMIN.GE.0.) IPOS=1
      IF(XMAX.LE.0.) IPOS=71
C SCALE TO PRINTER COORDINATES
      SCALE=70./XMAX
      IF(IPOS.EQ.36) SCALE=35./XMAX
      MARK='-'
      DO 2 I=1,NX
      RNDOFF=.5001
      IF(X(I).LT.0.) RNDOFF=-.5001
      N=IFIX(X(I)*SCALE+RNDOFF)+IPOS
      DO 5 J=1,71
5     ARR(J)=MARK
      ARR(IPOS)='|'
      ARR(N)='*'
      TYPE 100,I-1,ARR
      IF(NPRNT.EQ.1) PRINT 100,I-1,ARR
100   FORMAT(' ',I3,'-',71A1)
2     MARK=' '
      RETURN
      END


Program to test PLOT.FOR:

C TEST0 FORTRAN
      DIMENSION X(45)
1     TYPE*,' FOR SINE FCN - ENTER 0'
      TYPE*,' FOR COS FCN  - ENTER 1'
      ACCEPT*, ITYPE
      TYPE*,' ENTER FREQ (0<F<0.5)'
      ACCEPT*, F
      TYPE*,'PRINT GRAPH (1/0)'
      ACCEPT*, NPRNT
      PI=4.*ATAN(1.)
      DO 2 I=1,45
      ALPHA=2.*PI*F*FLOAT(I-1)
      IF(ITYPE.EQ.1) X(I)=COS(ALPHA)
      IF(ITYPE.EQ.0) X(I)=SIN(ALPHA)
2     TYPE 102,I-1,X(I)
102   FORMAT(' X(',I2,')=',F6.2)
```

```
      CALL PLOT(X,45,NPRNT)
      TYPE*,'GO AGAIN? (1/0)'
      ACCEPT*, IANS
      IF(IANS.EQ.1) GOTO 1
      STOP
      END

      SUBROUTINE FFT (LOG2N,XR,XI,NTYPE)
      DIMENSION XR(1),XI(1)
      SIGN=-1.
      IF(NTYPE.LT.0) SIGN=1.
      N=2**LOG2N
      NV2=N/2
      NM1=N-1
      J=1
      DO 7 I=1,NM1
      IF(I.GE.J) GOTO 5
      TR=XR(J)
      TI=XI(J)
      XR(J)=XR(I)
      XI(J)=XI(I)
      XR(I)=TR
      XI(I)=TI
5     K=NV2
6     IF(K.GE.J) GOTO 7
      J=J-K
      K=K/2
      GOTO 6
7     J=J+K
      PI=4.*ATAN(1.)
      DO 20 L=1,LOG2N
      LE=2**L
      LE1=LE/2
      UR=1.
      UI=0.
      WR=COS(PI/LE1)
      WI=SIGN*SIN(PI/LE1)
      DO 20 J=1,LE1
      DO 10 I=J,N,LE
      IP=I+LE1
      TR=XR(IP)*UR - XI(IP)*UI
      TI=XR(IP)*UI + XI(IP)*UR
      XR(IP)=XR(I) - TR
      XI(IP)=XI(I) - TI
      XR(I)=XR(I) + TR
10    XI(I)=XI(I) + TI
      TR=UR*WR - UI*WI
      TI=UR*WI + UI*WR
      UR=TR
20    UI=TI
      IF(NTYPE.GT.0) RETURN
      AIN=1./N
      DO 30 I=1,N
      XR(I)=XR(I)*AIN
30    XI(I)=XI(I)*AIN
      RETURN
      END
```

```
        SUBROUTINE PLOTMR(X,NPRNT)
        DIMENSION X(65)
        BYTE ARR(65),MARK
C FIND MAXIMUM OF X ARRAY - XMAX
        TYPE 200
        IF(NPRNT.EQ.1) PRINT 200
200 FORMAT(' MAGNITUDE RESPONSE')
        XMAX=0.
        DO 1 I=1,65
1       XMAX=AMAX1(XMAX,X(I))
        SCAL = XMAX/20.
        SCALD2=SCAL/2.
        DO 3 I=1,21
C LEVEL OF GRAPH LINE
        GLVL=FLOAT(21-I)*SCAL
        MARK=' '
        IF(I.EQ.1.OR.I.EQ.21) MARK='-'
        DO 4 J=1,65
4       ARR(J)=MARK
        ARR(1)='|'
        ARR(65)='|'
        DO 5 J=1,65
        A=ABS(X(J)-GLVL)
5       IF(A.LT.SCALD2) ARR(J)='*'
        IF(NPRNT.EQ.1) PRINT 100,GLVL,ARR
3       TYPE 100,GLVL,ARR
100 FORMAT(' 'F6.2,1X,65A1)
        RETURN
        END

        SUBROUTINE PLOTPR(X,NPRNT)
        DIMENSION X(65)
        BYTE ARR(65),MARK
        PI=4.*ATAN(1.)
        SCAL=PI/10.
        TYPE 101
        IF(NPRNT.EQ.1) PRINT 101
101 FORMAT(' PHASE RESPONSE')
        SCALD2=SCAL/2.
        DO 3 I=1,21
        PHLVL=FLOAT(11-I)*SCAL
        MARK=' '
        IF(I.EQ.1)  MARK='-'
        IF(I.EQ.11) MARK='-'
        IF(I.EQ.21) MARK='-'
        DO 4 J=1,65
        ARR(J)=MARK
        ARR(1)='|'
        ARR(65)='|'
        DO 5 J=1,65
        A=ABS(X(J)-PHLVL)
5       IF(A.LT.SCALD2) ARR(J)='*'
        IF(NPRNT.EQ.1) PRINT 100,PHLVL,ARR
3       TYPE 100,PHLVL,ARR
100 FORMAT(' 'F6.2,1X,65A1)
        RETURN
        END
```

```
        SUBROUTINE PLOTLM(X,DBSCAL,SPEC,
       +                  IF1,IF2,NPRNT)
        DIMENSION X(65),Y(65)
        BYTE ARR(65),MARK
        TYPE 101
        IF(NPRNT.EQ.1) PRINT 101
101 FORMAT(' LOG-MAGNITUDE RESPONSE')
        XMAX=-1000.
        DO 1 I=1,65
1       XMAX=AMAX1(XMAX,X(I))
C SCALE ARRAY BETWEEN 0 AND DBSCAL
        DO 2 I=1,65
        Y(I)=X(I)-XMAX
2       Y(I)=AMAX1(Y(I),DBSCAL)
        SCAL=-DBSCAL/20.
        SCALD2=SCAL/2.
        DO 3 I=1,21
        MARK=' '
        IF(I.EQ.1.OR.I.EQ.21) MARK='-'
        DO 4 J=1,65
4       ARR(J)=MARK
C INDICATE FREQUENCY MARKS AND LIMITS
        ARR(1)='|'
        IF(IF1.GT.0.)  ARR(IF1)='|'
        IF(IF2.GT.0.)  ARR(IF2)='|'
        ARR(65)='|'
C DB LEVEL OF GRAPH LINE
        DBINC=-FLOAT(I-1)*SCAL
        DO 5 J=1,65
C SPECIFICATION LINE
        A=ABS(SPEC-DBINC)
        IF(A.LT.SCALD2) ARR(J)='-'
C DATA POINT
        A=ABS(Y(J)-DBINC)
5       IF(A.LT.SCALD2) ARR(J)='*'
        IF(NPRNT.EQ.1)PRINT 100,DBINC,ARR
3       TYPE 100, DBINC,ARR
100 FORMAT(' 'F6.2,1X,65A1)
        RETURN
        END


Program to test FFT.FOR, PLOTMR.FOR,
PLOTPR.FOR and PLOTLM.FOR:

C   TEST1 FORTRAN
        DIMENSION X(128),Y(128)
        DO 1 I=1,128
        X(I)=0.
1       Y(I)=0.
        DO 2 I=1,8
2       X(I)=1.
        CALL FFT(7,X,Y,1)
        DO 3 I=1,65
        A=X(I)
        B=Y(I)
        X(I)=SQRT(A**2+B**2)
        PHS=0.
```

```
      IF(A.NE.0..AND.B.NE.0.) PHS=ATAN2(B,A)
3     Y(I)=PHS
      NPRNT=1
      CALL PLOTMR(X,NPRNT)
      CALL PLOTPR(Y,NPRNT)
      DBSCAL=-40
      SPEC=-20
      IF1=20
      IF2=30
      DO 4 I=1,65
      IF(X(I).LE.0.0001) X(I)=.0001
4     X(I)=20.*ALOG10(X(I))
      CALL PLOTLM(X,DBSCAL,
     +            SPEC,IF1,IF2,NPRNT)
      STOP
      END
```

Pascal programs.

```pascal
procedure plot(x:vector; nx,nprnt:integer;
               var dsk:text);
var
  i, j, zpt, ival, m, offset: integer;
  xmx, xmn, scale, rndoff: real;
  arr: array[1..71] of char;
  dot: char;
begin
writeln; if nprnt = 1 then writeln(dsk);
xmx := 0.; xmn := 0.;
for i := 1 to nx do begin
  if x[i] > xmx then xmx := x[i];
  if x[i] < xmn then xmn := x[i];
  end;
if xmx = 0. then xmx := 1.;
if xmn = 0. then begin
  {zero axis at left}
  zpt := 1; scale :=70./xmx; offset := 1;
  end;
if (xmx > 0.0) and (xmn < 0.0) then  begin
  {zero axis at middle}
  if -xmn > xmx then xmx := -xmn;
  zpt := 36; scale := 70./(2.*xmx);
  offset := 36;
  end;
write('maximum value=',xmx :6:2); writeln;
if nprnt=1 then begin
  write(dsk,'maximum value=',xmx :6:2);
  writeln(dsk);
  end;
for i := 1 to nx do begin
  dot :='-';
  write(i-1 :3,dot);
  if nprnt = 1 then
    write(dsk, i-1 :3,dot);
  rndoff := 0.5001;
```

```pascal
  if x[i] < 0 then rndoff := -0.5001;
  ival:=trunc(x[i]*scale+rndoff)+offset;
  for j := 1 to 71 do begin
    if i = 1 then dot := '-'
             else dot := ' ';
    if j = zpt then dot := '|';
    if j = ival then dot := '*';
    write(dot);
    if nprnt = 1 then write(dsk,dot);
    end;
  writeln; if nprnt = 1 then writeln(dsk);
  end;
writeln; if nprnt = 1 then writeln(dsk);
if nprnt = 1 then begin
  for j := 1 to 71 do write(dsk,dot);
  end; {empty print buffer}
end;{plot}
```

Program to test plot.pas:

```pascal
program test0(output);
type
  vector = array [1..80] of real;
var
  blnk: char;
  dsk: text;
  y: vector;
  x,f,an: real;
  n, nx, nprnt, sc: integer;

procedure fopen(var dsk:text);
var
  filnam:string[14];
begin
write('enter filename '); read(filnam);
writeln;
assign(dsk,filnam); rewrite(dsk);
end; {fopen}

{$i plot }

begin
write('f = (0<f<0.5) '); read(f);
write(' sine or cosine? (1/0) ');
read(sc); writeln;
f := 2.*pi*f;
for n := 1 to 45 do begin
  an := n-1;
  if sc = 0 then y[n] := cos(f*an);
  if sc = 1 then y[n] := sin(f*an);
  end;
write('print? (1/0)'); read(nprnt);
writeln;
if nprnt = 1 then fopen(dsk);
plot(y, 45, nprnt, dsk); blnk := ' ';
if nprnt=1 then
```

```
   for n:=1 to 160 do write(dsk,blnk);
end.


procedure fft(var xr,xi:vector;
              log2n,ntype:integer);
const
   pi = 3.1415926;
var
   n,n2,i,id,j,k,le,lel: integer;
   ur,ui,wr,wi,tr,ti,sign,inv,omega: real;
begin
sign := -1.;
if ntype < 0 then sign :=1.;
n := 1; for i := 1 to log2n do n := n*2;
n2 := n div 2;
j := 1; {time decimation}
for i := 1 to n-1 do begin
   if i < j then begin {switch}
      tr    := xr[j]; ti    := xi[j];
      xr[j] := xr[i]; xi[j] := xi[i];
      xr[i] := tr;    xi[i] := ti;
      end;
   k := n2;
   while k < j begin
      j := j-k; k := k div 2;
      end;
   j := j+k;
   end;
for k := 1 to log2n do begin
   {butterfly computations}
   le := 1; for i := 1 to k do le := le*2;
   lel := le div 2;
   ur := 1.; ui := 0.;
   omega := pi/lel;
   wr := cos(omega);
   wi := sign * sin(omega);
   for j := 1 to lel do begin
      i := j; while i <= n begin
         id := i + lel;
         tr := xr[id] * ur - xi[id] * ui;
         ti := xr[id] * ui + xi[id] * ur;
         xr[id] := xr[i] - tr;
         xi[id] := xi[i] - ti;
         xr[i] := xr[i] + tr;
         xi[i] := xi[i] + ti;
         i := i+le;
         end;
      tr := ur*wr - ui*wi;
      ti := ur*wi + ui*wr;
      ur := tr; ui := ti;
      end;
   end;
if ntype < 0 then begin
   inv := 1./n;
   for i := 1 to n do begin
      xr[i] := xr[i]*inv;
      xi[i] := xi[i]*inv;
      end;
   end;
end {fft};


procedure plotmr(x:vector; nprnt:integer;
                 var dsk:text);
const
    blnk = ´ ´;
    num = 65;
var
   dot, mark: char;
   i, j, zpt, ival, m, offset: integer;
   xmx, zero, scale, scale2, inc: real;
begin
writeln(´magnitude response´); writeln;
if nprnt = 1 then begin
   writeln(dsk,´magnitude response´);
   writeln(dsk);
   end;
xmx := 0.;
for i := 1 to num do begin
   if x[i] > xmx then xmx := x[i];
   end;
scale := xmx/20.; scale2 := scale/2.;
for i := 1 to 21 do begin
   inc := (21-i)*scale;
   if (i = 1) or (i = 21)
     then begin
       mark := ´-´;
       if i = 1 then begin
         write(blnk,xmx :6:2);
         if nprnt = 1 then
           write(dsk,blnk,xmx :6:2);
         end;
       if i = 21 then begin
         zero := 0.;
         write(blnk,zero :6:0);
         if nprnt = 1 then
           write(dsk,blnk,zero :6:0);
         end;
       end
   else begin
     mark := blnk;
     for j := 1 to 7 do write(blnk);
     if nprnt = 1 then
       for j := 1 to 7 do
         write(dsk,blnk);
     end;
   for j := 1 to num do begin
     if (j = 1) or (j = num) then begin
       dot := ´|´;
       if (i = 1) or (i = 21)
         then dot := ´+´;
       end;
     if abs(x[j]-inc) < scale2
```

```
      then dot := '*';
    write(dot);
    if nprnt = 1 then write(dsk,dot);
    dot := mark;
    end;
  writeln;
  if nprnt = 1 then writeln(dsk);
  end;
end; {plotmr}


procedure plotpr(x:vector; nprnt:integer;
  var dsk:text);
const
      blnk = ' ';
      num = 65;
var
    dot, mark: char;
    i, j, zpt, ival, m, offset: integer;
    xmx, pi, val, scale, scale2, inc: real;
begin
writeln('phase response'); writeln;
if nprnt = 1 then begin
  writeln(dsk,'phase response');
  writeln(dsk);
  end;
pi := 4.*arctan(1);
xmx := pi;
scale := xmx/10.; scale2 := scale/2.;
for i := 1 to 21 do begin
  inc := (21-i)*scale - pi;
  if (i = 1) or (i = 21) or (i=11)
    then begin
      mark := '-';
      val := xmx;
      if i = 11 then val := 0.;
      if i = 21 then val := -xmx;
      write(blnk,val :6:2);
      if nprnt = 1
        then write(dsk,blnk,val :6:2);
      end
    else begin
      mark := blnk;
      for j := 1 to 7 do write(blnk);
      if nprnt = 1 then
        for j := 1 to 7 do
          write(dsk,blnk);
      end;
    for j := 1 to num do begin
      if (j = 1) or (j = num) then begin
        dot := '|';
        if (i = 1) or (i = 21)
          then dot := '+';
        end;
        if abs(x[j]-inc) < scale2
          then dot := '*';
        write(dot);
```

```
          if nprnt = 1 then write(dsk,dot);
          dot := mark;
          end;
        writeln;
        if nprnt = 1 then writeln(dsk);
        end;
end; {plotpr}


procedure plotlm(x:vector;
                  dbscal, spec:real;
                  if1,if2,nprnt:integer;
                  var dsk:text);
var
  blnk,mark: char;
  arr: array [1..65] of char;
  i, j, k, zpt, ival, m, offset: integer;
  xmx, xmn, scale, scale2, dbinc: real;
begin {plotlm}
writeln('log-magnitude response');
writeln;
if nprnt = 1 then begin
  writeln(dsk,'log-magnitude response');
  writeln(dsk);
  end;
blnk := ' ';
m := 65;
{find max of x array}
xmx := -300;
for i := 1 to m do
  if x[i] > xmx then xmx := x[i];
{normalize to 0 dB}
for i := 1 to m do begin
  x[i] := x[i] - xmx;
  if x[i] < dbscal then x[i] := dbscal;
  end;
scale := dbscal/20.;
scale2 := abs(scale)/2.;
for i := 1 to 21 do begin
  mark := ' ';
  if (i = 1) or (i = 21) then mark := '-';
  for j := 1 to 65 do arr[j] := mark;
  arr[1] := '|';
  if if1 > 0 then arr[if1] := '|';
  if if2 > 0 then arr[if2] := '|';
  arr[65] := '|';
  dbinc := (i-1)*scale;
  for j := 1 to 65 do begin
    if abs(spec-dbinc) < scale2
      then arr[j] := '-';
    if abs(x[j]-dbinc) < scale2
      then arr[j] := '*';
    end;
  j := 5 * ((i-1) div 5);
  if j = i-1
    then begin
      write(blnk,dbinc :6:1);
```

```
      if arr[1] = '|' then arr[1] := '+';
      if nprnt = 1
        then write(dsk,blnk,dbinc :6:1);
      end
    else begin
      for k := 1 to 7 do write(blnk);
      if nprnt = 1
        then for k := 1 to 7
          do write(dsk,blnk);
      end;
    for j := 1 to 65 do write(arr[j]);
    if nprnt = 1 then begin
      for j := 1 to 65
        do write(dsk,arr[j]);
      writeln(dsk);
      end;
    writeln;
    end;
end; {plotlm}
```

Program to test fft.pas, plotmr.pas, plotpr.pas and plotlm.pas:

```
program test1 (input, output);
type
    vector = array [1..128] of real;
var
    dsk: text;
    blnk: char;
    x,y: vector;
    i,nprnt,n,if1,if2: integer;
    a,b,pi,dbscal,spec: real;

{$i fopen }
{$i fft }
{$i plotmr }
{$i plotpr }
{$i plotlm }
begin
write('print (1/0)? '); read(nprnt);
writeln;
if nprnt = 1 then fopen(dsk);
for i := 1 to 128 do begin
  x[i] := 0.; y[i] := 0.;
  end;
for i := 1 to 8 do x[i] := 1.;
fft(x,y,7,1);
a := 1.; pi := 4.*arctan(a);
for i := 1 to 65 do begin
  a := sqrt(sqr(x[i]) + sqr(y[i]));
  b := 0.;
  if x[i] <> 0.
    then b := arctan(y[i]/x[i]);
  if x[i] = 0. then begin
    if y[i] > 0. then b := pi/2.;
    if y[i] < 0. then b := -pi/2;
```

```
    end;
  if x[i] < 0 then begin
    if y[i] < 0. then b := b-pi;
    if y[i] >= 0. then b := b+pi;
    end;
  x[i] := a; y[i] := b;
  end;
plotmr(x,nprnt,dsk);
plotpr(y,nprnt,dsk);
if1 := 20; if2 := 30;
dbscal := -40.; spec := -20.;
for i := 1 to 65 do begin
  if x[i] <= 0.0001   then
    x[i] := 0.0001;
  x[i] := 8.68*ln(x[i]);
  end;
plotlm(x,dbscal,spec,if1,if2,nprnt,dsk);
blnk := ' ';
if nprnt = 1 then for i := 1 to 160 do
  write(dsk,blnk); {clear print buffer}
end.
```

C-language programs.

```c
void plot(x,nx,nprnt,fp)
FILE    *fp;
float   x[];
int     nx,nprnt;
{
int     i,j,ival,zpt,offset;
float   xmx,xmn,scale,rndoff;
char    dot;

xmx = 0.; xmn = 0.;
for (i = 0; i < nx; i++) {
  if (x[i] > xmx) xmx = x[i];
  if [x[i] < xmn) xmn = x[i];
  }
if (xmx < -xmn) xmx = -xmn;
if (xmx == 0.) xmx = 1.;
printf("max val= %6.2f \n",xmx);
if (nprnt == 1)
  fprintf(fp,"max val= %6.2f \n",xmx);
if (xmn == 0.) {
  zpt = 0; scale = 70./xmx;
  offset = 0;
  }
if (xmx > 0. && xmn < 0.) {
  if (-xmn > xmx) xmx = -xmn;
  zpt = 35; scale = 35./xmx;
  offset = 35;
  }
for (i = 0; i < nx; i++) {
  printf("%3d-",i);
```

```
if (nprnt == 1) fprintf(fp,"%3d-",i);
for (j = 0; j < 71; j++) {
   if (i == 0) dot = ´-´;
      else dot = ´ ´;
   if (j == zpt) dot = ´|´;
   rndoff = .5001;
   if (x[i] < 0.) rndoff = -.5001;
   ival = offset + (rndoff + x[i]*scale);
   if (ival == j) dot = ´*´;
   printf("%1c",dot);
   if (nprnt == 1) fprintf(fp,"%1c",dot);
   }
printf("\n");
if (nprnt == 1) fprintf(fp,"\n");
}
return;
}


Program to test plot.c:

/* test0.c */
#include <math.h>
#include <stdio.h>
#include <plot.c>
main()
{
   FILE    *fp;
   float   x[128],f;
           pi = 4.*atan(1.);
   int     ans,i,j,sc,nx,nprnt;
   void    plot();

   fp = fopen("test0.dat","w");
start:  printf("sin or cos? (1/0) ");
   scanf("%d",&sc);
   printf("f= (0<f<0.5) ");
   scanf("%f",&f);
   printf("print results? (1/0) ");
   scanf("%d",&nprnt);
   if (nprtn == 1) {
      if (sc == 1) fprintf(fp,"sine fcn ");
      if (sc == 0) fprintf(fp,"cos fcn ");
      fprintf(fp,"with freq %6.3f \n",f);
      }
   f = 2.*pi*f; nx = 45;
   for (i = 0; i < nx; i++) {
      if (sc == 1) x[i] = sin(f*i);
      if (sc == 0) x[i] = cos(f*i);
      }
   plot(x,nx,nprnt,fp);
   printf("go again? (1/0)");
   scanf("%d",&ans);
   if (ans == 1) goto start;
   fclose(fp);
}
```

```
void fft(log2n,xr,xi,ntype)
float   *xr, *xi;
int     log2n, ntype;
{
int     n,nl,n2,i,j,k,l,le,lel,id,ip
double  ur,ui,wr,wi,tr,ti,
        pi=4.*atan(1.);
float   fm,fl,sign,in;

sign = -1.; if (ntype < 0) sign =1;
fm = log2n; n = pow(2.,fm);
n2 = n/2; nl = n-1; j = 1;
for (i=1; i<=nl; i++) {
   ip = i-1;
   if (i<j) {
      id = j-1;
      tr = *(xr+id); ti = *(xi+id);
      *(xr+id) = *(xr+ip);
      *(xi+id) = *(xi+ip);
      *(xr+ip) = tr; *(xi+ip) = ti;
      }
   k=n2;
   while (k<j) {
      j = j-k; k = k/2;
      }
   j = j+k;
   }
for (k=1; k<=log2n, k++) {
   fl = k; le = pow(2.,fl);
   lel = le/2; ur = 1.; ui = 0.;
   wr = cos(pi/lel);
   wi = sign*sin(pi/lel);
   for (j=1; j<=lel; j++) {
      for (i=j; i<=n; i=i+le) {
         ip = i-1; id = ip+lel;
         tr = *(xr+id)*ur - *(xi+id)*ui;
         ti = *(xr+id)*ui + *(xi+id)*ur;
         *(xr+id) = *(xr+ip) - tr;
         *(xi+id) = *(xi+ip) - ti;
         *(xr+ip) = *(xr+ip) + tr;
         *(xi+ip) = *(xi+ip) + ti;
         }
      tr = ur*wr - ui*wi;
      ti = ur*wi + ui*wr;
      ur = tr; ui = ti;
      }
   }
if (ntype < 0) {
   in = 1./n;
   for (i=0; i<n; i++) {
      *(xr+i) = *(xr+i) * in;
      *(xi+n) = *(xi+n) * in;
      }
   }
return;
}
```

```
void plotmr(x,nprnt,fp)
FILE     *fp;
float    x[];
int      nprnt;
{
int      i,j,nx,ival,zpt,offset;
float    xmx,xmn,scale,rndoff;
char     dot;

nx = 65; xmx = 0.; xmn = 0.;
for (i = 0; i < nx; i++) {
  if (x[i] > xmx) xmx = x[i];
  if (x[i] < xmn) xmn = x[i];
  }
if (xmx < -xmn) xmx = -xmn;
if (xmx == 0.) xmx = 1.;
printf("magnitude response\n");
if (nprnt == 1)
  fprintf(fp,"magnitude response\n");
printf("max val= %6.2f \n",xmx);
if (nprnt == 1)
  fprintf(fp,"max val= %6.2f \n",xmx);
if (xmn == 0.) {
  zpt = 20; scale = 20./xmx; offset = 0;
  }
if (xmx > 0. && xmn < 0.) {
  if (-xmn > xmx) xmx = -xmn;
  zpt = 10; scale = 10./xmx; offset = 10;
  }
for (i = 0; i <= 20; i++) {
  for (j = 0; j < nx; j++) {
    if (j == 0 || j == nx-1)
      dot = '|'; else dot = ' ';
    if (i == 0 || i == 20) dot = '-';
    rndoff = .5001;
    ival = offset + (rndoff + x[j]*scale);
    if (ival == 20-i) dot = '*';
    printf("%1c",dot);
    if (nprnt == 1) fprintf(fp,"%1c",dot);
    }
  printf("\n");
  if (nprnt == 1) fprintf(fp,"\n");
  }
return;
}


void plotpr(x,nprnt,fp)
FILE     *fp;
float    x[];
int      nprnt;
{
int      i,j,nx,ival,zpt,offset;
float    xmx,scale,rndoff;
double   pi = 4.*atan(1.);
char     dot;
```

```
nx = 65; xmx = pi;
printf("phase response\n");
if (nprnt == 1)
  fprintf(fp,"phase response\n");
zpt = 10; scale = 10./xmx; offset = 10;
for (i = 0; i <= 20; i++) {
  for (j = 0; j < nx; j++) {
    if (j == 0 || j == nx-1)
      dot = '|'; else dot = ' ';
    if (i == 0 || i == 20 || i == 10)
      dot = '-';
    rndoff = .5001;
    if (x[j] < 0.) rndoff = -.5001;
    ival = offset + (rndoff + x[j]*scale);
    if (ival == 20-i) dot = '*';
    printf("%1c",dot);
    if (nprnt == 1) fprintf(fp,"%1c",dot);
    }
  printf("\n");
  if (nprnt == 1) fprintf(fp,"\n");
  }
return;
}


void plotlm
    (x,dbscal,spec,if1,if2,nprnt,fp)
FILE     *fp;
float    x[],dbscal,spec;
int      if1,if2,nprnt;
{
int      i,j,nx,zpt,offset;
float    xt[65],xmx,scale,scale2,
         rndoff,dbinc;
char     dot;

nx = 65; xmx = -1000.;
for (i = 0; i < nx; i++){
  if (x[i] > xmx) xmx = x[i];
  }
for (i = 0; i < nx; i++) {
  xt[i] = x[i] - xmx;
  if (xt[i] < dbscal) xt[i] = dbscal;
  }
printf("log-magnitude response   ");
printf("dbscal= %6.2f   ",dbscal);
printf("spec= %6.2f\n",spec);
if (nprnt == 1){
  fprintf(fp,"log-magnitude response ");
  fprintf(fp,"dbscal= %6.2f   ",dbscal);
  fprintf(fp,"spec= %6.2f\n",spec);
  }
scale = dbscal/20.;
scale2 = fabs(scale/2.);
for (i = 0; i <= 20; i++) {
  dbinc = i*scale;
  for (j = 0; j < nx; j++) {
```

```
      dot = ´ ´;
      if (j == 0 || j == nx-1
        || j == if1 || j == if2)
        dot = ´|´;
      if (i == 0 || i == 20) dot = ´-´;
      if (fabs(spec-dbinc) < scale2)
        dot = ´-´;
      if (fabs(xt[j]-dbinc) < scale2)
        dot = ´*´;
      printf("%1c",dot);
      if (nprnt == 1)
        fprintf(fp,"%1c",dot);
      }
    printf("\n");
    if (nprnt == 1) fprintf(fp,"\n");
    }
  return;
  }


Program to test fft.c, plotmr.c,
plotpr.c and plotlm.c:

/* test1.c */
#include <math.h>
#include <stdio.h>
#include <fft.c>
#include <plotmr.c>
#include <plotpr.c>
#include <plotlm.c>

main()
{
  FILE    *fp;
  float   x[128],y[128],dbscal,spec;
  double  a,b,f,
          pi = 4.*atan(1.);
  int     i,j,if1,if2,nprnt,nx,ntype,log2n;
  void    fft(),plotmr(),plotpr(),plotlm();

  fp = fopen("test1.dat","w");
  nx = 128; nprnt = 1;
  for (i = 0; i < nx; i++) {
    x[i] = 0.; y[i] = 0.;
    }
  for (i = 0; i < 8; i++) {
    x[i] = 1.;
    }
  log2n = 7; ntype = 1;
  fft(log2n,x,y,ntype);
  nx = 65;
  for (i = 0; i < nx; i++) {
    a = x[i]; b = y[i];
    x[i] = sqrt(a*a + b*b);
    if (a == 0. && b == 0.) y[i] = 0.;
      else y[i] = atan2(b,a);
    }
```

```
  plotmr(x,nprnt,fp);
  plotpr(y,nprnt,fp);
  for (i = 0; i < nx; i++) {
    if (x[i] < 0.0001) x[i] = 0.0001;
    x[i] = 20.*log10(x[i]);
    }
  dbscal = -40.; spec = -20.;
  if1 = 19; if2 = 29;
  plotlm(x,dbscal,spec,if1,if2,nprnt,fp);
  fclose(fp);
  }
```

Results produced by test0:

maximum value= 1.00

```
  0------------------------------------*--------------------------------
  1-                                   |     *
  2-                                   |        *
  3-                                   |           *
  4-                                   |             *
  5-                                   |               *
  6-                                   |                *
  7-                                   |                 *
  8-                                   |                  *
  9-                                   |                  *
 10-                                   |                 *
 11-                                   |               *
 12-                                   |            *
 13-                                   |         *
 14-                                   |       *
 15-                                   |    *
 16-                                *  |
 17-                             *     |
 18-                          *        |
 19-                      *            |
 20-                   *               |
 21-                *                  |
 22-             *                     |
 23-           *                       |
 24-          *                        |
 25-*                                  |
 26-         *                         |
 27-          *                        |
 28-           *                       |
 29-             *                     |
 30-                *                  |
 31-                   *               |
 32-                      *            |
 33-                          *        |
 34-                             *     |
 35-                                *  |
 36-                                   |   *
 37-                                   |      *
 38-                                   |        *
 39-                                   |          *
 40-                                   |           *
 41-                                   |            *
 42-                                   |            *
 43-                                   |           *
 44-                                   |         *
```

Test1 results:

magnitude response

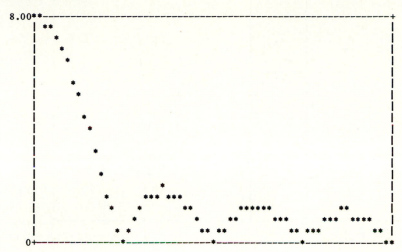

phase response

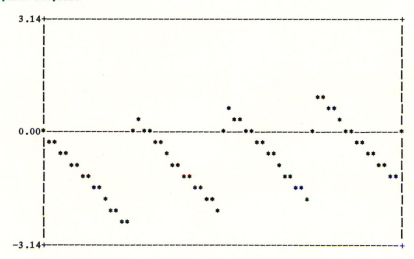

log-magnitude response

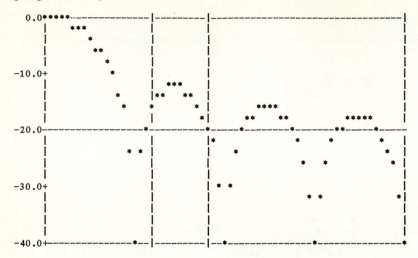

APPENDIX B

FILTER SPECIFICATION FOR PROJECTS TABLE B-1

Frequency values for the filter specification shown in Fig. B.1. The frequency values below are given in terms of k, the 128-point DFT bin number. The value $k = 1$ corresponds to $\omega = 0$. The frequency is then $\omega_k = 2\pi(k - 1)/128$. The lowpass and highpass filters are specified with the pair of values k_1 and k_2 that define ω_{k_1} and ω_{k_2}. The bandpass and bandreject filters are specified with one set of values for k_1, k_2, k_3 and k_4.

Lowpass and highpass frequency values								Bandpass and bandreject frequency values							
k_1	k_2	k_1	k_2	k_1	k_2	k_1	k_2	k_1	k_2	k_3	k_4	k_1	k_2	k_3	k_4
4	21	33	48	6	19	31	46	4	21	33	48	6	19	31	46
5	22	34	49	7	20	32	47	5	22	34	49	7	20	32	47
6	23	35	50	8	21	33	48	6	23	35	50	8	21	33	48
7	24	36	51	9	22	34	49	7	24	36	51	9	22	34	49
8	25	37	52	10	23	35	50	8	25	37	52	10	23	35	50
9	26	38	53	11	24	36	51	9	26	38	53	11	24	36	51
10	27	39	54	12	25	37	52	10	27	39	54	12	25	37	52
11	28	40	55	13	26	38	53	11	28	40	55	13	26	38	53
12	29	41	56	14	27	39	54	12	29	41	56	14	27	39	54
13	30	42	57	15	28	40	55	13	30	42	57	15	28	40	55
14	31	43	58	16	29	41	56	14	31	43	58	16	29	41	56

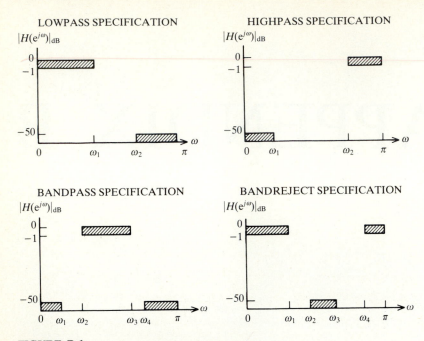

FIGURE B-1
Lowpass, highpass, bandpass and bandreject specifications for the projects.

SUBJECT INDEX

A

ADC (see analog-to-digital conversion)
Adder component, 3, 41, 169
Advance component, 3, 41, 169
Aliasing
  frequency domain, 103
  time domain, 125, 127, 341, 433, 438
Allpass filter, 310
Amplitude function, 75
Analog filter design methods, 263
  Butterworth filter, 264
  Chebyshev filter, 270
  elliptic filter, 275
Analog-to-digital conversion (ADC), 12
  Nyquist criterion, 102
  quantization effects, 395
  sampling of anlog signals, 103, 280
Anti-aliasing filter, 105, 264
Antisymmetric sequences, 86, 89
  (also see Symmetric sequences)

B

Bartlett window, 328
Bilinear transform, 288
Binary point, 391
Blackman window, 335
Butterfly computation, 147
Butterworth filter, 264

C

Cascade second-order section, 215
Cascade structure of filters, 164, 215

C

Causal system, 40
Chebyshev filter, 270
Circular convolution, 95, 129, 130, 321
Coefficient quantization, 399
  FIR filter, 400
  IIR filter, 402
Comb filter, 224, 364
Conversion of decimal to binary, 392
Convolution
  circular, 95, 129, 130, 321
  linear, 31, 132
  sum evaluation, 32
CSOS (see cascade second-order section)
Cutoff frequency, 236, 264

D

DAC (see digital-to-analog conversion)
Decimation-in-time FFT algorithm, 142
Delay component, 3, 41, 169
DFT (see Discrete Fourier transform)
DFT program, 135
Difference equation, 41, 162
Digital filter (see FIR or IIR filter)
Digital-to-analog conversion, 12, 106
Direct form structures, 210
Discrete Fourier transform (DFT)
  computation of, 123
  definition, 119, 123
  evaluating with fast Fourier transform, 142
  Goertzel algorithm, 139
  linear convolution with, 132
  of infinite duration sequences, 127
  of noncausal sequences, 125
  program (DFT), 135

Discrete Fourier transform (DFT) (*contd.*)
  properties, 129
  reducing computation time, 136
  relationship to Fourier transform, 119
Discrete Hilbert transform
  application to filter design, 436
  definition, 433
  program (DHT), 452
Discrete-time sequences
  complex exponential, 23
  unit-sample response, 27
  unit-sample, 21
  sinusoidal, 22
Discrete-time systems
  descriptions
    time-domain:
      convolution, 31
      difference equation, 41, 162
      unit-sample response, 21
    frequency-domain:
      transfer function, 67, 83
    z-domain:
      system function, 162
  linear time-invariant (LTI) system
    causality, 40
    stability, 38
Dominant singularity, 187

E

Elliptic filter, 275
Energy evolution factor, 427
Euler's identities, 25
Even sequence, 86, 188, 189

F

Fast Fourier transform, 142
  butterfly pattern, 147
  decimation-in-time algorithm, 142
  overlap-and-add method, 379
FFT (see fast Fourier transform)
Filter implementation
  from difference equation, 41
  from system function, 169
Filter structures
  cascade, 164
  cascade second-order section, 215
  direct forms, 210
  frequency-sampling, 221
  parallel, 167
  parallel second-order section, 218
Finite geometric sum formula, 36

Finite-impulse response (FIR) filter, 209
  approximation to inverse filter, 432
  design techniques, 318
    DFT method, 341
    DFT/windowing method, 348, 354
    frequency-sampling method, 361, 372
    windowing method, 319
  finite-precision effects, 400, 409
  frequency-sampling structure, 221
  implementation with FFT, 379
  linear-phase structures, 220
Finite-precision arithmetic effects, 391
FIR (see Finite impulse response filter)
Fixed-point
  addition, 405
  number representation, 391
  multiplication, 406
Floating-point
  addition, 407
  number representation, 394
  multiplication, 408
Fourier transform
  definition, 67
  linear-phase form, 75
  of continuous-time signals, 99
  of real-valued sequences, 72
  of symmetric sequences, 86
  of delayed sequences, 85
  of product of two sequences, 94
  program (FT), 81
  properties, 69
  relationship to z-transform, 184, 232
Frequency-sampling
  filter structure, 221
  FIR filter design method, 361
Frequency transformation for lowpass filters,
    295
FT program, 81

G

Gain constant, 241
Geometric evaluation of z-transform, 183
Gibbs phenomenon, 326
Goertzel algorithm, 139

H

Hamming window, 332
Hanning window, 330
Hibert transform (see Discrete Hilbert
    transform)

I

IDFT (see inverse discrete Fourier transform)
IIR (see Infinite-impulse response filter)
Impulse-invariance method, 280
Infinite-impulse response filter, 209
  design techniques:
    impulse-invariance method, 280
    bilinear-transform method, 288
    transformations of lowpass filter, 295
  finite-precision effects, 402, 410
  limit cycle, 410
Infinite geometric sum formula, 36
Interactive filter design, 247, 256
Interpolation, 108, 264
Inverse filter, 422
Inverse Fourier transform, 92
Inverse discrete Fourier transform, 123
Inverse z-transform, 196

L

Limit cycle, 410
Linear difference equations, 41
Linear convolution, 32
Linear-phase filter
  energy evolution of, 427
  FIR filter structure, 220
  unit-sample response of, 220
  z plane pole/zero pattern of, 195
Linear-phase form of transfer function, 75
Linear-phase response, 243, 349, 426
Linear-phase system, 426
Linear system, 26
  causality, 40
  stability, 38
Log-magnitude response
  definition, 238
  program (LOGMAG), 240
LOGMAG program, 240
Lowpass frequency transformations, 295
  to bandpass, 301
  to bandreject, 306
  to highpass, 298
  to lowpass, 296

M

Magnitude response, 71, 236
Magnitude-phase program (MAGPHS), 81
MAGPHS program, 81
Maximum-phase system, 426
Minimum-phase system, 426, 433

N

Multiplication
  fixed-point, 406
  floating-point, 408
Multiplier component, 3, 41, 169

N

Noncausal filter
  definition, 40, 41
  filter structure, 169
  implementation of, 54
Nonrecursive filter, 5, 210
  program (NONREC), 48
NONREC program, 48
NULL program, 45
Number representations
  Fixed-point, 391
  Floating-point, 394
Nyquist criterion, 102

O

Order, 5
Overflow, 405
Overlap-and-add method, 379

P

Padding with zeros, 123
Parallel second-order section, 218
Parallel structure, 167, 218
Periodic extension
  of spectra, 102
  of time sequences, 120, 125, 127, 341, 432
Phase distortion, 244, 440
Phase response, 71, 243
  jumps in, 76
  of minimum-phase system, 426
  of maximum-phase system, 426
  principal value, 73
PLOTLM program, 240
POLE program, 181
Pole/zero pattern in z plane, 170
  of linear-phase systems, 195
  of maximum-phase sequences, 426
  of minimum-phase sequences, 426
  of symmetric sequences, 189
  to z-transform, 235
Prewarping, 290
Program descriptions:
  DFT, 135
  FT, 81
  LOGMAG, 240

Program descriptions: (*contd.*)
  MAGPHS, 81
  NONREC, 48
  NULL, 45
  PLOTLM, 240
  POLE, 181
  QUANT, 412
  REC, 49
  SCALE, 413
  SHIFT, 54
Programming style, 45
PSOS (see parallel second-order section)

Q

Quantization
  in ADC, 395
  of filter coefficients, 399
  simulation of, 412
Quantizer characteristic, 395
QUANT program, 412

R

Raised-cosine windows, 330
Random number generator, 414
Rectangular window, 322
Reconstruction filter, 105
REC program, 49
Recursive filter, 5, 209
  program (REC), 49
Relevant singularity, 246
Region of convergence, 174
Resonator, 224
Round-off, 392, 406

S

Sampling analog signals, 2, 96
Second-order sections, 215, 218, 400, 402
SCALE program, 413
SHFTRG program, 47
SHIFT program, 54
Shift register program (SHFTRG), 47
Spectrum, 67, 71
Stability criterion, 38
Sum-of-products program (SUMP)
SUMP program, 47
Superposition principle, 26, 381, 398
Symmetric sequences, 86
  Fourier transform of, 86
  pole/zero pattern of, 189
  z-transform of, 188
System function, 162

T

Time-invariant system, 31
Transfer function, 67, 83
Transition region, 368
Triangular window, 328
Truncation of arithmetic results, 392, 406, 409
Truncation of sequences, 321
Two's-complement notation, 393

U

Unit circle, 189
Unit-sample response, 27
Unit-sample sequence, 21
Unit-step sequence, 22

V

Vector notation of complex numbers, 23

W

Window sequences, 321
  Bartlett, 328
  Blackman, 335
  comparison, 340
  FIR filter design with, 319
  rectangular, 322
  Hamming, 332
  Hanning, 330
  main-lobe width, 325, 340
  raised-cosine, 330
  side-lobe magnitude, 325, 340
  triangular, 328

Z

ZERO program, 181
Z plane, 169
Z-transform
  definition, 158
  filter implementation from, 167
  from difference equation, 162
  inverse, 196
  of symmetric sequences, 188
  pole/zero pattern, 170, 235
  properties, 160
  region of convergence, 174
  relation to Fourier transform, 184